AF756800

Advances in
NANO-OPTICS AND NANO-PHOTONICS

TIP ENHANCEMENT

Series Editors

Satoshi Kawata
Department of Applied Physics
Osaka University, Japan

Vladimir M. Shalaev
Purdue University
School of Electrical and Computer Engineering
West Lafayette, IN, USA

Advances in
NANO-OPTICS AND NANO-PHOTONICS

TIP ENHANCEMENT

Edited by

SATOSHI KAWATA
Department of Applied Physics
Osaka University, Japan

VLADIMIR M. SHALAEV
Purdue University
School of Electrical and Computer Engineering
West Lafayette, IN, USA

ELSEVIER

AMSTERDAM • BOSTON • HEIDELBERG • LONDON • NEW YORK • OXFORD
PARIS • SAN DIEGO • SAN FRANCISCO • SINGAPORE • SYDNEY • TOKYO

Elsevier
Radarweg 29, PO Box 211, 1000 AE Amsterdam, The Netherlands
The Boulevard, Langford Lane, Kidlington, Oxford OX5 1GB, UK

First edition 2007

Copyright © 2007 Elsevier B.V. All rights reserved

No part of this publication may be reproduced, stored in a retrieval system or transmitted in any form or by any means electronic, mechanical, photocopying, recording or otherwise without the prior written permission of the publisher

Permissions may be sought directly from Elsevier's Science & Technology Rights Department in Oxford, UK: phone (+44) (0) 1865 843830; fax (+44) (0) 1865 853333; email: permissions@elsevier.com. Alternatively you can submit your request online by visiting the Elsevier web site at http://elsevier.com/locate/permissions, and selecting *Obtaining permission to use Elsevier material*

Notice
No responsibility is assumed by the publisher for any injury and/or damage to persons or property as a matter of products liability, negligence or otherwise, or from any use or operation of any methods, products, instructions or ideas contained in the material herein. Because of rapid advances in the medical sciences, in particular, independent verification of diagnoses and drug dosages should be made

Library of Congress Cataloging-in-Publication Data
A catalog record for this book is available from the Library of Congress

British Library Cataloguing in Publication Data
A catalogue record for this book is available from the British Library

ISBN-13: 978-0-444-52058-6
ISBN-10: 0-444-52058-9
ISSN: 1871-0018

For information on all Elsevier publications
visit our website at books.elsevier.com

Printed and bound in the United Kingdom
Transferred to Digital Print 2010

Working together to grow
libraries in developing countries
www.elsevier.com | www.bookaid.org | www.sabre.org
ELSEVIER BOOK AID International Sabre Foundation

List of Contributors

Neil Anderson	University of Rochester, The Institute of Optics, Rochester, New York, USA
Renaud Bachelot	Nanotechnology and Optical Instrumentation Laboratory, Delaunay Institute, University of Technology of Troyes (UTT), Troyes, France
Jon A. Dieringer	Department of Chemistry, Northwestern University, Evanston, Illinois, USA
Hans-Jürgen Eisler	Nano-Optics Group, National Center of Competence for Research in Nanoscale Science, Institute of Physics, University of Basel, Klingelbergstr, Basel, Switzerland
Javad N. Farahani	Nano-Optics Group, National Center of Competence for Research in Nanoscale Science, Institute of Physics, University of Basel, Klingelbergstr, Basel, Switzerland
M. D. Foster	Department of Polymer Science, The University of Akron, Akron, Ohio, USA
M. Futamata	Nanoarchitechtonics Research Center, National Institute of Advanced Industrial Science and Technology, Higashi, Tsukuba, Japan
Achim Hartschuh	Physikalische und Theoretische Chemie, Universität Tübingen, Tübingen, Germany
R. D. Hartschuh	Department of Polymer Science, The University of Akron, Akron, Ohio, USA
Bert Hecht	Nano-Optics Group, National Center of Competence for Research in Nanoscale Science, Institute of Physics, University of Basel, Klingelbergstr, Basel, Switzerland

Taro Ichimura Graduate School of Frontier Biosciences, Osaka University, Osaka, Japan

Yasushi Inouye Graduate School of Frontier Biosciences, Osaka University, Osaka, Japan
RIKEN, Japan

Satoshi Kawata Graduate School of Engineering, Osaka University, Osaka, Japan
RIKEN, Japan

A. Kisliuk Department of Polymer Science, The University of Akron, Akron, Ohio, USA

N. Lee Department of Polymer Science, The University of Akron, Akron, Ohio, USA

J. F. Maguire Materials and Manufacturing Directorate/ MLBP, Wright-Patterson AFB, OH, USA

Y. Maruyama Tsukuba Research Laboratory, Hamamatsu Photonics K. K., Tsukuba, Japan

D. Mehtani Department of Polymer Science, The University of Akron, Akron, OH, USA

Alfred J. Meixner Physikalische und Theoretische Chemie, Universität Tübingen, Tübingen, Germany

Peter Mühlschlegel Nano-Optics Group, National Center of Competence for Research in Nanoscale Science, Institute of Physics, University of Basel, Klingelbergstr, Basel, Switzerland

Lukas Novotny University of Rochester, The Institute of Optics, Rochester, New York, USA

Dieter W. Pohl Nano-Optics Group, National Center of Competence for Research in Nanoscale Science, Institute of Physics, University of Basel, Klingelbergstr, Basel, Switzerland

Huihong Qian Physikalische und Theoretische Chemie, Universität Tübingen, Tübingen, Germany

Thomas Schmid Department of Chemistry and Applied Biosciences, ETH Zurich, Zurich, Switzerland

A. P. Sokolov Department of Polymer Science, The University of Akron, Akron, Ohio, USA

Richard P. Van Duyne	Department of Chemistry, Northwestern University, Evanston, Illinois, USA
Prabhat Verma	Graduate School of Frontier Biosciences, Osaka University, Osaka, Japan Graduate School of Engineering, Osaka University, Osaka, Japan
Boon-Siang Yeo	Department of Chemistry and Applied Biosciences, ETH Zurich, Zurich, Switzerland
Matthew A. Young	Department of Chemistry, Northwestern University, Evanston, Illinois, USA
Anatoly V. Zayats	Nano-Optics and Near-Field Spectroscopy Group, Centre for Nanostructured Media, IRCEP, UK
Renato Zenobi	Department of Chemistry and Applied Biosciences, ETH Zurich, Zurich, Switzerland
Weihua Zhang	Department of Chemistry and Applied Biosciences, ETH Zurich, Zurich, Switzerland

Preface

With the increased advancement in the fields of nano-sciences, optics and photonics have been utilized to investigate, fabricate and manipulate nanomaterials. As it turns out, this area has developed rapidly with plenty of interesting results, demanding for a need of a well-organized documentation to follow the scientific achievements. Sensing the need of the time, we have decided to produce this book series, *Advances in Nano-Optics & Nano-Photonics*, dedicated to the new and interesting results obtained by leading scientists in the area.

As the first volume of this book series, tip enhancement has been chosen. Tip enhancement effect, which was first proposed in 1999, made a great breakthrough in Raman spectroscopic analysis. Within just a few years, it has been significantly used, in particular, for nano-imaging and nano-analysis, not only because of its high factor of enhancement, but also because of the high-resolution capability down to 10 nm due to the localization of the field/hot spot of the resonant plasmon polaritons in a structured nano-metal.

In this volume, we invited distinct scientists who have made great contributions to tip-enhancement science including theoretical development, numerical studies, instrumentation and experiments.

In the first two chapters, the authors start a discussion on the mechanism of surface enhancement, which builds up the basic science involved in tip-enhancement, and examine the possibilities of single molecular sensitivity in surface-enhanced Raman scattering. They further discuss the tip-enhancement and the plasmonic materials suitable for the same. Then the next chapter is devoted to near-field effects in tip-enhancement, where the authors examine several nanomaterials with both linear and nonlinear Raman scatterings. Next three chapters are focused on the applications in the fields of chemical, biological and material sciences, with special emphasis on single wall carbon nanotubes and semiconductor materials. The contributors in the next chapter discuss the optical structuring and manipulation based on field enhancement near metallic nanostructures, while the near-field microscopy based on second-harmonic generation is discussed in the following chapter. Finally, the last chapter discusses the optical antennas and single emitters based on the localized resonant enhancement of optical field.

We, as the Editors of this volume, show our heartfelt gratitude to all the contributors for their enthusiasm to share some of their recent results in the framework of this volume and for the efforts they have given in bringing this volume to its present shape.

Satoshi Kawata and Vladimir M. Shalaev

Contents

Chapter 1

Plasmonic materials for surface-enhanced and tip-enhanced Raman spectroscopy

by

M.A. Young, J.A. Dieringer, R.P. Van Duyne

Department of Chemistry, Northwestern University, 2145 Sheridan Road, Evanston, Illinois 60208

Tip Enhancement
Advances in Nano-Optics and Nano-Photonics
ISSN: 1871-0018

S. Kawata and V.M. Shalaev (Editors)
© *2007 Elsevier B.V. All rights reserved.*
DOI: 10.1016/S1871-0018(06)01001-6

Contents

§ 1. Introduction

Plasmonics is an emerging branch of nano-optics and nano-photonics that examines the properties of the collective electronic excitations in noble metal films or nanoparticles known colloquially as surface plasmons. The unusual dispersion properties of these metals near the plasmon resonance enable excitation of plasmons in nanostructures that access a very large range of wavevectors over a narrow frequency range, and accordingly, resonant plasmon excitation allows for light localization in ultra-small volumes. Near-field localization of light combined with quantitative electromagnetic simulation methods and a broad array of established and emerging nanofabrication methods creates the conditions for dramatic scientific progress as well as a new class of subwavelength optical components. The excitement of plasmonics lies in its potential to achieve highly miniaturized and sensitive photonic devices by controlling, manipulating, and amplifying light on the nanometer length scale (Maier et al., 2001, 2003; Barnes et al., 2003; Haynes et al., 2003; Andrew and Barnes, 2004; Ozbay, 2006).

To date, a variety of passive plasmonic devices have been demonstrated, including filters (Barnes et al., 2003), waveguides (Barnes et al., 2003; Maier et al., 2003), polarizers (Haynes and Van Duyne, 2003a), Bragg reflectors (Barnes et al., 2003), and nanoscopic light sources (Lezec et al., 2002). On the horizon are active plasmonic devices, such as light output enhancers for organic light-emitting diodes (Smith et al., 2004; Wedge et al., 2004) as well as switches and modulators (Andrew and Barnes, 2004). Furthermore, our rapidly improving understanding of the interactions between adsorbed molecules and plasmonic nanostructures (i.e., molecular plasmonics) (Van Duyne, 2004) is having a significant impact on a broad spectrum of other applications, including nanoscale optical spectroscopy (Anderson, 2000; Hayazawa et al., 2000; Stockle et al., 2000; Pettinger et al., 2002, 2004; Ichimura et al., 2004a, 2004b; Ren et al., 2004; Mehtani et al., 2005; Saito et al., 2005; Anderson et al., 2005a; Rasmussen and Deckert, 2006), surface-enhanced Raman spectroscopy (SERS) (Campion and Kambhampati, 1998; Schatz and Van

Duyne, 2002; Moskovits, 2005; Tian, 2005; Haynes et al., 2005a, 2005b), surface plasmon resonance sensing (Brockman et al., 2000; Haes and Van Duyne, 2004), and nanolithography (Halteen and Van Duyne, 1995; Srituravanich et al., 2004; Felidj et al., 2005).

There are two types of surface plasmon resonance – localized and propagating. This chapter will mostly be concerned with the former, termed localized surface plasmon resonance (LSPR). This occurs in noble metal nanoparticles in the 10–200 nm size range and results in amplification of the electric field **E** near the particle surfaces such that the $|\mathbf{E}|^2$ can be 100–10,000 times greater in intensity than the incident field. The field has a spatial range on the order of 10–50 nm and is strongly dependent on nanoparticle size, shape, and local dielectric environment. Propagating plasmons, also called surface plasmon polaritons (SPP), are associated with smooth, thin films of silver and gold with thicknesses in the 10–200 nm range. Propagating plasmons lead to smaller field enhancements (10–100) and have a larger spatial range normal to the plane of propagation ($\sim 1,000$ nm).

This chapter will focus on the fabrication and characterization of plasmonic materials that show promise in chemical/biological sensing and surface-enhanced spectroscopy applications. This includes tip-enhanced Raman spectroscopy (TERS), a recently discovered near-field technique for which plasmonics will be a critical feature of its continued development toward becoming an important analytical technique. In the first part, the simple, massively parallel method of nanosphere lithography (NSL) and its use in the fabrication of size- and shape-controlled nanostructures is briefly reviewed. Also, the essential physics of the LSPR and the theoretical methods used to understand it are described, and key results concerning the short- and long-range distance dependencies of the electromagnetic fields surrounding the nanoparticles are summarized. In the second part of this chapter, we briefly introduce SERS and TERS. We focus on the relationship between LSPR spectroscopy and SERS, as revealed by surface-enhanced Raman excitation spectroscopy (SERES). SERES provides a systematic, reproducible way to optimize the signal intensity in SERS experiments. These experiments are directly relevant to optimization of TERS experiments, where the electromagnetic enhancement mechanism is expected to play a crucial role. We also discuss the distance dependence of SERS, probed via atomic layer deposition (ALD) experiments, and we review a 2D correlation method for single-molecule SERS (SMSERS). Finally, we present some initial work on force dependence in TERS. The chapter concludes with our perspective on the future of molecular plasmonics as it relates to the new and very promising field of TERS.

§ 2. Nanosphere lithography

NSL (Hulteen and Van Duyne, 1995) is a surprisingly powerful yet simple approach to the fabrication of nanoparticle arrays with precisely controlled shape, size, and interparticle spacing.

NSL (fig. 1) begins with the self-assembly of monodisperse polystyrene or SiO_2 nanospheres of diameter D to form a single- or double-layer

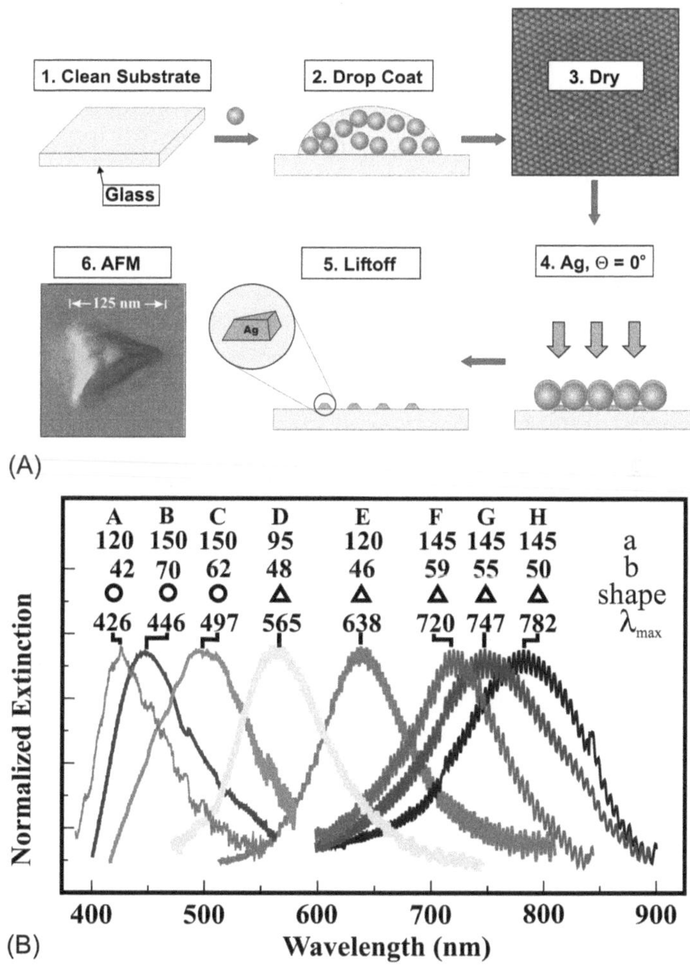

Fig. 1. (A) Schematic representation of the NSL fabrication process. The AFM image in step 3 is 5 μm × 5 μm. (B) Size- and shape-tunable LSPR spectra of various Ag nanoparticles (labeled A–H) fabricated by NSL. The wavelength of maximum extinction, λ_{max}, is changed by varying the in-plane width a and out-of-plane height b of the nanoparticles.

colloidal crystal mask for material deposition. A substrate (fig. 1(A)) is prepared so that the nanospheres move freely until they reach their lowest energy configuration. This is achieved by chemically modifying the nanosphere surface with a negative charge that is electrostatically repelled by a negatively charged substrate such as mica or chemically treated glass. As the solvent (water) evaporates, capillary forces draw the nanospheres together, and they crystallize into a hexagonally close-packed pattern on the substrate. As is the case in all naturally occurring crystals, nanosphere masks include a variety of defects that arise as a result of nanosphere polydispersity, site randomness, point defects (vacancies), line defects (slip dislocations), and polycrystalline domains. Typical defect-free domain sizes are in the 10–100 μm range. Following self-assembly of the nanosphere mask, a metal or other material is then deposited by physical vapor deposition from a collimated source normal to the substrate through the nanosphere mask to a controlled thickness. The resulting surface is referred to as a metal (e.g., Ag) "film over nanosphere" (FON) surface. AgFON surfaces are robust plasmonic materials for SERS applications (Litorja et al., 2001; Dick et al., 2002). If the nanosphere mask is removed, typically by sonicating the entire sample in a solvent, surface-confined nanoparticles are left behind that have a triangular footprint. In a typical NSL process, deposition of 50 nm of Ag over a single-layer mask self-assembled from nanospheres with $D = 400$ nm produces nanotriangles with in-plane width $a \approx 100$ nm, out-of-plane height $b \approx 50$ nm, and an interparticle separation distance $d_{\mathrm{ip}} \approx 230$ nm.

§ 3. Size- and shape-tunable localized surface plasmon resonance spectra

NSL-derived nanoparticles exhibit intense UV-visible extinction (i.e., the sum of absorption and scattering) bands that are not present in the spectrum of the bulk metal. Figure 1(B) shows that the LSPR spectra can easily be tuned all the way from the near-UV through the visible spectrum (Jensen et al., 2000a) and even into the mid-IR (Jensen et al., 2000b) by changing the size or shape (triangle or hemisphere) of the nanoparticles. Note that the LSPR bandwidth does not change significantly as the wavelength at peak maximum, $\lambda_{\max}$, is tuned. Additionally, several other surprising LSPR optical properties have been discovered for NSL-derived Ag nanoparticles: (1) $\lambda_{\max}$ shifts by 2–6 nm per 1 nm variation in nanoparticle width or height; (Jensen et al., 2000a) (2) the molar decadic extinction coefficient is $\varepsilon = 3 \times 10^{11}\ \mathrm{M}^{-1}\,\mathrm{cm}^{-1}$; (Jensen et al., 2000a) (3) the LSPR oscillator strength per atom is equivalent to that of atomic silver in gas or liquid phases; (Jensen et al., 2000a) (4) resonant Rayleigh scattering (Michaels et al., 1999; Schultz et al., 2000) occurs with an efficiency equivalent to that of 10^6 fluorophors (Yguerabide and Yguerabide, 1998), and (5) local electromagnetic fields are amplified by factors of $|\mathbf{E}|^2 \approx 10^4$, leading to intense signals in all surface-enhanced spectroscopies (Schatz and Van Duyne, 2002).

§ 4. Fundamentals of localized surface plasmon resonance spectroscopy

The simplest theoretical approach available for modeling the optical properties of nanoparticles is classical electrodynamics (i.e., solving Maxwell's equations with the metal dielectric constant taken from bulk measurements). For spherical particles, this leads to the following (Mie theory) expression for the extinction coefficient $E(\lambda)$ in the long-wavelength limit (Kreibig and Vollmer, 1995):

$$E(\lambda) = \frac{24\pi N_A a^3 \varepsilon_m^{3/2}}{\lambda \ln(10)} \left[\frac{\varepsilon_i}{(\varepsilon_r + 2\varepsilon_m)^2 + \varepsilon_i^2} \right], \tag{4.1}$$

where N_A is the areal density of nanoparticles, a the radius of the metallic nanosphere, ε_m the dielectric constant of the medium surrounding the nanosphere (assumed to be a positive, real number), λ the wavelength, and ε_r and ε_i are the real and imaginary parts of the metal dielectric function. This formula predicts a resonant peak when $\varepsilon_r = -2\varepsilon_m$, which for silver and gold occurs in the visible portion of the spectrum. In addition, any change in the dielectric constant of the medium (e.g., when molecules adsorb on the particle) leads to a change in the resonance wavelength.

When one considers spheroidally shaped particles, the term $\varepsilon_r + 2\varepsilon_m$ in the denominator in eq. (4.1) is replaced by $\varepsilon_r + \chi\varepsilon_m$, where χ is a parameter that depends on the shape of the spheroid, increasing from 2 for a sphere to 17 for a spheroid with an aspect ratio of 5:1. This leads to strong dependence of λ_{max} on particle shape, and there is also a strong size dependence that arises from electrodynamic effects that are not contained in eq. (4.1). In addition, many of the samples considered in this review contain an ensemble of nanoparticles that are supported on a substrate, leading to a dependence on interparticle spacing (Haynes et al., 2003) and substrate dielectric constant (Haynes and Van Duyne, 2001).

§ 5. Electrodynamic calculations

Equation (4.1) provides useful insight, but it only applies to spherical particles. In order to describe particles like those pictured in fig. 1(A), it is

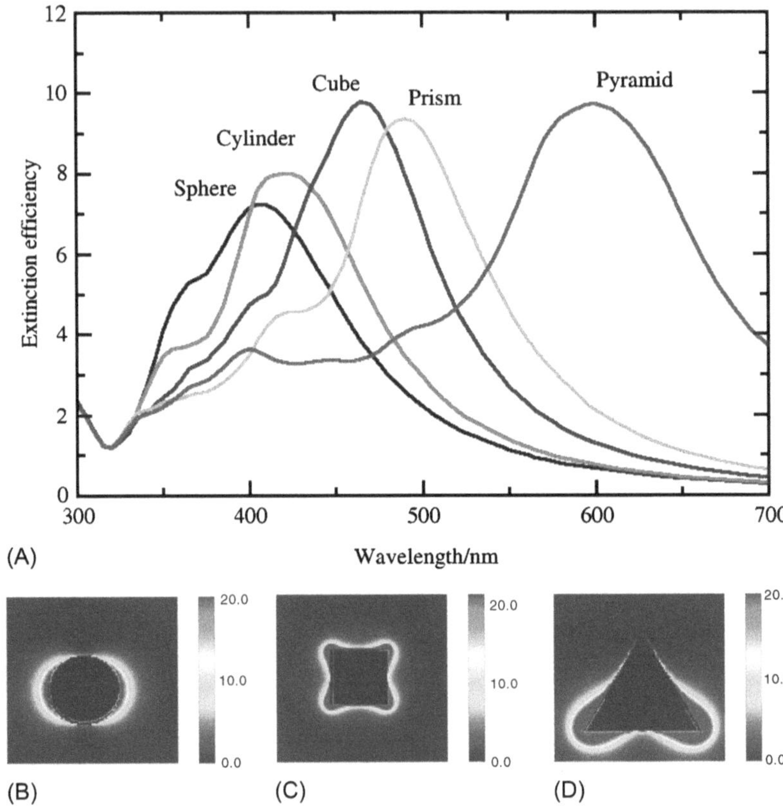

Fig. 2. (A) Extinction efficiency (ratio of cross section to effective area) of silver nanoparticles in vacuum having the shapes indicated. Each particle has the same volume, taken to be that of a sphere with a radius of 50 nm. $|\mathbf{E}|^2$ contours ($\mathbf{E}$ is electric field) for a (B) sphere, (C) cube, and (D) pyramid, plotted for wavelengths corresponding to the plasmon peak in (A), with peak $|\mathbf{E}|^2$ values of 54, 745, and 9770, respectively.

necessary to use a numerical method. Several methods have been developed for solving Maxwell's equations by finite-element-based approaches, and one that we have found to be particularly useful is the discrete dipole approximation (DDA) (Flatau et al., 1990; Goodman et al., 1991; Yang et al., 1995). In this method, the particle is subdivided into an array of cubical, polarizable elements. When a plane wave field is applied to the particle, the resulting induced polarizations in the elements are calculated, and from these the extinction and local fields are determined. Figure 2(A) shows extinction cross sections that have been calculated with this approach for several particle shapes, all for silver particles with the same volume. This shows that a sphere has a plasmon near the blue end of the spectrum (fig. 2(B)), and the particles with points (cubes, prisms, and pyramids) are nearer the red end of the spectrum (figs. 2(C) and 2(D)).

The electromagnetic (EM) mechanism of SERS (Schatz and Van Duyne, 2002) predicts an enhancement factor proportional to $|\mathbf{E}(\omega)|^2|\mathbf{E}(\omega')|^2$, where ω and ω' are the incident and Stokes shifted frequencies, respectively. To estimate the enhancement factor, in figs. 2(B)–2(D) we show contours of $|\mathbf{E}|^2$ around three of the particles for wavelengths corresponding to $\lambda_{\max}$ and for polarizations that lead to the largest $|\mathbf{E}|^2$. These figures show that the peak field for a sphere is on the order of 10^2, while that for the truncated tetrahedron is 10^4. If we approximate the enhancement factor as $|\mathbf{E}(\omega)|^4$, we see that the highest enhancements are on the order of 10^8, which as noted later is about what is found for NSL-derived Ag nanoparticles.

§ 6. The distance dependence of the localized surface plasmon resonance

The electrodynamics results in fig. 2 show that the electromagnetic fields surrounding Ag nanoparticles excited at $\lambda_{\max}$ drop off quickly as one moves (a few nm) away from the particle surface. How can the range of these fields be measured experimentally? Scanning near-field optical microscopy is certainly a possibility; however, the resolution currently obtainable ($\sim$ 10 nm) would only provide a rough picture.

We have pursued an alternative strategy to experimentally measure the range of the electromagnetic fields. It is apparent from eq. (4.1) that $\lambda_{\max}$ of noble metal nanoparticles is highly dependent on the dielectric properties of the surrounding environment. For NSL-derived Ag nanoparticles, it has been demonstrated experimentally that $\lambda_{\max}$ is a linear function of the solvent refractive index, n_{m} (where the external dielectric constant, $\varepsilon_{\mathrm{m}} = n_{\mathrm{m}}^2$) with a slope of approximately 200 nm per refractive index unit (RIU) (Jensen et al., 1999). In addition, $\lambda_{\max}$ is responsive to molecule-induced changes in the local dielectric environment. The LSPR wavelength shift response, $\Delta\lambda_{\max}$, of Ag nanoparticles fabricated by NSL has been used to develop a new class of nanoscale optical biosensors (Haes and Van Duyne, 2002; Riboh et al., 2003; Yonzon et al., 2004; Haes et al., 2005). On the most elementary level, the LSPR wavelength shift response of these sensors can be understood using a model of the refractive-index response of propagating surface plasmons on a planar noble metal surface (Jung et al., 1998)

$$\Delta\lambda_{\max} = m\Delta n[1 - \exp(-2d/l_{\mathrm{d}})], \tag{6.1}$$

where $\Delta\lambda_{\max}$ is the wavelength shift, m the refractive-index sensitivity, Δn the change in refractive index induced by an adsorbate, d the effective adsorbate layer thickness, and l_d the characteristic electromagnetic field decay length. This model assumes a single exponential decay of the electromagnetic field normal to the planar surface, which is accurate for a propagating surface plasmon but is undoubtedly an oversimplification for the electromagnetic fields associated with noble metal nanoparticles. While this oversimplified model does not quantitatively capture all aspects of the LSPR nanosensor response, it does provide some guidance for sensor optimization. In particular, eq. (6.1) highlights the importance of distance dependence as described by the electromagnetic field decay length, l_d.

Haes and co-workers explored the long-range distance dependence of the LSPR nanosensor using self-assembled monolayers (SAMs) of 11-mecaptoundecanoic acid (11-MUD) and Cu^{2+} ions adsorbed on arrays of noble metal nanoparticles with various sizes, shapes, and compositions (Haes et al., 2004a). The spatial resolution of this experiment was limited to the thickness of the 11-MUD/Cu^{2+} monolayer which was at least 1.6 nm. Several interesting characteristics of the long-range behavior were found including (1) the LSPR shift vs. SAM thickness is nonlinear; (2) Ag nanoparticles are more sensitive than Au nanoparticles; (3) nanotriangles have larger sensing distances than nanohemispheroids; (4) increasing the nanoparticle in-plane width results in larger sensing distances; and (5) decreasing nanoparticle out-of-plane height results in larger sensing distances (Haes et al., 2004a). Semiquantitative theoretical calculations revealed that the plasmon resonance shift is controlled by the average electromagnetic field over the nanoparticle surface (Haes et al., 2004a).

Similarly, the short-range distance dependence (0–3 nm) of the LSPR nanosensor has been studied using alkanethiol, $CH_3\text{-}(CH_2)_xSH$ ($x = 2-11$, $13-15$, and 17), monolayers (Malinsky et al., 2001; Haes et al., 2004b). It was found that eq. (6.1) does a remarkably good job of accounting for the short-range LSPR response if one assumes a value $l_d = 5-6$ nm. In addition, the dependence of $\Delta\lambda_{max}$ on the chain length of the alkanethiol monolayer was found to be linear, with a large slope of 3.1–3.3 nm per CH_2 unit (Haes et al., 2004b). In addition, it has recently become possible to use dark-field resonant Rayleigh scattering to perform the $\Delta\lambda_{max}$ measurement for isolated single particles, and it was found that for chemically synthesized Ag triangular prisms, $\Delta\lambda_{max}$ is a linear function of the refractive index of the external medium with a slope of $\sim$ 200 nm/RIU (McFarland and Van Duyne, 2003). Also, the same LSPR shift of ~ 40 nm as seen for the NSL particle array was observed for a monolayer of hexadecanethiol on a single particle; more importantly, the dependence of $\Delta\lambda_{max}$ on the chain length of the alkanethiol monolayer was found to be linear, with a large slope of 3.5 nm per CH_2 unit (McFarland and Van Duyne, 2003).

Even though much important information was obtained from these previous long- and short-range distance dependence studies, it was hypothesized that new information could be obtained if it were possible to deposit single layers of a material with thicknesses of ~ 1 Å. Furthermore, while the refractive index of the bulk SAM molecule is known, the refractive index of the SAM is not known making an accurate theoretical model of the experiment difficult (Haes et al., 2004a).

ALD is just such a fabrication method that produces highly uniform and controlled thin films. Precursor gases are alternately pulsed through the reactor and purged away resulting in a self-limiting growth process that constructs a film one monolayer at a time (Ritala and Leskela, 2002). Highly uniform monolayers of Al_2O_3 can be deposited with ~ 1 Å thickness resolution and a refractive index of 1.57 (Ott et al., 1997; Groner et al., 2004). This work utilized ALD of Al_2O_3 to probe both the long- and short-range LSPR distance dependences of Ag nanoparticles in one integrated experiment. The nucleation and growth of Al_2O_3 on Ag surfaces was studied using a quartz crystal microbalance, variable angle spectroscopic ellipsometry, and X-ray photoelectron spectroscopy measurements. These measurements revealed that the Al_2O_3 deposits on the LSPR Ag surface in a layer-by-layer fashion and the initial nucleation may proceed via the thermal decomposition of trimethylaluminum (TMA) on Ag (Whitney et al., 2005).

Figure 3 depicts LSPR extinction spectra for triangular Ag nanoparticles with an in-plane width (a) of 90 nm and out-of-plane height (b) of 40 nm. The UV–vis spectra for Ag nanoparticles with 0–450 cycles of TMA and water are presented. As subsequent ALD Al_2O_3 layers are completed, the LSPR λ_{max} position shifts to longer wavelengths which is consistent with previous work (Haes et al., 2004a). Importantly, these results clearly demonstrate that the LSPR nanosensor has sufficient sensitivity to detect the deposition of each successive Al_2O_3 monolayer. For nanoparticles with $a = 90$ and $b = 40$ nm, a 5-nm LSPR λ_{max} shift is observed with 2 Å of Al_2O_3. Furthermore, by extrapolation it should be possible to detect submonolayers of material. ALD of Al_2O_3 allows for the fabrication of angstrom-thick monolayers which gives this work a 10× increase in the spatial resolution compared to previous work (Haes et al., 2004a).

Figure 4 depicts plots of the LSPR λ_{max} shift vs. Al_2O_3 film thickness for triangular nanoparticles with $a = 90$ nm and $b = 30$ nm (1), 40 nm (2), 51 nm (4), and hemispheroidal nanoparticles with $a = 104$ nm and $b = 54$ nm (3).

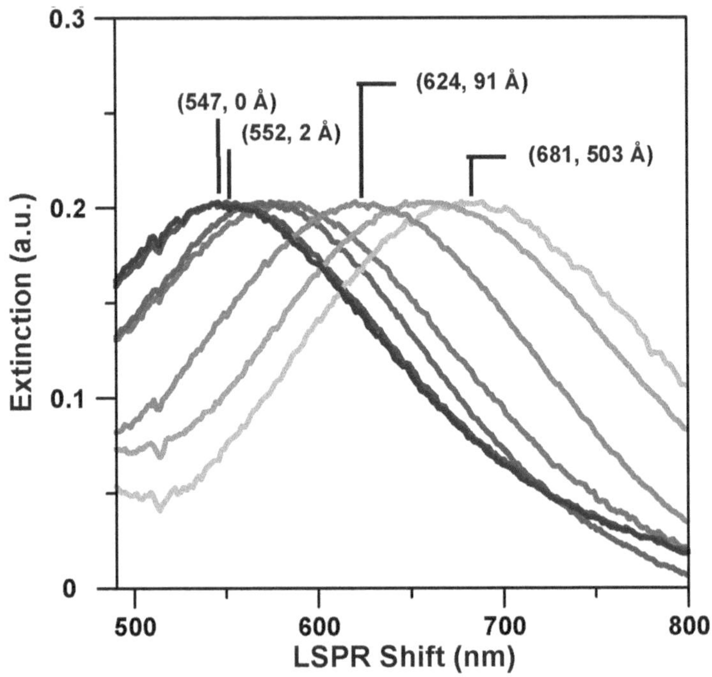

Fig. 3. LSPR spectroscopy of Ag nanoparticles ($a = 90$ and $b = 40$ nm) for 0–450 AB cycles of TMA and water.

In fig. 4(A), the LSPR λ_{max} shift vs. Al_2O_3 film thickness response is shown for 0–600 AB cycles. At short distances from the nanoparticle surface, the LSPR λ_{max} shift follows a steep linear slope, but as the distance from the nanoparticle increases, the curve levels off as the nanoparticle reaches its saturation point. As the nanoparticle height decreases and the in-plane width remains constant, the LSPR λ_{max} shift increases. The short-range distance dependence is highlighted in fig. 4(B) which depicts the λ_{max} shift vs. Al_2O_3 film thickness response for the first 0–20 AB cycles. After each cycle, a LSPR extinction spectrum was collected which presents a highly detailed view of the short-range distance dependence of the LSPR of these nanoparticles. Because Al_2O_3 layers deposited by ALD are ~ 1.1 Å in thickness (Ott et al., 1997), this is the first time that extremely detailed pictures of both the short- and long-range distance dependences of the LSPR nanosensor have been obtained in a single integrated experiment. In fact, these results show that at short distances from the nanoparticle surface, the LSPR λ_{max} shift vs. layer thickness follows a steep linear trend compared with the moderate slope at a larger distance from the nanoparticle surface. Nanoparticles with fixed in-plane widths and decreasing out-of-plane heights yield larger sensing

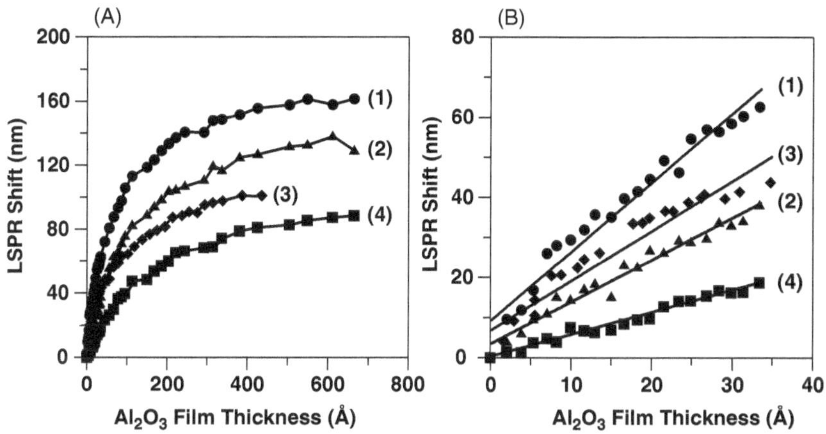

Fig. 4. LSPR shift vs. Al_2O_3 film thickness. (A) Out-of-plane height dependence on the long- and short-range distance dependence for Ag triangular nanoparticles $a = 90$ and $b = (1)$ 30, (2) 40, and (4) 51 nm and Ag hemispherical nanoparticles $a = 104$ and $b = 52$ nm (3). Data presented for 0–600 AB cycles of TMA and water. (B) Out-of-plane height dependence of the short-range distance dependence for the Ag triangular nanoparticles $a = 90$ and $b = (1)$ 30, (2) 40, and (4) 51 nm, and Ag hemispherical nanoparticles $a = 104$ and $b = 52$ nm (3). Data presented for 0–20 AB cycles of TMA and water. Linear regression was used to fit the data to lines described by the following equations: $y = 1.7x + 9.1$, $R^2 = 0.9602$ (1); $1.0x + 3.5$, $R^2 = 0.9689$ (2); $1.2x + 7.0$; $R^2 = 0.9293$ (3); and $0.5x + 0.1$, $R^2 = 0.9744$ (4).

distances (fig. 4(B)-1). The scatter in these plots results from slight misalignments resulting from repositioning the LSPR samples between measurements as the samples were moved between the ALD reactor and the spectrometer. The scatter would be expected to be greatly reduced if the extinction spectra were recorded in situ, which is the focus of future work. Nevertheless, the data in figs. 3 and 4 clearly demonstrate that the LSPR sensors possess angstrom-level sensitivity to the Al_2O_3 thickness changes. Previous results have shown that hemispheroidal nanoparticles have a smaller sensing distance than triangular nanoparticles of similar volume (Haes et al., 2004a). Unexpectedly, fig. 4(B)-3 shows that the hemispheroidal particles ($a = 104$ and $b = 52$ nm) presented here have a larger LSPR λ_{max} shift at long distances compared with triangular nanoparticles with $a = 90$ and $b = 51$ nm (fig. 4(A)-4). Also, at short distances from the nanoparticle surface, the hemispheroidal particles have larger LSPR λ_{max} shifts than both triangular nanoparticles with $a = 90$ and $b = 40$ and 51 nm.

The following results were obtained in this study: (1) a detailed study of the long-range distance dependence of the LSPR sensor with 10× increased spatial resolution afforded by ALD in comparison with earlier experiments using SAMs; (2) LSPR nanosensors were shown to have single Al_2O_3 layer detection capabilities; and (3) at short range, triangular nanoparticles with fixed widths and smaller out-of-plane heights have larger LSPR sensing distances.

§ 7. Surface-enhanced Raman spectroscopy

The local electromagnetic fields that accompany photon excitation of the LSPR are a key factor leading to the intense signals observed in all surface-enhanced spectroscopies (Schatz and Van Duyne, 2002), as well as for TERS. SERS is characterized by an ensemble-averaged intensity enhancement factor, EF, of $\sim 10^6$ for analytes bound to noble metal surfaces that possess random roughness (Jeanmaire and Van Duyne, 1977) or EF $\approx 10^7 - 10^8$ for surfaces with intentionally nanofabricated feature sizes in the $\sim$ 100 nm range (Haynes and Van Duyne, 2003b). Reports of single-molecule detection (Kneipp et al., 1997; Nie and Emory, 1997) using SERS on Ag nanoparticle clusters have rejuvenated interest in this widely used analytical technique. A clear understanding of the mechanism responsible for the enormous enhancement factors ($\sim 10^{14} - 10^{15}$) observed in SM SERS remains elusive.

Since Raman intensities scale as the product of incident field intensity and polarizability derivative, it comes as no surprise that there are two commonly considered mechanisms for SERS, one of which involves enhancements in the field intensity as a result of plasmon resonance excitation, and the other the enhancement in polarizability due to chemical effects such as charge transfer excited states (Moskovits, 1979; Gersten and Nitzan, 1980; Liao et al., 1981; Philpott, 1983; Metiu and Das, 1984; Schatz, 1984; Kerker, 1987). The first of these is the well-known electromagnetic (EM) enhancement mechanism, and in this mechanism the enhancement factor EF at each molecule is (approximately) given by

$$\mathrm{EF} = |\mathbf{E}(\omega)|^2 |\mathbf{E}(\omega')|^2, \tag{7.1}$$

where $\mathbf{E}(\omega)$ is the local electric field enhancement factor at the incident frequency ω and $\mathbf{E}(\omega')$ the corresponding factor at the Stokes shifted frequency ω'. The EM mechanism is the predominant mechanism in SERS, as well as in TERS. The EM mechanism of SERS predicts that there is a well-defined relationship between the LSPR spectrum and the surface-enhanced Raman excitation spectrum (SERES) of a SERS active surface. In particular, very specific quantitative predictions for the magnitude of the enhancement factor on nanoparticle size, shape, and local dielectric environment are made. Owing to the significant experimental challenges, many EM predictions remain purely in the domain of theory. The plasmonic materials that are now readily available from NSL and electron-beam lithography (EBL) provide a new platform for the detailed study of the electromagnetic field enhancement mechanism associated with SERS. Consequently, there is renewed experimental effort in SERES seeking to verify various aspects of the EM theory, as well as to elucidate the conditions necessary for optimization of surface-enhanced spectroscopies. For example, the EM mechanism primarily gives rise to TERS, and thus the information obtained is expected to be applicable to the important task of optimizing TERS experiments.

§ 8. Wavelength-scanned surface-enhanced Raman excitation spectroscopy

Wavelength-scanned SERES (WS SERES) involves the measurement of SERS enhancement for several laser excitation wavelengths, λ_{ex}. This technique was recognized as a useful tool for probing the EM mechanism immediately following the discovery of SERS. An obvious limitation of this technique is that the number of data points is determined by the tunability of the excitation laser and detection system. These substantial instrumental requirements have led to the majority of SERES publications suffering from low data point density and/or limited spectral coverage (Blatchford et al., 1982; Van Duyne et al., 1993; Vlckova et al., 1997; Gregory et al., 2001; Felidj et al., 2003). These limitations prevent the establishment of conclusive generalizations from SERES data. Additionally, most SERES experiments have been performed using surface-enhancing substrates with an unknown or poorly characterized distribution of roughness features. In the few cases where the surfaces are carefully characterized, it was shown that there were a wide distribution of roughness feature sizes (Von Raben et al., 1984; Van Duyne et al., 1993). Other studies did not include optical characterization of the substrate (Vlckova et al., 1997; Gregory et al., 2001), which prevents any direct comparison of the excitation profiles to the spectral location of the LSPR λ_{max}. The most common substrates historically employed in SERES experiments are Ag island films and Ag colloidal solutions. In these cases, the majority of the SERS excitation profiles peak at an excitation wavelength of maximum enhancement ($\lambda_{ex,max}$) near 500–600 nm (Weitz et al., 1982; Kerker et al., 1984; Von Raben et al., 1984; Feilchenfeld and Siiman, 1986; Fornasiero and Grieser, 1987). The peaks of the excitation profiles have been shown to shift to longer wavelengths with increased aggregation (Blatchford et al., 1982; Kerker et al., 1984; Von Raben et al.,

1984; Fornasiero and Grieser, 1987), which is a qualitative result predicted by the EM mechanism. With these substrates it is difficult to make a direct comparison between the LSPR of the substrate and the SERS excitation profile because the LSPR of the substrate is a superposition of a wide variety of LSPR wavelengths corresponding to the different roughness features.

Two exceptions to the above statements regarding roughness features are the well-known experiments by Liao and co-workers on microlithographically prepared Ag posts (Blatchford et al., 1982) and recent work by Felidj and co-workers on e-beam lithographically produced arrays of gold elongated nanoparticles (Felidj et al., 2003). The former groundbreaking work demonstrated excitation profiles where $\lambda_{ex,max}$ shifts to longer wavelengths with increased particle aspect ratio and with increased dielectric constant of the medium surrounding the particles. These results qualitatively agree with the EM mechanism, but the LSPR of these substrates was not characterized for a direct comparison. In the latter work, the SERS enhancement was shown to peak at the midpoint between the excitation wavelength and the wavelength of the Raman-scattered photon. This experiment was the first observation of precisely what is predicted by the EM mechanism. Unfortunately, this result was obtained on one sample with a profile consisting of three data points.

The limitation of laser and detection tunability has been circumvented by several researchers using a unique approach that involves investigating substrates with variations in the spectral location of the LSPR λ_{max} (Oldenburg et al., 1999; Weimer and Dyer, 2001; Haynes and Van Duyne, 2003b). These variations allow investigation of the relationship between the LSPR and SERS enhancement using a single excitation wavelength. Our work using plasmon-sampled SERES (PS SERES) on well-defined arrays of nanoparticles was the first systematic study using this technique. PS SERES is a particularly attractive technique from a practical standpoint due to the fact that varying the LSPR λ_{max} of nanoparticles is typically far easier (and more cost effective) than accessing a variety of laser wavelengths. The conclusion of that study was that the condition for maximum enhancement occurred when the peak extinction wavelength of the LSPR, λ_{max}, is located near the midpoint between the energy of laser excitation and the energy of the Raman-scattered photons. This conclusion supports the EM mechanism, which predicts that maximum SERS intensity is achieved when the LSPR strongly enhances both the incident and scattered photon intensities. WS SERES allows for a more thorough study of the relationship between the LSPR and the Raman enhancement, and it affords the advantage of insuring that particle size and shape do not change throughout a given excitation spectrum.

The work reviewed below utilized a broadly tunable Raman system to measure excitation profiles with the greatest data point density ever demonstrated in a WS SERES experiment (McFarland et al., 2005). A broadly tunable laser system, a versatile detection system, and a well-characterized surface-enhancing substrate were all employed in order to overcome the traditional shortcomings of WS SERES experiments. The use of a CW-modelocked Ti:Sapphire and its harmonics allow for continuous tunability over the spectral ranges 350–500 nm and 700–1,000 nm. The visible region not covered by the Ti:Sapphire system was augmented with the use of a solid-state laser and a tunable dye laser. A triple spectrograph equipped with a CCD camera allows for rapid, multichannel spectral acquisition with efficient rejection of Rayleigh-scattered photons. The SERS substrates used in this work are triangular nanoparticle arrays fabricated by NSL. These substrates present a significant advantage over many of the traditional SERS substrates for SERES studies because NSL-fabricated triangular nanoparticles exhibit extremely narrow size distributions, making them an indispensable tool for probing the fundamental characteristics of SERS. Even though the surface coverage of these nanoparticles is $\sim 7\%$, strong SERS intensities are observed from analytes adsorbed to these substrates due to the strong enhancement (EF $\sim 10^8$) NSL-fabricated arrays exhibit (Haynes and Van Duyne, 2003b).

For this detailed set of WS SERES experiments performed on optically and topographically characterized SERS substrates, the relative SERS enhancement of the substrates has been shown to vary by three orders of magnitude over the spectral range investigated. It is worth noting that this was not principally a study on the practical application of SERS for chemical analysis. Factors such as spectrograph throughput, detector efficiency, and the ν^4 scattering dependence of Raman photons play an important role in the practice of Raman spectroscopy. Instead, this study was designed to contribute fundamental insights into the origins of the SERS effect and to test various aspects of the EM mechanism not previously studied. However, the conclusions reached are extremely important for the optimization of a surface-enhanced spectroscopy.

Figure 5 shows a schematic of the instrumentation used for the SERES experiments. All optical measurements were performed using an inverted microscope equipped with a $20\times$ objective (NA = 0.5). The light scattered by the samples was analyzed with a three-stage spectrograph equipped with a liquid nitrogen-cooled, deep-depletion CCD detector.

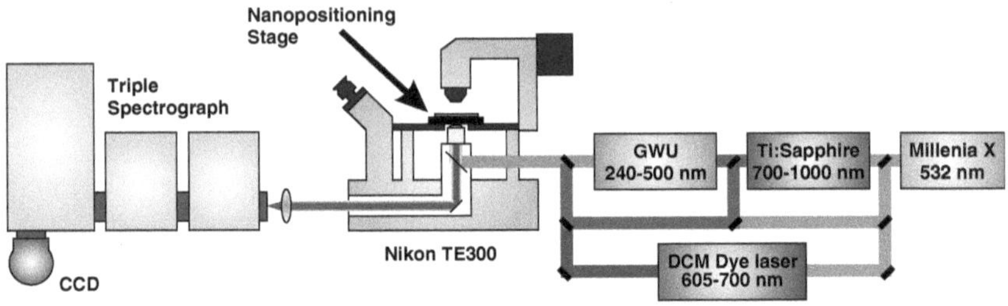

Fig. 5. Schematic diagram of the WS SERES apparatus.

For the NSL-fabricated triangular nanoparticles, in situ measurement of the LSPR spectrum was achieved by illuminating the sample with the microscope lamp and analyzing the transmitted light with a fiber-optically coupled miniature spectrometer. It is worthwhile to note that all illumination powers reported in this section were the laser powers incident on the microscope beamsplitter, not the power incident on the sample. Based on experimental measurements, approximately 5–10% of the reported power is incident on the sample; however, because of the intensity standard the absolute power at the sample is not a critical measurement.

In order to correct for any variation of the SERS intensity not due to the enhancement by the substrate, the $1,444\,\text{cm}^{-1}$ normal Raman scattering band of neat cyclohexane was used as an intensity standard. This standard was used to correct for the inherent ν^4 behavior of Raman scattering, spectral dependence of the detection system, and differences in the illumination power.

A representative SERS spectrum of benzenethiol on a Ag nanoparticle array is shown in fig. 6. An AFM image of the sample from which this spectrum was taken is shown in the inset. This array was fabricated by depositing 55 nm of Ag through a mask formed with 450 nm diameter nanospheres. Figure 7 shows four excitation profiles for the $1,575\,\text{cm}^{-1}$ peak of benzenethiol, each with an LSPR $\lambda_{\max}$ at a distinctly different location. The SERES profile in fig. 7(A) consists of 13 data points measured over the spectral range 420–500 nm. Because the formation of a monolayer of benzenethiol on these nanoparticle arrays results in a significant red shift in the position of the LSPR $\lambda_{\max}$, it was necessary to anneal this sample at 300°C for 1 h prior to benzenethiol addition in order to achieve a final LSPR $\lambda_{\max}$ at a wavelength shorter than 500 nm. It has been previously shown that annealing NSL-derived samples results in a large blue shift of the LSPR due to changing the shape of the nanoparticles (Jensen et al., 2000a). The LSPR $\lambda_{\max}$ of this substrate was

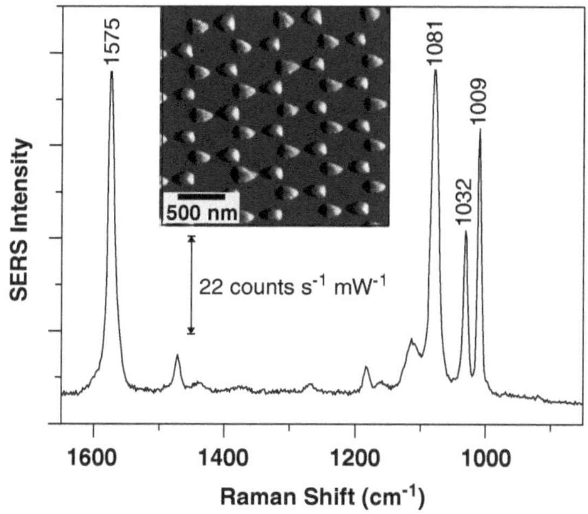

Fig. 6. Representative SERS spectrum of benzenethiol-dosed NSL substrate. $\lambda_{ex} = 620\,nm$, $P = 3.0\,mW$, acquisition time = 150 s. An atomic force micrograph of the sample is shown in the inset.

measured to be 489 nm ($20,450\,cm^{-1}$). The largest SERS enhancement occurs at $\lambda_{ex} = 485\,nm$. Fitting a Gaussian lineshape to the data reveals that the peak of the excitation profile, $\lambda_{ex,max}$, is 480 nm ($20,833\,cm^{-1}$). The peak EF value for this sample was calculated to be 5.5×10^5. This value is low in comparison to the values determined for the other samples because the shape of the nanoparticles is made more ellipsoidal by annealing. In addition to shifting the LSPR λ_{max} to shorter wavelengths, this change decreases the intensity of the electromagnetic fields at the nanoparticle surfaces.

The SERES profile in figs. 6 and 7(B) consists of 14 data points measured over the spectral range 532–690 nm. The LSPR λ_{max} of this substrate was measured to be 663 nm ($15,083\,cm^{-1}$). The largest SERS enhancement occurs for $\lambda_{ex} = 625\,nm$. The maximum of a Gaussian lineshape fit to the data is 625 nm ($16,000\,cm^{-1}$). The peak EF value for this sample is 1.2×10^7. The SERES profile in fig. 7(C) consists of 15 data points measured over the spectral range 532–740 nm. The LSPR λ_{max} of this substrate was measured to be 699 nm ($14,306\,cm^{-1}$). The largest SERS enhancement occurs for $\lambda_{ex} = 670\,nm$. The maximum of a Gaussian lineshape fit to the data is 671 nm ($14,903\,cm^{-1}$). The peak EF value for this sample is 1.4×10^7. The SERES profile in fig. 7(D) consists of 15 data points measured over the spectral range 630–800 nm. The LSPR λ_{max} of this substrate was measured to be 810 nm ($12,346\,cm^{-1}$). The largest SERS enhancement occurs for $\lambda_{ex} = 770\,nm$. The maximum of a

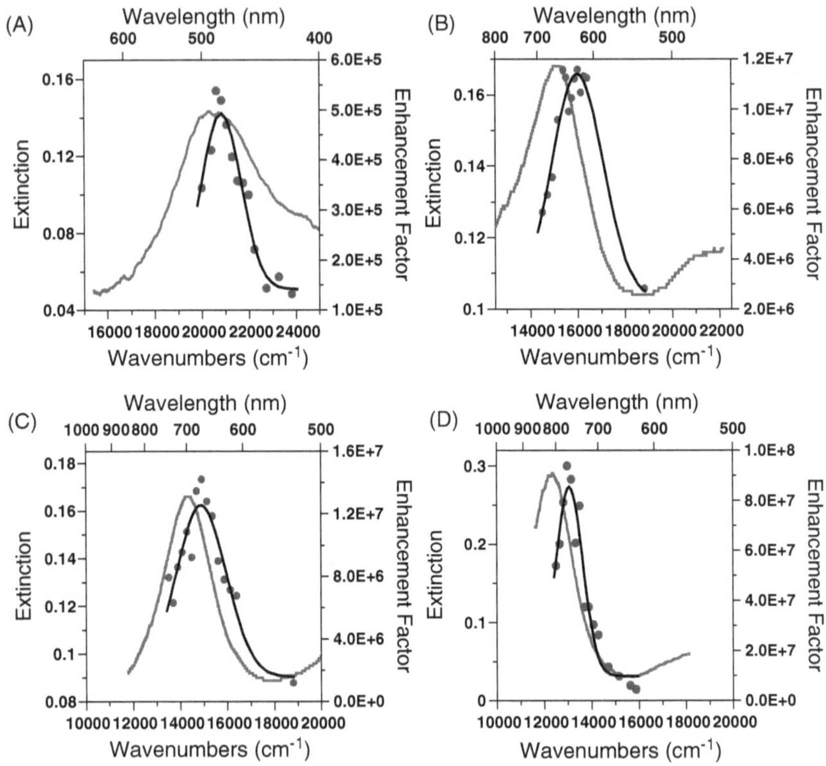

Fig. 7. Surface-enhanced Raman excitation spectra of the $1,575\,\text{cm}^{-1}$ peak of benzenethiol with cyclohexane as intensity standard. (A) Substrate annealed at 300°C for 1 h. LSPR $\lambda_{max} = 489$ nm, profile fit maximum at $\lambda_{ex,max} = 480$ nm. (B) LSPR $\lambda_{max} = 663$ nm, profile fit maximum at $\lambda_{ex,max} = 625$ nm. (C) LSPR $\lambda_{max} = 699$ nm, profile fit maximum at $\lambda_{ex,max} = 671$ nm. (D) LSPR $\lambda_{max} = 810$ nm, profile fit maximum at $\lambda_{ex,max} = 765$ nm.

Gaussian lineshape fit to the data is 765 nm ($13,072\,\text{cm}^{-1}$). The peak EF value for this sample is 9.3×10^7.

In order to verify that this behavior can be generalized, two SERES experiments were undertaken in which a different benzenethiol band ($1,081\,\text{cm}^{-1}$) and intensity standard were monitored. In this case, the intensity standard was the $520\,\text{cm}^{-1}$ phonon mode of silicon. The wavelength-dependent absorptivity of silicon requires that the measured Raman intensities must be corrected for differences in laser penetration depth. The penetration depth was calculated at all of the excitation wavelengths using the silicon absorptivities measured by Aspnes and Studna (Aspnes and Studna, 1983). The silicon spectra were then normalized so that the intensities were representative of equivalent probe volumes. In addition, a correction was performed to account for the fact that the 520 cm^{-1} band of Si scatters at a significantly different wavelength than the 1,081 cm^{-1} band of benzenethiol, particularly at redder excitation wavelengths. No correction was performed to account for variation in the Raman scattering cross section of silicon because over the range of excitation wavelengths utilized in this work, the differences in the experimentally determined values of the polarizability of silicon are negligible (Grimsditch and Cardona, 1980). The excitation spectra are shown in fig. 8. The SERES profile in fig. 8(A) consists of 13 data points

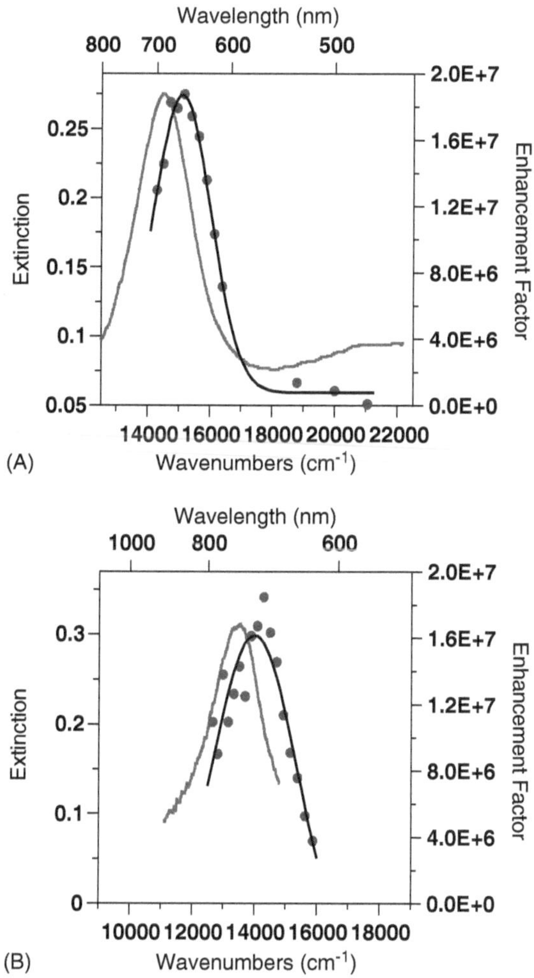

Fig. 8. Surface-enhanced Raman excitation spectra of the 1,081 cm^{-1} peak of benzenethiol with Si as intensity standard. (A) LSPR $\lambda_{max} = 690$ nm, profile fit maximum at $\lambda_{ex,max} = 662$ nm. (B) LSPR $\lambda_{max} = 744$ nm, profile fit maximum at $\lambda_{ex,max} = 715$ nm.

measured over the spectral range 475–700 nm. The LSPR λ_{max} of this substrate was measured to be 690 nm (14,493 cm^{-1}). The largest SERS enhancement occurs for $\lambda_{ex} = 660$ nm. The maximum of a Gaussian lineshape fit to the data is 662 nm (15,106 cm^{-1}). The peak EF value for this sample is 1.9×10^7. The SERES profile in fig. 8(B) consists of 17 data points measured over the spectral range 630–790 nm. The LSPR λ_{max} of this substrate was measured to be 744 nm (13,441 cm^{-1}). The largest relative SERS intensity occurs for $\lambda_{ex} = 700$ nm. The maximum of a Gaussian lineshape fit to the data is 715 nm (13,986 cm^{-1}). The peak EF value for this sample is 1.8×10^7.

Each substrate exhibits a SERES profile that has a similar lineshape to the extinction spectrum of the substrate. Also, the $\lambda_{ex,max}$ for the NSL-fabricated substrates is consistently a shorter wavelength than the LSPR λ_{max}. In all cases, the maximum SERS enhancement occurs when the substrate LSPR λ_{max} is located between λ_{ex} and λ_{vib}. Under these conditions, both the incident and scattered photons experience enhancement by the LSPR. This data is in accordance with the EM mechanism of SERS and the experimental work performed previously using PS SERES.

If the peak in the SERS enhancement occurs when the LSPR λ_{max} of the sample is approximately equal to $(\lambda_{ex} + \lambda_{vib})/2$, then $\lambda_{ex,max}$ should be different for the various Raman bands of benzenethiol on a single sample. It is expected that $\lambda_{ex,max}$ will have a larger separation from the LSPR λ_{max} for a large Raman shift than for a small shift. Excitation profiles for three benzenethiol peaks on a single substrate are shown in fig. 9. For this substrate, the LSPR λ_{max} is 729 nm. Figure 9(A) shows the SERS excitation profile for the 1,575 cm^{-1} peak of benzenethiol, normalized to the 1,444 cm^{-1} peak of liquid cyclohexane. The separation in wavenumbers between the LSPR λ_{max} and $\lambda_{ex,max}$ is 734 cm^{-1}. In fig. 9(B), the excitation profile for the 1,081 cm^{-1} benzenethiol peak (normalized to the 1,028 cm^{-1} peak of cyclohexane) is shown. The separation in wavenumbers between the LSPR λ_{max} and $\lambda_{ex,max}$ is 569 cm^{-1}. Finally, in fig. 9(C), the excitation profile for the 1,009 cm^{-1} benzenethiol peak (normalized to the 1,028 cm^{-1} peak of cyclohexane) is shown, and the separation in wavenumbers between the LSPR λ_{max} and $\lambda_{ex,max}$ is 488 cm^{-1}. This data demonstrates the qualitative trend whereby the $\lambda_{ex,max}$ in the excitation spectra of larger Raman shifted bands yield a larger separation from the LSPR λ_{max} than those of smaller Raman shifted bands, and this once again lends support to the EM mechanism.

As described above, the spectral location of the LSPR is extremely sensitive to the presence of molecular adsorbates (Malinsky et al., 2001; Haes

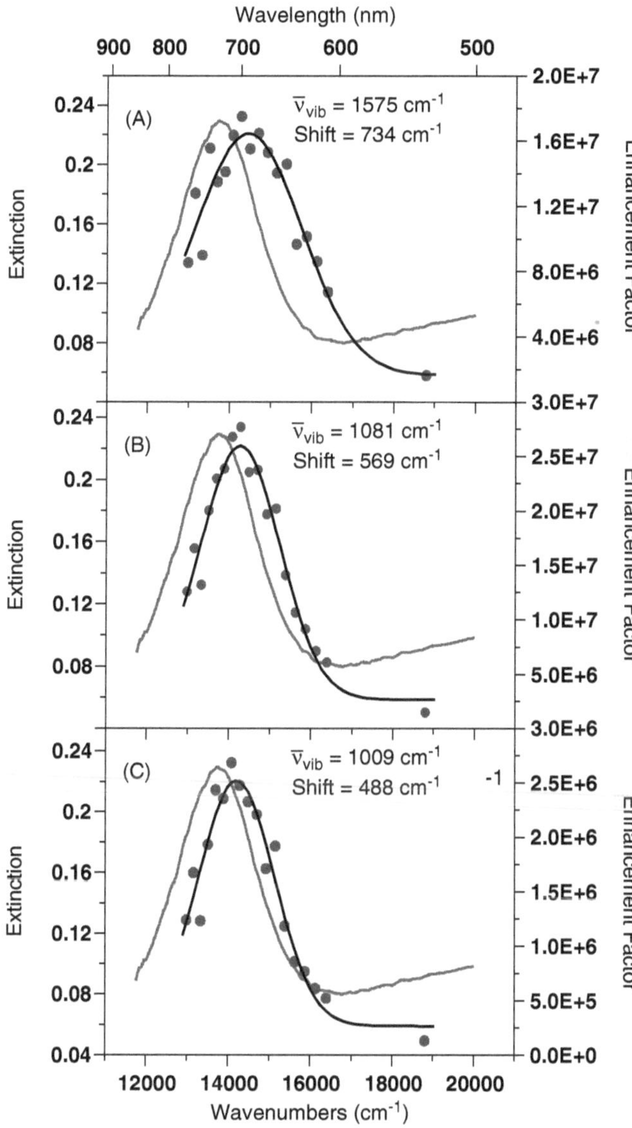

Fig. 9. Effect of Stokes Raman shift. (A) Profile of the $1,575\,\mathrm{cm}^{-1}$ vibrational mode of benzenethiol. Distance between LSPR $\lambda_{\max}$ and excitation profile fit line $\lambda_{\mathrm{ex,max}} = 734\,\mathrm{cm}^{-1}$. $\mathrm{EF} = 1.8 \times 10^7$. (B) $1,081\,\mathrm{cm}^{-1}$ vibrational mode, shift $= 569\,\mathrm{cm}^{-1}$, $\mathrm{EF} = 2.8 \times 10^7$. (C) $1,009\,\mathrm{cm}^{-1}$ vibrational mode, shift $= 488\,\mathrm{cm}^{-1}$, and $\mathrm{EF} = 2.7 \times 10^6$.

and Van Duyne, 2002; McFarland and Van Duyne, 2003). Therefore, it is important to note that the relationship between the LSPR spectra and SERES profiles depicted in figs. 7–9 pertain to LSPR spectra measured after adsorption of the analyte molecule. Figure 10 demonstrates the

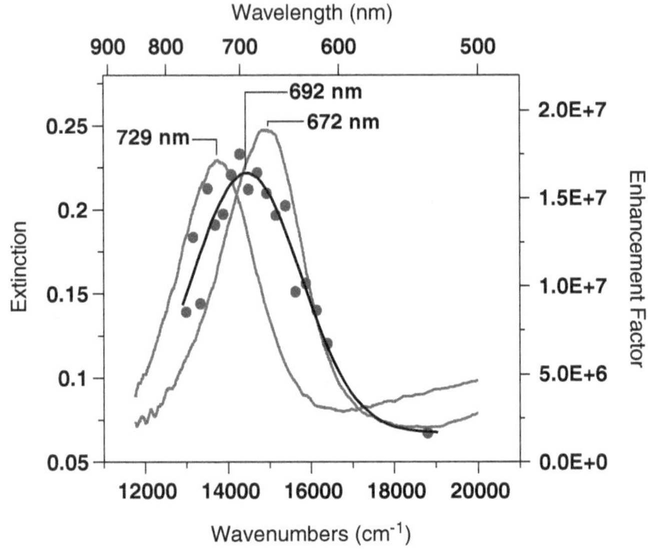

Fig. 10. LSPR shift and SERES profile for the $1,575\,\text{cm}^{-1}$ peak of benzenethiol. The line with $\lambda_{max} = 672\,\text{nm}$ is the LSPR extinction of the bare nanoparticle array. The line with $\lambda_{max} = 729\,\text{nm}$ is the LSPR extinction of the nanoparticle array with an adsorbed monolayer of benzenethiol. The line with $\lambda_{ex,max} = 692\,\text{nm}$ is the best fit to the SERES data points.

importance of considering this point. For a bare nanoparticle array, the LSPR λ_{max} was measured to be 672 nm. After incubation in 1 mM benzenethiol for more than 3 h, thorough rinsing with methanol, and drying, the LSPR λ_{max} was observed to have red-shifted by 57 nm to 729 nm. Measurement of the WS SERES profile yields $\lambda_{ex,max} = 692\,\text{nm}$. This is blue-shifted with respect to the LSPR λ_{max} of the adsorbate-covered sample, as observed for the other samples used in this study, but red-shifted with respect to the LSPR λ_{max} of the bare nanoparticle array. This demonstrates that it is critical to characterize the LSPR of a SERS substrate after analyte adsorption in order to choose the appropriate laser excitation wavelength for maximizing EF or to draw any conclusions about the fundamental mechanism of the SERS effect.

This work demonstrates the most thorough WS SERES experiments ever performed on optically and topographically characterized SERS substrates (McFarland et al., 2005). The experimental apparatus utilized has proven effective for the measurement of relative SERS enhancements that vary by three orders of magnitude. This work demonstrates that the relationship between the substrate LSPR and the SERES profile for size-homogenous nanoparticles is consistent throughout the visible range. In all cases, the experimentally observed behavior is consistent with that predicted by the EM mechanism. Specifically, the strongest SERS enhancement occurs under conditions where the incident and Raman-scattered photons are both strongly enhanced. The largest EF measured was $\sim 10^8$ for the triangular nanoparticle arrays studied. Ultimately, refinement of the experimental apparatus and optimization of SERS enhancement will allow SERES to be performed using single nanoparticle substrates. This level provides the best possible case in terms of reducing sample heterogeneity. These experiments are expected to provide key information to validate the EM mechanism of SERS and will present an additional technique that can be used to study the SMSERS effect.

§ 9. SERS enhancement factor calculation

SERS EF values were calculated by comparing the intensity of the appropriate benzenethiol peak (1,081 or $1,575\,\mathrm{cm}^{-1}$) measured in the SERS experiments to the corresponding peak measured from liquid benzenethiol. The normal Raman spectrum was measured at $\lambda_{\mathrm{ex}} = 480$, 532, and 670 nm from a 37-μm-thick cell filled with benzenethiol. When the intensities are normalized for laser power and acquisition time, under the same collection conditions, the SERS EF is given by

$$\mathrm{EF} = \frac{N_{\mathrm{vol}} I_{\mathrm{surf}}}{N_{\mathrm{surf}} I_{\mathrm{vol}}}, \tag{9.1}$$

where N_{vol} and N_{surf} are the number of molecules probed in the liquid sample and on the SERS substrates, respectively, and I_{vol} and I_{surf} the corresponding normal Raman and SERS intensities. Based on scanning silicon knife edge measurements, the beam waist of the 20× objective is 4.0 μm with $\lambda_{\mathrm{ex}} = 532\,\mathrm{nm}$. Assuming a benzenethiol packing density of 6.8×10^{14} molecules cm^{-2} and the 7.4% surface coverage of nanoparticles, approximately 6.3×10^6 molecules are probed on the triangular nanoparticle array substrates (N_{surf}). For the normal Raman experiment, the probe volume was approximated as a cylinder with a radius of 2.0 μm and a height of 37 μm, resulting in 2.7×10^{12} molecules being probed (N_{vol}). The values for N_{surf} and N_{vol}, along with the SERS intensities for the 1,081 and $1,575\,\mathrm{cm}^{-1}$ peaks (measured with one of the excitation wavelengths listed above), were substituted into eq. (9.1) to calculate the EFs. The peak EF values were calculated by compensating for the differences in the relative SERS intensity measured at $\lambda_{\mathrm{ex,max}}$ vs. an excitation wavelength at which a liquid benzenethiol spectrum was measured. Figure 11 shows the spectra necessary for calculating the peak EF for a

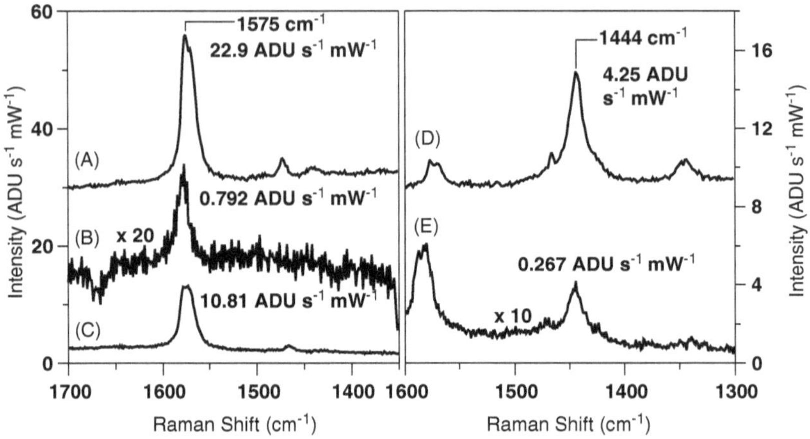

Fig. 11. Example of EF calculation. (A) SERS spectrum of benzenethiol on a nanoparticle array at $\lambda_{ex} = 670\,nm$. (B) Raman spectrum of neat benzenethiol at $\lambda_{ex} = 670\,nm$. (C) SERS spectrum of benzenethiol on the same sample as (A) at $\lambda_{ex} = 770$ (the peak of the excitation profile). (D) Raman spectrum of cyclohexane at $\lambda_{ex} = 670\,nm$. (E) Raman spectrum of cyclohexane at $\lambda_{ex} = 770\,nm$.

single SERE spectrum. This example pertains to the spectrum where $\lambda_{max} = 810\,nm$ and $\lambda_{ex,max} = 765\,nm$ (fig. 7(D)). Figure 11(A) shows the SERS spectrum of benzenethiol on a nanoparticle array taken at $\lambda_{ex} = 670\,nm$. The $1,575\,cm^{-1}$ peak intensity is $22.9\,ADU\,s^{-1}\,mW^{-1}$ (I_{surf}). Figure 11(B) shows the normal Raman spectrum of benzenethiol in the 37.0-μm-thick cell at $\lambda_{ex} = 670\,nm$. The $1,575\,cm^{-1}$ peak intensity is $0.792\,ADU\,s^{-1}\,mW^{-1}$ (I_{vol}). When substituted into the EF equation, these values yield an EF of 1.24×10^7. This EF is for $\lambda_{ex} = 670$, but because the maximum normalized SERS intensity for this profile was $\lambda_{ex} = 770\,nm$, this value needs to be multiplied by an appropriate scale factor to calculate the peak EF. The normalized SERS intensity at $\lambda_{ex} = 670\,nm$ is calculated by dividing the SERS intensity of the $1,575\,cm^{-1}$ benzenethiol peak in fig. 11(A) ($22.9\,ADU\,s^{-1}\,mW^{-1}$) by the normal Raman intensity of the $1,444\,cm^{-1}$ peak of cylcohexane at the same excitation wavelength, fig. 11(D) ($4.25\,ADU\,s^{-1}\,mW^{-1}$), yielding a relative intensity of 5.39. The normalized SERS intensity at $\lambda_{ex} = 770\,nm$ is calculated by dividing the intensity in fig. 11(C) ($10.8\,ADU\,s^{-1}\,mW^{-1}$) by that in fig. 11(E) ($0.267\,ADU\,s^{-1}\,mW^{-1}$), which yields a relative intensity of 40.4. Therefore, the normalized SERS intensity at $\lambda_{ex} = 770\,nm$ is larger than that at $\lambda_{ex} = 670$ by a factor of $40.4/5.39 = 7.50$. Multiplying the EF value at $\lambda_{ex} = 670$ by this factor yields a peak EF of 9.3×10^7. This value, the largest peak EF in this study, was measured from the substrate with the longest-wavelength LSPR λ_{max}. The values for EF reported in this paper are conservative estimates because the probe volume assumed for the normal Raman experiment is likely to be an underestimation of the actual value. Also, the calculated number of adsorbate molecules is most likely an overestimation because the calculation assumes perfect array formation. The nanoparticle arrays contain a small number of defects, which can be assumed to contribute little to the observed characteristics of the SERES profiles. If Raman scattering from the defects were contributing significantly, it is unlikely that the systematic trends demonstrated by this work would be observed. Even with consideration of these points, the peak EF calculated for the triangular nanoparticle arrays is consistent with the average values reported in other experimental work using these same substrates (Haynes and Van Duyne, 2003b) and theoretical modeling of triangular Ag nanoparticles (Hao and Schatz, 2004).

§ 10. SERS distance dependence by atomic layer deposition

The distance dependence of SERS is both a critical mechanistic and practical issue. The electromagnetic mechanism predicts that SERS is a long-range effect that does not require the adsorbate to be in direct contact with the surface; but, does require it to be within a few nanometers of the surface. In contrast, the chemical mechanism predicts that SERS is a short-range effect that requires the adsorbate to be chemisorbed directly on the surface. From a practical perspective, there are certain experiments, such as those involving surface-immobilized biological molecules, where direct contact between the adsorbate of interest and the surface must be prevented for the experiment to succeed. Additionally, in many cases TERS experiments rely solely on EM enhancement and thus the distance dependence of the enhancement is of immediate interest. In all of the SERS distance dependence studies published to date, randomly roughened surfaces (viz., island films, cold-deposited films in UHV, etc.) were used and the Raman-active adsorbate was spaced from the surface using a polymer film, cold-condensed molecular layer, or self-assembled monolayer. Each of these spacer methods has its limitations. The ideal distance dependence experiment would be one in which the thickness of the spacer layer could be easily varied in thickness from a few angstroms to a hundred nanometers. Furthermore, the spacers would be conformal to handle roughened and nanostructured surfaces, pinhole free, and chemically uniform. ALD is just such a spacer fabrication method that produces highly uniform and controlled thin films, as described above.

In this work we have deposited Al_2O_3 multilayers onto AgFON surfaces to probe the distance dependence of SERS. In previous work, the distance dependence of SERS has been theoretically approximated as

$$I = \left(1 + \frac{r}{a}\right)^{-10}, \tag{10.1}$$

where I is the Raman intensity, a the average size of the field-enhancing features on the surface and r the distance from the surface to the adsorbate. Figure 12(A) shows the SER spectra for pyridine adsorbed on AgFON surfaces coated with four different thicknesses of ALD deposited Al_2O_3. A plot of the relative intensity of the $1,594\,cm^{-1}$ band as a function of Al_2O_3 thickness is shown in fig. 12(B). Fitting the experimental data to eq. (10.1) leads to the average size of the enhancing particle $a = 12.0\,nm$. The term d_{10} defines the surface to molecule distance required to decrease the SERS intensity by a factor of 10. The data presented in this work clearly show that SERS is a long-range effect with a d_{10} value for this particular surface nanostructure of 2.8 nm.

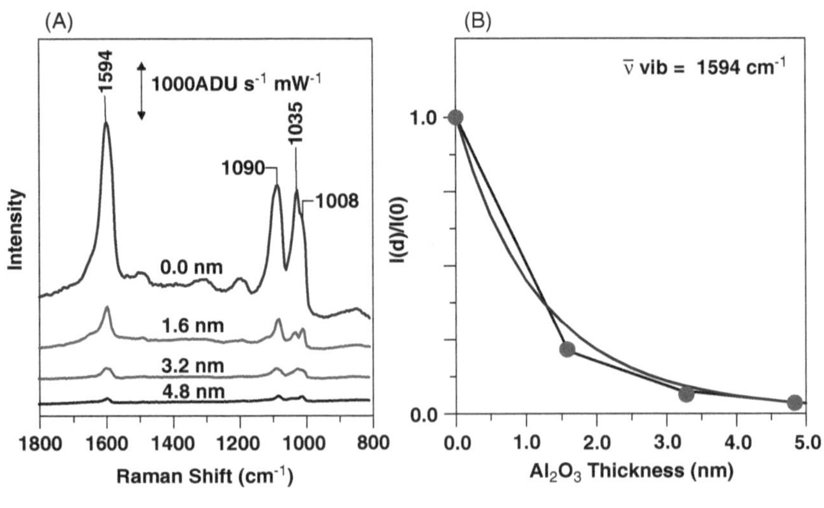

Fig. 12. (A) SERS spectra of pyridine adsorbed to silver film over nanosphere (AgFON) samples treated with various thicknesses of alumina (0.0, 1.6, 3.2, and 4.8 nm). $\lambda_{ex} = 532\,nm$, $P = 1.0\,mW$, acquisition time = 300 s. (B) Plot of SERS intensity as a function of alumina thickness for the $1,594\,cm^{-1}$ band (filled circles and straight line segments). The solid curved line is a fit of this data to eq. (9.1).

§ 11. 2D correlation analysis of SMSERS and single nanoparticle SERS data

The SERS community has been immersed in debate concerning the mechanism of SMSERS since the first reports of the phenomenon in 1997 (Kneipp et al., 1997; Nie and Emory, 1997). Correlated LSPR spectroscopy (Michaels et al., 1999), polarization measurements (Bosnick et al., 2002), and nanoparticle size-dependent experiments (Emory et al., 1998) have all been carried out in order to understand SMSERS better, but none have uncovered the elusive details of the mechanism that leads to the enormous EFs of $10^{14} - 10^{15}$ that are observed. General opinion is that nanoparticle junctions may provide the extraordinary field enhancement required to detect single molecules, but even this theory has not been proven conclusively. Figure 13(A) shows the spectral variations observed in six SMSERS spectra collected from Rhodamine 6G (R6G) bound to chemically synthesized Ag nanoparticles. This data demonstrates how the

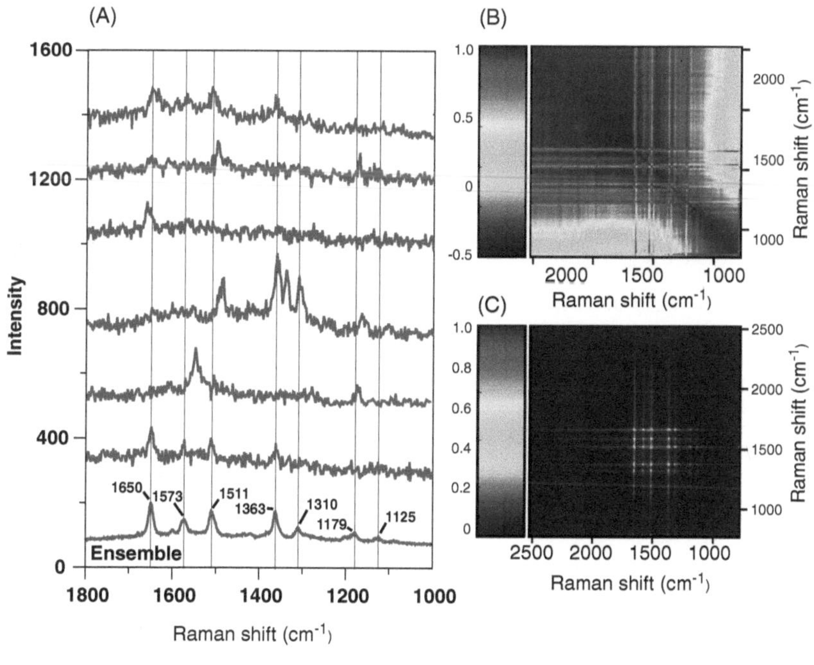

Fig. 13. (A) Temporal fluctuation of the spectral features for SMSERS of R6G on chemically synthesized Ag nanoparticles. The ensemble SERS spectrum is included to indicate the prominent Raman peaks of R6G. (B) Correlation plot of SMSERS of R6G. (C) Correlation plot of SERS of R6G at higher coverage (>100 molecules nanoparticle^{-1}) on single-Ag nanoparticles.

relative intensities of the individual vibrational peaks of R6G vary dramatically between spectra. This temporal fluctuation of the scattering spectrum has been considered a signature of single-molecule detection. In certain instances only a subset of the R6G peaks is visible, and in others spurious peaks are present. These spurious peaks may be due to extreme spectral wandering of the R6G peaks, but is much more likely due to SERS of photodegradation products of R6G (Emory et al., 2001). It is important to note that under the experimental conditions used, the excitation wavelength ($\lambda_{ex} = 514.5\,nm$) is located within the electronic absorption band of R6G. Resonant excitation could account for up to three orders of magnitude enhancement in the observed SMSERS intensity and also facilitate the photodecomposition of R6G. For comparison, SERS of R6G adsorbed onto Ag nanoparticles at relatively high coverage (>100 molecules nanoparticle^{-1}) was also investigated. Even under these conditions, temporal fluctuation of the scattering spectrum was observed and analysis confirmed the detection of R6G. The differences between the behaviors observed for SMSERS and that for single-nanoparticle SERS is best demonstrated by correlation analysis of the two data sets (Nie and Emory, 1997). The correlation diagrams for these two experiments are shown in figs. 13(B) and 13(C). The off-diagonal features represent the degree of correlation between individual Raman modes throughout the temporal fluctuation of the spectrum. In these diagrams, a correlation value of 1 signifies strong correlation, a value of zero signifies no correlation, and a value of (-1) signifies strong anticorrelation. In terms of the spectral data, strong correlation indicates that two Raman peaks tend to appear and disappear together, no correlation indicates that there is no relationship between the appearance or disappearance of the two peaks, and anticorrelation indicates that features do not appear simultaneously. The lack of strong correlation between peaks in the SMSERS data (i.e., the absence of well-defined features off-diagonal) indicates that the R6G peaks fluctuate independently of one another. This suggests that different molecular orientations of adsorption or different hot spot geometries may favor SERS of different modes. This behavior supports the conclusion that the observed spectra are due to the SERS of single-R6G molecules. In contrast, the correlation analysis for the single-nanoparticle data indicates strong correlation among all the R6G spectral features. This analysis reveals that there are clear differences between the temporal fluctuations observed in SMSERS spectra and those obtained from single nanoparticles with higher analyte coverage. We believe that a more detailed analysis of the correlation behavior of SMSERS spectral features may provide further insight into the mechanism underlying SMSERS.

§ 12. Tip-enhanced Raman scattering

One of the most severe restrictions in the application of SERS to a wide variety of problems in ultrahigh vacuum (UHV) single-crystal surface science, electrochemistry, heterogeneous catalysis, microelectronics, and tribology is the requirement that the surface be roughened or nanostructured Ag, Au, or Cu. Weaver broadened the applicability of SERS using a transition metal overlayer approach for studying heterogeneous catalysis. The overlayer approach involves creating nanoscale surface roughness features on Au and subsequently electrodepositing a thin layer of catalytic metal on top of these features (Zou and Weaver, 1998; Wasileski et al., 2000; Park et al., 2002). The overlayer provides the catalytic activity while the underlying substrate maintains the SERS activity. Systems studied using these substrates include methanol oxidation on rhodium and palladium as well as formic acid electrooxidation on platinum group metals. Van Duyne used a similar approach involving pulsed laser deposition of platinum over AgFON surfaces in order to broaden the molecular generality of SERS (Freunscht et al., 1997).

A new technique, TERS (Anderson, 2000; Hayazawa et al., 2000, 2002, 2003; Stockle et al., 2000; Pettinger et al., 2002, 2004; Ren et al., 2004; Watanabe et al., 2004; Mehtani et al., 2005; Anderson et al., 2005a, 2005b; Rasmussen and Deckert, 2006), has recently been developed that can, in principle, circumvent this limitation and achieve complete substrate generality for SERS. TERS uses a scanning probe microscopy (SPM) tip as an external enhancing unit. Because the enhancement originates from an external source, this technique makes significantly enhanced Raman signals accessible for any adsorbate on any surface. When a noble metal SPM tip is irradiated with light that excites the tip LSPR, electromagnetic field enhancement takes place near the tip apex. This enhancement can be used to increase the Raman scattering intensity from molecules in the vicinity of the tip, even though they are not directly adsorbed to it, due to the long-range nature of the interaction between the molecules and the enhanced electromagnetic field. SPM tips made from or coated with Ag or Au have been shown to enhance Raman scattering intensities of molecules near the tip by 3 to 6 orders of magnitude (Pettinger et al., 2002, 2004; Watanabe et al., 2004).

An additional advantage afforded by TERS is its nanometer scale spatial resolution. TERS combines the high sensitivity and rich chemical information content of SERS with the excellent spatial resolution provided by SPM. TERS can provide Raman spectra with a spatial resolution near the radius of curvature of the SPM tip, which is far better than the diffraction limit typically encountered in optical microscopy. For this reason, TERS has been referred to as apertureless near-field Raman spectroscopy. TERS has been used to provide high-resolution images of nanoscale samples, and since the light detected is Raman scattering, chemical information is built in.

TERS is a relatively new technique and its substantial potential is largely untapped. TERS shows promise as an important analytical technique in the areas of nanoscale chemical analysis, nanotribology, molecular electronics, heterogeneous catalysis, and SM SERS studies. We are beginning to explore the utility of TERS in the field of nanotribology – the study of friction, wear, and pressure at the nanoscale. Below we present the first study of the force dependence of TER spectra from SAMs. From the tribology perspective, the SAM is viewed as a model lubricant. The role of lubricants in determining mechanical properties is critical for modern technologies including hard disk drives, microelectromechanical systems (MEMS), and many others.

Figure 14 shows a schematic diagram of the TERS apparatus used for the results reported here. The laser excitation light is delivered using the inverted microscope objective, and the Raman-scattered light is collected through the same objective. The AFM scan head is mounted directly on

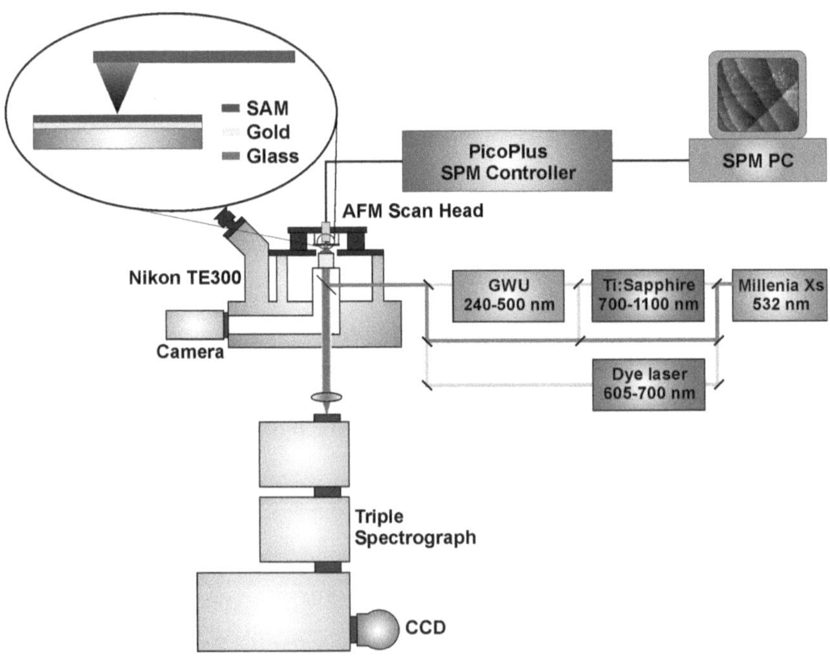

Fig. 14. Schematic diagram of the TERS setup.

the inverted microscope, and contact mode feedback is used to bring the tip into contact with the sample. The Raman-scattered light is collected through a triple spectrograph in order to block the incident laser and the AFM laser. The SAM samples were formed on a thin (8 nm) layer of Au deposited on a glass cover slip.

In fig. 15, representative TER spectra are shown for benzenethiol adsorbed on an 8 nm Au island film that is very weakly SERS active at $\lambda_{ex} = 532$ nm. In fig. 15(A), the AFM tip, coated with 40 nm Ag, is engaged and makes contact with the sample yielding a high signal-to-noise ratio TER spectrum. In fig. 15(B), the tip is withdrawn from the sample by 100 μm, with the laser spot still focused on the sample. Figure 15(C) shows the same high signal-to-noise ratio TER spectrum when the AFM tip is re-engaged in contact mode. Finally, fig. 15(D) shows a spectrum with the laser focused on the AFM tip after the sample has been removed. This spectrum was acquired after several engage–disengage cycles and serves to show that benzenethiol molecules do not transfer from the Au film to the AFM tip. Thus, the majority of the signal originates from molecules that are not themselves adsorbed to the tip. Taking the difference in surface areas between the laser spot, $\sim 1\ \mu m^2$, and the contact area of the AFM tip, $\sim 300\ nm^2$, into account, the average enhancement factor for each of the Raman modes is estimated to be $> 10^4$. These high signal-to-noise spectra unambiguously demonstrate significant enhancement brought on by the presence of the Ag-coated AFM tip.

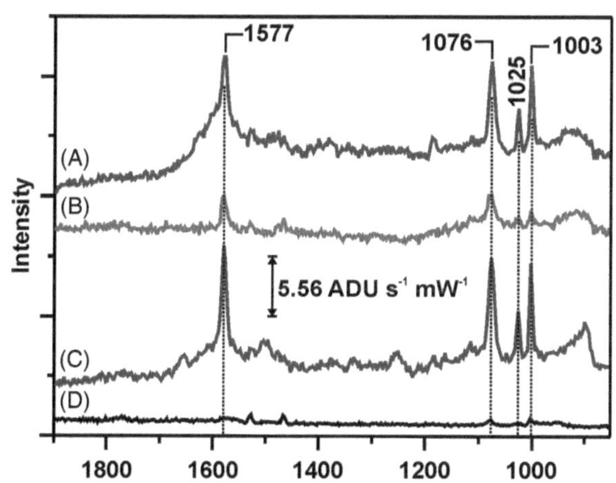

Fig. 15. TERS of a SAM of benzenethiol adsorbed to an 8-nm Au film on glass. AFM tip is coated with 40 nm Ag. Excitation $\lambda = 532$ nm. (A) AFM tip engaged in contact mode. (B) AFM tip withdrawn by 100 μm. (C) AFM tip re-engaged in contact mode. (D) Laser focused solely on the AFM tip after the sample has been removed.

§ 13. TERS force dependence using AFM

Figure 16 presents a data set that shows the effect of tip force on the TER spectra. An AFM tip, coated with 15 nm of Ag, was brought into contact with a SAM of benzenethiol, again on an 8 nm Au island film on glass. The tip was scanned over the monolayer at a constant speed, and the force was increased from a starting value of 19 nN to a final tip force of 391 nN. Throughout this force range, the TERS spectra acquired are of high signal-to-noise ratio. The TERS intensities show a weak dependence of force. Over the range 0–50 nN, the TERS intensity decreases by 10–15% of its low force value. This is followed by the 50–400 nN range where the TERS intensity is approximately force independent. It is also observed that there are no significant changes in vibrational frequency over the entire 0–400 nN range. Clearly much more work is needed to

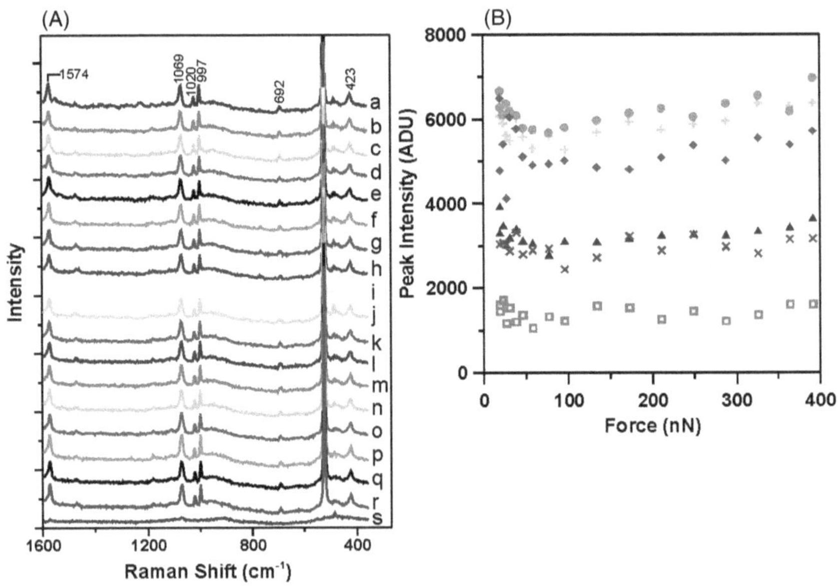

Fig. 16. (A) Force dependence of TERS spectra of a SAM of benzenethiol with the AFM tip scanning in contact mode at a constant speed (2.0 lines s^{-1}) and varying force. The very intense peak at 520 cm^{-1} comes from the Si AFM tip. (a) tip in contact, not scanning, (b) tip scanning, force = 19 nN, (c) 23 nN, (d) 27 nN, (e) 31 nN, (f) 38 nN, (g) 46 nN, (h) 58 nN, (i) 77 nN, (j) 96 nN, (k) 134 nN, (l) 173 nN, (m) 211 nN, (n) 249 nN, (o) 288 nN, (p) 326 nN, (q) 364 nN, (r) 391 nN, (s) tip withdrawn 100 μm. (B) Graph of the peak intensity of each of the benzenethiol Raman modes as a function of tip force. • denotes the 1,069 cm^{-1} peak, + denotes the 997 cm^{-1} peak, ◆ denotes the 1,574 cm^{-1} peak, ▲ denotes the 423 cm^{-1} peak, × denotes the 1,020 cm^{-1} peak, and □ denotes the 692 cm^{-1} peak. The intensity of each of the peaks remains relatively constant over this force range.

further investigate the force dependence of TERS. At present our working hypothesis is that the decrease in TERS intensity in the 0–50 nN range is due to reversible orientation changes of the benzenethiol molecules in the vicinity of the tip.

§ 14. Conclusion and outlook

We have detailed the fabrication and characterization of plasmonic materials that show promise in sensing as well as surface- and tip-enhanced Raman spectroscopy. NSL is a simple approach for making nanoparticle arrays that are very well-defined in shape and size, and have continuously tunable LSPRs throughout the visible spectrum. Investigation of the distance dependence of the LSPR has shown that the $\Delta\lambda_{max}$ for NSL-derived nanoparticles is sensitive to a single layer of Al_2O_3 deposited by ALD. At short distances from the Ag nanoparticle surface, there is a much larger LSPR λ_{max} red shift than is seen at distances further from the nanoparticle surface. In addition, nanoparticles with fixed in-plane widths and decreasing out-of-plane heights have larger sensing distances, which is the same trend that was observed for alkanethiol layers at larger distances from the nanoparticle surface.

The WS SERES results reviewed herein are the most detailed ever performed on optically and topographically characterized SERS substrates. The SERS enhancement factor has been shown to vary significantly as a function of laser wavelength for substrates with LSPR λ_{max} values throughout the visible spectrum. The consistent conclusion, verifying the electromagnetic mechanism prediction, is that the SERS enhancement factor is maximized when both the incident laser and Raman-scattered photons are strongly enhanced. This occurs when the incident laser is on the higher energy side of λ_{max}, and the Raman shift is on the lower energy side. Additionally, the electromagnetic mechanism prediction concerning the effect of the Stokes shift on SERES has been verified. Specifically, the data demonstrates that a peak with a smaller Raman shift shows a maximum enhancement closer in energy to the LSPR λ_{max} than a peak with a larger Raman shift. The conditions for maximum SERS enhancement are determined after the addition of the analyte molecule because of the significant shift in the LSPR λ_{max} caused by analyte adsorption. The adsorbate-induced LSPR shift and SERES Stokes shift are in opposite directions. The largest EF measured was $\sim 10^8$ for the triangular nanoparticle arrays studied, which is in accordance with theory and previous experimental work.

The distance dependence of SERS on a AgFON has been probed using ALD, and the distance at which the SERS intensity decreased by a factor of 10 was found to be 2.8 nm. TERS is an ideal tool for probing the distance dependence of the EM component of SERS, and this will be the subject of future studies.

Two-dimensional correlation analysis for SMSERS and ensemble averaged SERS was reviewed. The data showed a lack of correlation among the various R6G Raman modes for SMSERS, and strong correlation among all the R6G spectral features for the ensemble averaged case. This kind of analysis is a powerful check on whether SMSERS is indeed being achieved, and it is expected to provide further insight into the mechanism underlying SMSERS.

TERS has great potential to provide material, surface and crystallographic generality for SERS. TERS has a currently demonstrated spatial resolution of $\sim 15\,\mathrm{nm}$ (Anderson et al., 2005a, 2005b). Our work demonstrates high signal-to-noise ratio TER spectra for a benzenethiol monolayer on a Au island film using a Ag-coated AFM tip. An $EF_{\mathrm{TERS}} > 10^4$ is demonstrated. The force dependence of TERS was studied for the benzenethiol/Au island film system. Over the range 0–50 nN the TERS intensity decreased by 10–15% and was force independent from 50 to 400 nN. No force dependent vibrational frequency shifts were observed.

In order for TERS to become a routine technique for near-field chemical analysis, it will be necessary to optically characterize the tips used for TERS. According to the WS SERES experiments reviewed here, the best case for a TERS experiment will be when the LSPR $\lambda_{\max}$ is located in between the excitation laser wavelength and the wavelength of the Stokes Raman shift of interest. TERS is expected to be an important area in the emerging field of molecular plasmonics with important applications in analytical nanoscale analysis. In particular, TERS has great potential for enabling new applications in UHV surface science, heterogeneous catalysis, electrochemistry, tribology, silicon microelectronics, molecular electronics, as well as many other areas.

Acknowledgments

The authors gratefully acknowledge support from the Air Force Office of Scientific Research MURI program (Grant F49620-02-1-0381), the National Science Foundation (EEC-0118025, DMR-0076097, DMR-0520513, CHE-0414554, DGE-0114429), and the U.S. Department of Energy (DE-FG02-03ER15457).

References

Anderson, M.S., 2000, Appl. Phys. Lett. **76**, 3130.

Anderson, N., Hartschuh, A., Cronin, S., Novotny, L., 2005a, J. Am. Chem. Soc. **127**, 2533.

Anderson, N., Hartschuh, A., Novotny, L., 2005b, Mater. Today (Oxford United Kingdom) **8**, 50.

Andrew, P., Barnes, W.L., 2004, Science (Washington, DC, United States) **306**, 1002.

Aspnes, D.E., Studna, A.A., 1983, Phys. Rev. B **27**, 985.

Barnes, W.L., Dereux, A., Ebbesen, T.W., 2003, Nature (London, United Kingdom) **424**, 824.

Blatchford, C.G., Campbell, J.R., Creighton, J.A., 1982, Surface Sci. **120**, 435.

Bosnick, K.A., Jiang, J., Brus, L.E., 2002, J. Phys. Chem. B **106**, 8096.

Brockman, J.M., Nelson, B.P., Corn, R.M., 2000, Ann. Rev. Phys. Chem. **51**, 41.

Campion, A., Kambhampati, P., 1998, Chem. Soc. Reviews **27**, 241.

Dick, L.A., McFarland, A.D., Haynes, C.L., Van Duyne, R.P., 2002, J. Phys. Chem. B **106**, 853.

Emory, S.R., Ambrose, W.P., Goodwin, P.M., Keller, R.A., 2001, Proc. SPIE-The Int. Soc. Opt. Eng. **4258**, 63.

Emory, S.R., Haskins, W.E., Nie, S., 1998, J. Am. Chem. Soc. **120**, 8009.

Feilchenfeld, H., Siiman, O., 1986, J. Phys. Chem. **90**, 2163.

Felidj, N., Aubard, J., Levi, G., Krenn, J.R., Hohenau, A., Schider, G., Leitner, A., Aussenegg, F.R., 2003, Appl. Phys. Lett. **82**, 3095.

Felidj, N., Laurent, G., Aubard, J., Levi, G., Hohenau, A., Krenn, J.R., Aussenegg, F.R., 2005, J. Chem. Phys. **123**, 221103/1.

Flatau, P.J., Stephens, G.L., Draine, B.T., 1990, J. Opt. Soc. Am. A **7**, 593.

Fornasiero, D., Grieser, F., 1987, J. Chem. Phys. **87**, 3213.

Freunscht, P., Van Duyne, R.P., Schneider, S., 1997, Chem. Phys. Lett. **281**, 372.

Gersten, J., Nitzan, A., 1980, J. Chem. Phys. **73**, 3023.

Goodman, J.J., Draine, B.T., Flatau, P.J., 1991, Opt. Lett. **16**, 1198.

Gregory, B.W., Clark, B.K., Standard, J.M., Avila, A., 2001, J. Phys. Chem. B **105**, 4684.

Grimsditch, M., Cardona, M., 1980, Phys. Status Solidi B **102**, 155.

Groner, M.D., Fabreguette, F.H., Elam, J.W., George, S.M., 2004, Chem. Mater. **16**, 639.

Haes, A.J., Chang, L., Klein, W.L., Van Duyne, R.P., 2005, J. Am. Chem. Soc. **127**, 2264.

Haes, A.J., Van Duyne, R.P., 2002, J. Am. Chem. Soc. **124**, 10596.

Haes, A.J., Van Duyne, R.P., 2004, Anal. Bioanal. Chem. **379**, 920.

Haes, A.J., Zou, S., Schatz, G.C., Van Duyne, R.P., 2004a, J. Phys. Chem. B **108**, 109.

Haes, A.J., Zou, S., Schatz, G.C., Van Duyne, R.P., 2004b, J. Phys. Chem. B **108**, 6961.

Halteen, J.C., Van Duyne, R.P., 1995, J. Vac. Sci. Technol. A **13**, 1553.

Hao, E., Schatz, G.C., 2004, J. Chem. Phys. **120**, 357.

Hayazawa, N., Inouye, Y., Sekkat, Z., Kawata, S., 2000, Optics Commun. **183**, 333.

Hayazawa, N., Inouye, Y., Sekkat, Z., Kawata, S., 2002, J. Chem. Phys. **117**, 1296.

Hayazawa, N., Yano, T., Watanabe, H., Inouye, Y., Kawata, S., 2003, Chem. Phys. Lett. **376**, 174.

Haynes, C.L., McFarland, A.D., Van Duyne, R.P., 2005b, Anal. Chem. **77**, 338A.

Haynes, C.L., McFarland, A.D., Zhao, L., Van Duyne, R.P., Schatz, G.C., Gunnarsson, L., Prikulis, J., Kasemo, B., Kaell, M., 2003, J. Phys. Chem. B **107**, 7337.

Haynes, C.L., Van Duyne, R.P., 2001, J. Phys. Chem. B **105**, 5599.

Haynes, C.L., Van Duyne, R.P., 2003a, Nano Lett. **3**, 939.

Haynes, C.L., Van Duyne, R.P., 2003b, J. Phys. Chem. B **107**, 7426.

Haynes, C.L., Yonzon, C.R., Zhang, X., Van Duyne, R.P., 2005a, J. Raman Spectrosc. **36**, 471.

Hulteen, J.C., Van Duyne, R.P., 1995, J. Vac. Sci. Technol. A **13**, 1553.

Ichimura, T., Hayazawa, N., Hashimoto, M., Inouye, Y., Kawata, S., 2004a, Appl. Phys. Lett. **84**, 1768.
Ichimura, T., Hayazawa, N., Hashimoto, M., Inouye, Y., Kawata, S., 2004b, Phys. Rev. Lett. **92**, 220801/1.
Jeanmaire, D.L., Van Duyne, R.P., 1977, J. Electroanal. Chem. Interfacial Electrochem. **84**, 1.
Jensen, T.R., Duval, M.L., Kelly, K.L., Lazarides, A.A., Schatz, G.C., Van Duyne, R.P., 1999, J. Phys. Chem. B **103**, 9846.
Jensen, T.R., Malinsky, M.D., Haynes, C.L., vanDuyne, R.P., 2000a, J. Phys. Chem. B **104**, 10549.
Jensen, T.R., Van Duyne, R.P., Johnson, S.A., Maroni, V.A., 2000b, Appl. Spectrosc. **54**, 371.
Jung, L.S., Campbell, C.T., Chinowsky, T.M., Mar, M.N., Yee, S.S., 1998, Langmuir **14**, 5636.
Kerker, M., 1987, Stud, Phys. Theor. Chem. **45**, 3.
Kerker, M., Siiman, O., Wang, D.S., 1984, J. Phys. Chem. **88**, 3168.
Kneipp, K., Wang, Y., Kneipp, H., Perelman, L.T., Itzkan, I., Dasari, R.R., Feld, M.S., 1997, Phys. Rev. Lett. **78**, 1667.
Kreibig, U., Vollmer, M., 1995. Optical Properties of Metal Clusters. Springer Series in Materials Science 25. Springer, Heidelberg, 552 pp.
Lezec, H.J., Degiron, A., Devaux, E., Linke, R.A., Martin-Moreno, L., Garcia-Vidal, F.J., Ebbesen, T.W., 2002, Science **297**, 820.
Liao, P.F., Bergman, J.G., Chemla, D.S., Wokaun, A., Melngailis, J., Hawryluk, A.M., Economou, N.P., 1981, Chem. Phys. Lett. **82**, 355.
Litorja, M., Haynes, C.L., Haes, A.J., Jensen, T.R., Van Duyne, R.P., 2001, J. Phys. Chem. B **105**, 6907.
Maier, S.A., Brongersma, M.L., Kik, P.G., Meltzer, S., Requicha, A.A.G., Atwater, H.A., 2001, Advanced Mater. (Weinheim, Germany) **13**, 1501.
Maier, S.A., Kik, P.G., Atwater, H.A., Meltzer, S., Harel, E., Koel, B.E., Requicha, A.A.G., 2003, Nature Mater. **2**, 229.
Malinsky, M.D., Kelly, K.L., Schatz, G.C., Van Duyne, R.P., 2001, J. Am. Chem. Soc. **123**, 1471.
McFarland, A.D., Van Duyne, R.P., 2003, Nano Lett. **3**, 1057.
McFarland, A.D., Young, M.A., Dieringer, J.A., Van Duyne, R.P., 2005, J. Phys. Chem. B **109**, 11279.
Mehtani, D., Lee, N., Hartschuh, R.D., Kisliuk, A., Foster, M.D., Sokolov, A.P., Maguire, J.F., 2005, J. Raman Spectrosc. **36**, 1068.
Metiu, H., Das, P., 1984, Annu. Rev. Phys. Chem. **35**, 507.
Michaels, A.M., Nirmal, M., Brus, L.E., 1999, J. Am. Chem. Soc. **121**, 9932.
Moskovits, M., 1979, Solid State Commun. **32**, 59.
Moskovits, M., 2005, J. Raman Spectrosc. **36**, 485.
Nie, S., Emory, S.R., 1997, Science (Washington, D. C.) **275**, 1102.
Oldenburg, S.J., Westcott, S.L., Averitt, R.D., Halas, N.J., 1999, J. Chem. Phys. **111**, 4729.
Ott, A.W., Klaus, J.W., Johnson, J.M., George, S.M., 1997, Thin Solid Films **292**, 135.
Ozbay, E., 2006, Science (Washington, DC, United States) **311**, 189.
Park, S., Yang, P., Corredor, P., Weaver, M.J., 2002, J. Am. Chem. Soc. **124**, 2428.
Pettinger, B., Picardi, G., Schuster, R., Ertl, G., 2002, Single Mol. **3**, 285.
Pettinger, B., Ren, B., Picardi, G., Schuster, R., Ertl, G., 2004, Phys. Rev. Lett. **92**, 096101/1.
Philpott, M.R., 1983. J. Phy. Colloque 295.
Rasmussen, A., Deckert, V., 2006, J. Raman Spectrosc. **37**, 311.
Ren, B., Picardi, G., Pettinger, B., Schuster, R., Ertl, G., 2004, Angewandte Chemie Int. Ed. **44**, 139.

Riboh, J.C., Haes, A.J., McFarland, A.D., Yonzon, C.R., Van Duyne, R.P., 2003, J. Phys. Chem. B **107**, 1772.
Ritala, M., Leskela, M., 2002, Handbook of Thin Film Mater. **1**, 103.
Saito, Y., Hayazawa, N., Kataura, H., Murakami, T., Tsukagoshi, K., Inouye, Y., Kawata, S., 2005, Chem. Phys. Lett. **410**, 136.
Schatz, G.C., 1984, Acc. Chem. Res. **17**, 370.
Schatz, G.C., Van Duyne, R.P., 2002, Electromagnetic mechanism of surface-enhanced spectroscopy, in: Chalmers, J.M., Griffiths, P.R. (Eds.), Handbook of Vibrational Spectroscopy, vol. 1, Wiley, New York, p. 759.
Schultz, S., Smith, D.R., Mock, J.J., Schultz, D.A., 2000, Proc. Natl. Acad. Sci. USA **97**, 996.
Smith, L.H., Wasey, J.A.E., Barnes, W.L., 2004, Appl. Phys. Lett. **84**, 2986.
Srituravanich, W., Fang, N., Sun, C., Luo, Q., Zhang, X., 2004, Nano Lett. **4**, 1085.
Stockle, R.M., Suh, Y.D., Deckert, V., Zenobi, R., 2000, Chem. Phys. Lett. **318**, 131.
Tian, Z.Q., 2005, J. Raman Spectrosc. **36**, 466.
Van Duyne, R.P., 2004, Science (Washington, DC, United States) **306**, 985.
Van Duyne, R.P., Hulteen, j.C., Treichel, D.A., 1993, J. Chem. Phys. **99**, 2101.
Vlckova, B., Gu, X.J., Moskovits, M., 1997, J. Phys. Chem. B **101**, 1588.
Von Raben, K.U., Chang, R.K., Laube, B.L., Barber, P.W., 1984, J. Phys. Chem. **88**, 5290.
Wasileski, S.A., Zou, S., Weaver, M.J., 2000, Appl. Spectrosc. **54**, 761.
Watanabe, H., Ishida, Y., Hayazawa, N., Inouye, Y., Kawata, S., 2004, Phys. Rev. B **69**, 155418/1.
Wedge, S., Wasey, J.A.E., Barnes, W.L., Sage, I., 2004, Appl. Phys. Lett. **85**, 182.
Weimer, W.A., Dyer, M.J., 2001, Appl. Phys. Lett. **79**, 3164.
Weitz, D.A., Garoff, S., Gramila, T.J., 1982, Opt. Lett. **7**, 168.
Whitney, A.V., Elam, J.W., Zou, S., Zinovev, A.V., Stair, P.C., Schatz, G.C., Van Duyne, R.P., 2005, J. Phys. Chem. B **109**, 20522.
Yang, W.-H., Schatz, G.C., Van Duyne, R.P., 1995, J. Chem. Phys. **103**, 869.
Yguerabide, J., Yguerabide, E.E., 1998, Anal. Biochem. **262**, 157.
Yonzon, C.R., Jeoung, E., Zou, S., Schatz, G.C., Mrksich, M., Van Duyne, R.P., 2004, J. Am. Chem. Soc. **126**, 12669.
Zou, S., Weaver, M.J., 1998, Anal. Chem. **70**, 2387.

Chapter 2

Towards single molecule sensitivity in surface-enhanced Raman scattering

by

M. Futamata

Nanoarchitectonics Research Center, National Institute of Advanced Industrial Science and Technology, 1-1-1 Higashi, Tsukuba 305-8562, Japan

Y. Maruyama

Tsukuba Research Laboratory, Hamamatsu Photonics K.K., Tsukuba 300-2635, Japan

Tip Enhancement
Advances in Nano-Optics and Nano-Photonics
ISSN: 1871-0018

S. Kawata and V.M. Shalaev (Editors)
© 2007 Elsevier B.V. All rights reserved.
DOI: 10.1016/S1871-0018(06)01002-8

Contents

§ 1. Introduction

It is crucial to elucidate the nature of constituents at an atomic or molecular scale in order to realize a bottom-up nanotechnology, e.g. a single molecule device or a nano-chemical factory. Surface-enhanced Raman scattering (SERS) available under ambient conditions even at the solid/liquid interfaces attracts increasing interest according to the potentialities of single molecule sensitivity (SMS) and also nanoscale spatial resolution when combined with near-field microscopy. SERS is a well-known phenomenon for about 30 years (Kerker, 1990; Otto et al., 1992) that Raman signal from adsorbates on roughened metal surfaces is enhanced by a factor of 10^4–10^6 due to excitation of surface plasmon polariton (SPP) on roughened metals and/or due to the 'first layer' enhancement including charge transfer resonance between adsorbates and metals. The enhancement factor obtained in the early stage of SERS history was a population-averaged value for enormous molecules sitting on numerous Ag particles with various shapes and sizes, and thus obviously insufficient to yield single molecule detection (SMD) in SERS. Recent progress in scanning probe microscopy (SPM) in addition to a highly sensitive charge-coupled device (CCD) detector enables us to observe extremely weak spectral signal from an individual metal particle or even from single molecule. Accordingly, several groups reported vast enhancement in SERS, e.g. 10^{14}, corresponding to SMS by probing only 'hot' particles with appropriate shapes and sizes (Kneipp et al., 1999; Krug et al., 1999; Maruyama et al., 2000; Michaels et al., 2000; Xu et al., 2000), whereas mechanism of the enormous enhancement is still obscure. Obviously this is because there are no straightforward methods to elucidate electronic and vibrational spectra of individual molecules at local environment on metal nanostructures, e.g. spatial resolution of near-field spectroscopy is still at ca. 10 nm relevant for such purposes. Some groups claimed the importance of junction for SMS-SERS based on the fact that only touching metal nanoparticles give enormous enhancement and 'blinking' of the signal that intensity suddenly and repeatedly alters with time as predicted by theoretical simulation for the local electric field. In contrast, Nie et al. observed even isolated Ag or Au nanoparticles showed the blinking, which were immobilized on the substrate and followed by activation with halide ions (Emory and Nie, 1998; Doering and Nie, 2003).

As is well known, valuable information is extracted from Raman spectra such as a molecular structure, orientation or interaction with neighboring species. In contrast, fluorescence spectroscopy precedent as an SMD method yields less informative broad spectra albeit useful to identify individual molecules. Thus, it is challenging to establish the SMD method with SERS in terms of application to various fields, such as elucidation of elementary reaction process at solid/liquid interfaces. The followings are commonly observed concerning the vast enhancement in SERS (Kneipp et al., 1999; Krug et al., 1999; Maruyama et al., 2000; Michaels et al., 2000; Xu et al., 2000): (1) only a small number of 'hot particles' emerge with prominent enhancement when surface coverage of adsorbates is between ca. several tens and one hundred per each Ag particle. (2) At lower surface coverage $<$ ca. 1 molecule/particle, the blinking of Raman signal is observed for merely trace amounts of Ag particles with dye or other biomolecules like hemoglobin or DNA bases. The blinking is believed to be a single molecule phenomenon, since it is not plausible for many molecules to move or to interchange their orientation at the same time. Although there are no direct methods to prove it, various supporting evidences were reported. For instance, Nie et al. reported intermittent properties, i.e. complete ON/OFF features of the blinking rationalized only by motion of individual molecules (Emory and Nie, 1998; Doering and Nie, 2003). Kneipp et al. found Poisson distribution of the SERS intensity for extremely low surface coverage (Kneipp et al., 1999), which is quantized intensity corresponding to a discrete number of adsorbed molecules in the sampled volume, e.g. null, one or two molecules, in contrast to Gaussian distribution at much higher coverage. Concerning this, Etchegoin and coworkers addressed more direct evidences were found that the SERS signal from Ag nanoparticles is dominated by a few molecules (Le Ru et al., 2006). A large number of dye molecules, i.e. ca. 1,200 mixed molecules per particle, assured most of sites are fully occupied in their experiments. Nevertheless, SERS spectra observed from the Ag particles diffusing in solution were reproduced with the Poisson distribution assuming an average number of 4 molecules/signal. Similar but significantly different approach has been continued by Aorca et al. using Langmuir Blodgett monolayer on evaporated Ag island films, in which perylene dye (PD) dispersed in arachidic acid (AA) with sufficiently low concentration ratio, showed the blinking features (Constantino et al., 2002; Goulet et al., 2005). At concentration between 10^3:1 and 10^5:1 for AA:PD, the breakdown of ensemble averaging was observed.

Surprisingly, the enhancement, most probably electromagnetic enhancement, is only 3.64×10^7 (Otto, 2006). Details on contribution of the resonance effect is not always clear, although the value itself agrees with that estimated based on Shalaev's theory, 10^6–10^7 with probability of 0.003 (Otto, 2006). These works remind us the number of active sites per Ag particle is 3.3, which was estimated by adsorption isotherm (Hildebrandt and Stockburger, 1984; Nie and Emory, 1997; Jiang et al., 2003; Sharaabi et al., 2005). It is essentially associated with the origin of the enormous enhancement, particularly to the charge transfer interaction at hot sites, further study is prospected to get deeper insight.

The anti-Stokes scattering of dyes showed much larger intensity than that expected under thermal equilibrium, which was first attributed to optical pumping of vibrational state due to enormously large local electric field (Kneipp et al., 1997). Maher et al. recently reported that: (1) the pumping was actually observed at 10 K, at which thermal excitation of vibrational modes above 600 cm^{-1} was suppressed, based on the fact that anti-Stokes bands at 610 cm^{-1} and 1,650 cm^{-1} have comparable intensity in contrast to extremely large difference in Boltzmann factors for these modes (Maher et al., 2006a). In addition, (2) the vibrational pumping was also ensured for the Raman bands above 1,000 cm^{-1} at higher laser power (Maher et al., 2006b). The latter point seems to be consistent with the former one, since the temperature range, 100–350 K, allows to monitor the pumping for such higher frequency modes by suppressing thermal excitation (Maher et al., 2006a).

It should be noted that apparent SERS intensity fluctuation may occur during photochemical reaction of organic adsorbates using even rather weak laser irradiation under electronic resonances (Kudelski and Pettinger, 2000). Therefore, experiments of the blinking in SERS signal should be performed under precisely controlled conditions to suppress such phenomena. Recently, Weiss and Haran (2001) reported that the rate of spectral fluctuation is proportional to laser intensity and suggested a nonthermal process such as photo-induced desorption of adsorbates. Bosnick et al. (2002) reported the SERS scattering intensity fluctuates due to motions of the molecule in and out of the hot spot and is highly localized around it. Also, thermal activation of the blinking was evidenced by the temperature dependence that the intensity fluctuation is suppressed at 77 K as described in Section 3.3 (Maruyama et al., 2004), which is compatible with the photo-induced molecular motion for particular molecules with appropriate electronic state. In addition, local electric field intensity evaluated by a numerical method, e.g. finite difference time domain (FDTD) method, yields vast enhancement of $>10^{10}$

at the junction of touching Ag particles with various sizes and shapes as in Section 3.2 (Futamata et al., 2003) in contrast to modest enhancement of 10^4–10^5 at other ordinary sites or on isolated particles. Similar results were obtained with using different numerical simulations, e.g. finite element method (Kottmann et al., 2001), multipole expansion and boundary charge method (Xu et al., 2000), discrete dipole analysis (Hao and Schatz, 2004). Thus, the observation of frozen molecules at low temperature indicates that the blinking is due to thermal process, most probably due to thermal diffusion of adsorbate molecules from the junctions to other ordinary sites on Ag particles.

Furthermore, in order to verify the above attributions, it is crucial to specify the adsorbed site of individual dye molecules on Ag nanoparticles. Recently, significant spectral changes were observed in elastic scattering during inactivation process of SERS from hot or blinking Ag particles (Itoh et al., 2003; Futamata et al., 2004a).[1] Adsorbed molecules sitting at the junction of Ag particles could account for vast enhancement and distinct spectral variation in elastic scattering. In the former paper (Futamata et al., 2004a), two-dimensional FDTD method (FDTD-2D) was utilized to evaluate the effect of dye adsorption to an Ag nanowire in the elastic scattering spectra. Although the essential features were obtained with such calculations, 3D analysis provides more accurate electromagnetic distribution, especially precise wavelength dependence in the near- and far-field for the complicate nanostructures. Accordingly, FDTD-3D adopted here in Section 3.4 yields detailed information on electromagnetic coupling between dye and localized surface plasmon (LSP) of Ag nanoparticles that determines the excitation profile of SERS.

At last, we would like to introduce the fabrication of metal nanostructures for SMS-SERS, which might be crucial to utilize this novel technique. In our knowledge, silver or gold nanoparticles prepared by chemical reduction of $AgNO_3$ or $HAuCl_4$ have been exploited for most of SMD or vast enhancement in SERS experiments (Kerker, 1990; Emory and Nie, 1998; Kneipp et al., 1999; Krug et al., 1999; Maruyama et al., 2000, 2004; Michaels et al., 2000; Xu et al., 2000; Weiss and Haran, 2001; Bosnick et al., 2002; Doering and Nie, 2003). This is probably due to experimental feasibility to explore the optimum metal nanoparticles with particular morphology or sizes. However, it has not been established to fabricate such optimum nanoparticles with sufficiently high yield by the chemical reduction, since the process is only macroscopically controlled

[1] Similar spectral changes for much larger amount of molecules on several tens of Ag nanoparticles were reported in aqueous media by Itoh et al. (2003).

through the reaction temperature or mixing speed. Indeed, variety of silver particles with different shapes and sizes are formed by means of citrate salt as a reducing chemical, while $NaBH_4$ provide homogeneous, isolated spherical silver particles with modest enhancement. Obviously, only scarce metal nanoparticles show the blinking among huge number of particles prepared by the citrate method (Kerker, 1990; Emory and Nie, 1998; Kneipp et al., 1999; Krug et al., 1999; Maruyama et al., 2000; Michaels et al., 2000; Xu et al., 2000; Constantino et al., 2002; Doering and Nie, 2003; Goulet et al., 2005; Otto, 2006). Chemically etched metal surfaces could provide much higher occurrence for the blinking in SERS as suggested by Doering and Nie (2003). On the other hand, SERS spectrum is strongly dependent on the nanoscale morphology at the active sites that are not precisely controlled with the etching conditions. Therefore, more efficient fabrication methods for the metal nanostructures with optimum morphology for SMS-SERS are requisite. As an alternative way, the nanosphere lithography (NSL) was pioneered by Van Duyne et al. (Jensen et al., 1999; Dick et al., 2000, 2002; Haynes and Van Duyne, 2001, 2003b) to control the LSP resonance with respect to the efficient SERS-active substrates and optical sensors. Of late, Van Duyne group reported detailed excitation profiles of SERS under plasmon resonances for the trigonal metal nanostructures, e.g. pronounced enhancement of ca. 10^9 observed for $Fe(bpy)_3$-$(PF_6)_2$ or other molecules with/without using electronic resonance effect (Haynes and Van Duyne, 2003a; Dieringer et al., 2006), which proves the prevalence of NSL. In addition, the trigonal nanostructure, in principle, has sharp edges which possibly yield enormous electric field under the LSP resonance as anticipated by the numerical simulation (Xu et al., 2000; Kottmann et al., 2001; Futamata et al., 2003). If the preparation condition is optimized to give sufficiently sharp nanostructures, it can be quite valuable to fabricate nanodevices with SMS-SERS (Yamaguchi et al., 2005). Another approach to collect SERS active Ag colloidal particles has been performed in 2D or 3D by Svedberg and Käll (Svedberg and Käll, 2006). The Ag naoparticles were optically trapped with the near IR laser (830 nm and 200 mW) using the axial gradient force dominant the dissipative force, generating a potential-well that is deep enough to overcome thermal energy. They succeeded to monitor gradual growth of SERS signal from adsorbed thiol monolayer with the accumulation of Ag nanoparticles (Svedberg and Käll, 2006). This method could provide a new insight into the electromagnetic coupling of metal nanoparticles, while it will also be useful to characterize dynamical properties of biomolecules such as protein at various positions in a living cell. Electron beam lithography has been installed to construct metal nanostructures as a 2D array with optimized morphology for SERS applications. In this method, a metal layer is evaporated on the structured resist, which was assembled by the electron beam irradiation followed by chemical development, and the resist is finally removed by using organic solvent to leave metal nanostructures. At present, the spatial resolution is still at about 10 nm, which is not sufficient to fabricate closely adjacent nanostructures or sharp-edged particles. Accordingly, the LSP of isolated structures or weakly coupled one are utilized, while convincing their SERS enhancement or LSP resonances predicted by numerical simulations. For instances, Félidj et al. fabricated oblate Au nanoparticles (aspect ratio of 10, $a = b = 200$ nm, $c = 20$ nm with a spacing of 200 nm) that yield SERS enhancement of 10^8 for pyridine-like molecules at the excitation wavelength of 674 nm (Félidj et al., 2004). More precisely controlled metal nanostructures will be formed in the near future with using improved spatial resolution. Another notable result was reported by Dickson and coworkers that Ag_2–Ag_8 nanoclusters (NC), consisting of 2 or 8 Ag atoms, which are formed and stabilized in dendrimer or oligopeptide molecules, provide scaffold-specific Raman scattering on strong Ag_n emission with intermittent nature (Peyser-Capadona et al., 2005; Zheng and Dickson, 2002). Antibunching enabled them to conclude the emission originated from a single AgNC. Simultaneously, it enhances Raman signal from the single molecule. Details on the electronic interaction between host molecules and AgNC have not been unambiguously characterized. However, it can be an alternative method to yield SMS in Raman scattering, if various molecules show the identical properties.

§ 2. Experiments and numerical analysis

2.1. Experimental setup for SERS measurement

2.1.1. Ag nanoparticles preparation

$AgNO_3$ was chemically reduced using excess amount of sodium citrate (Lee and Meisel, 1982), and mixed with adsorbates such as rhodamine 6G (R6G) or DNA bases and NaCl in aqueous solution with particular concentration to control the surface coverage. Then, the Ag particles were immobilized onto 3-aminopropyl trimethoxysilane (APTMS) covered Si substrate using a spin-coater. Here NaCl substitutes residual citrate anions on the Ag particles with Cl^- anions, which is essential for cationic dye to adsorb onto the Ag particles with electrostatic force. It was evidenced by an occasional appearance of SERS bands from citrate anions instead of dye at modest concentration of NaCl. Only a countable number of Ag particles are located in each sampled area on the Si substrate, which are separated with a grid of 50 μm × 50 μm, by adjusting a concentration of Ag-dispersed solution. It enables us to detect Raman spectra from individual Ag particles, since a separation (ca. several micrometer) between neighboring Ag particles is much larger than the laser beam diameter at the sample position (ϕ ca. 1 μm). Additional optics for external configuration in Raman spectroscopy, a highly sensitive CCD camera with an image intensifier, a notch filter to observe Raman images and a precisely adjustable X–Y mechanical stage (with an accuracy of 1 μm in both directions) were installed to a conventional microscope (Renishaw© Ramascope, see fig. 1). These are actually crucial to detect SERS images and SERS spectra from individual Ag particles between room temperature (RT) and 77 K, while compensating mechanical drift of sampled positions. Elastic scattering images and spectra of the individual Ag particles were observed with a CCD camera using white light source (Xe lamp) and a mask for dark field illumination as shown with the dotted line in fig. 1, whereas Raman images were obtained under an external geometry using Ar^+ laser at 488 nm (full line). Back-scattering geometry was used to measure Raman spectra with the same Ar^+ ion laser through a microscope objective and a polychromator containing a set of notch filters and a CCD detector (see broken line).

Sufficiently weak laser power, i.e. 1 μW μm^{-2} for Raman imaging and 70 μW μm^{-2} (recently, we found even 1 μW μm^{-2} provides equivalent results) for Raman spectra was used, which did not yield any indication of

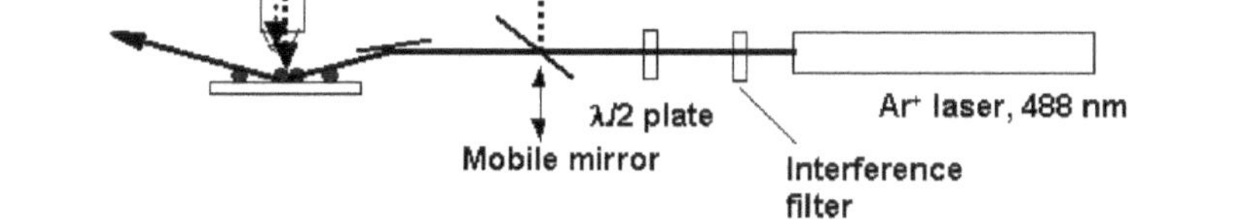

Fig. 1. Experimental setup for optical imaging, LSP extinction spectra, Raman imaging and Raman spectral measurement based on a Raman microscope. Ag particles with adsorbates are immobilized onto the APTMS-coated glass or Si substrates located under the objective (× 50).

photochemical reaction of constituents. With using a grid-like marker, we can correlate each Ag particle in the atomic force microscope (AFM) images, optical (LSP Raman and scattering) images and spectra, which enables us to explore the optimum Ag nanostructures.

2.2. *Numerical analysis of the local electric field and elastic scattering spectra for metal nanostructures*

Maxwell equations are solved for spheres or ellipsoids with or without substrate (Aravind and Metiu, 1983; Liver et al., 1984; Chew and Kerker, 1985; Kerker, 1990). However, analytical solutions have not been obtained for other complicated structures such as triangular, tetrahedral particles with/without unsymmetrical protrusions or pit, because isolation of variables in the differential equations is substantially difficult. Therefore, the numerical simulation such as the FDTD method is valuable, which transfers the differential equations to difference equations. The stationary solutions can be obtained at given positions for the initial electromagnetic field with various wavelengths and propagating directions. In the FDTD method, the nanostructure is surrounded by virtual boundaries with an appropriate size, and inside of the area is separated into small rectangular meshes with a particular size ($\Delta x \times \Delta y$ meshes, see fig. 2, here 2D expression is given to simplify the explanation). Namely, the metal particles and surroundings are depicted as a collection of these small meshes with a particular size and dielectric properties (Yee, 1966). In this system, curl in the Maxwell equations is given by

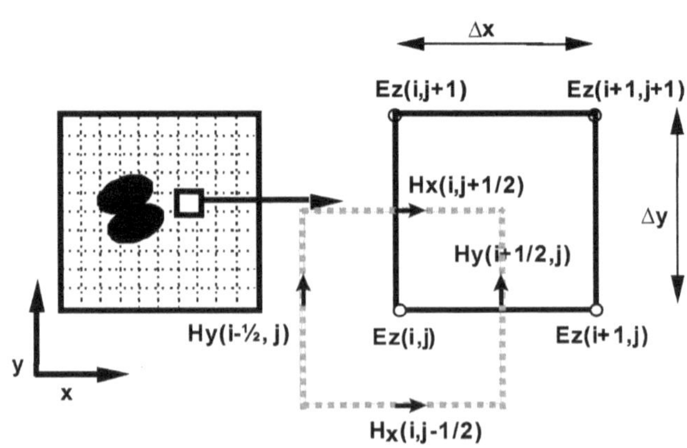

Fig. 2. Schematic image of a sample system and coordinate used in the FDTD calculation.

the following equations for TM (transverse magnetic) field (Taflove, 2000).

$$E_z^n(i,j) = C_{EZ}(i,j)E_z^{n-1}(i,j) + C_{EZLX}(i,j)\{H_y^{n-1/2}(i+1/2,j) - H_y^{n-1/2}(i-1/2,j)\} - C_{EZLY}(i,j)\{H_x^{n-1/2}(i,j+1/2) - H_y^{n-1/2}(i,j-1/2)\},$$

here

$$C_{EZ}(i,j) = \frac{1-(\sigma(i,j)\Delta t)/(2\varepsilon(i,j))}{1+(\sigma(i,j)\Delta t)/(2\varepsilon(i,j)},$$

$$C_{EZLX}(i,j) = \frac{\Delta t/\varepsilon(i,j)}{1+(\sigma(i,j)\Delta t)/(2\varepsilon(i,j))}\frac{1}{\Delta x}$$

$$C_{EZLY}(i,j) = \frac{\Delta t/\varepsilon(i,j)}{1+(\sigma(i,j)\Delta t)/(2\varepsilon(i,j))}\frac{1}{\Delta y}$$

$$H_x^{n+1/2}(i,j+1/2) = H_x^{n-1/2}(i,j+1/2) - C_{HXLY}(i,j+1/2)\{E_z^n(i,j+1) - E_z^n(i,j)\}$$

$$H_y^{n+1/2}(i+1/2,j) = H_y^{n-1/2}(i+1/2,j) - C_{HYLX}(i+1/2,j)\{E_z^n(i+1,j) - E_z^n(i,j)\}$$

$$C_{HXLY}(i,j+1/2) = \frac{\Delta t}{\mu(i,j+1/2)}\frac{1}{\Delta y},$$

$$C_{HYLX}(i+1/2,j) = \frac{\Delta t}{\mu(i+1/2,j)}\frac{1}{\Delta x}.$$

where $\sigma(i,j)$, $\varepsilon(i,j)$ and μ are conductivity, dielectric constant, and magnetic permeability, respectively. According to these equations, the local electromagnetic field E^{n-1} and $H^{n-1/2}$ at orthogonal coordinates $x = x(i)$, $y = y(j)$ and at time $t = t^{n-1}$, $t^{n-1/2}$ are calculated, and then the $H^{n+1/2}$ and E^n as a sequential and time-evolutional response to the incident electromagnetic field. The first order Mur scheme or PML (perfectly matched layer) was used as an absorbing boundary condition in conjunction with a recursive convolution method for metals with prominent dielectric dispersion in wavelength region studied here (Taflove, 2000).

In the FTDTD-2D simulation in Section 3.2, metal nanostructures were supposed to be a nanowire with infinite height (length) perpendicular to the cross section with various shapes, e.g. circle, triangle or ellipsoid. Local electric field was evaluated using the maximum field intensity on metal surfaces with using a sufficiently small mesh size, e.g. 0.25 nm × 0.25 nm or 0.1 nm × 0.1 nm, where variation of the field is rapid at junctions or surfaces of particles, contrary to less dense mesh size for outside the particles with 1 nm × 1 nm, typically. Such inhomogeneous meshes are quite useful and efficient to obtain the accurate values with rather short time especially in 3D evaluations. Actual calculation was performed for metal nanoparticles with various sizes, shapes, and ordered structures. We noted that the accurate wavelength dependence of the LSP resonance could not be reproduced in the 2D calculation, as the height of the nanostructures is not contained. However, it was also confirmed that accurate local field intensity is obtained for parallel polarization to the X–Y plane with detailed mesh sizes in much shorter processing time compared to the 3D simulation. Accordingly, FDTD-2D provide essential results that closely adjacent metal nanoparticles provide enormous field enhancement at SMS level accordingly the LSP coupling, while only modest enhanced was obtained for various isolated particles. In order to get deeper insight into the origin of the additional peak in the elastic scattering under SERS active state as well as enormous enhancement in the adsorbed layer, 3D calculation should be utilized.

Therefore, scattering cross section, far-field response, in Section 3.4 from Ag nanoparticles with or without R6G adsorbates was evaluated at various wavelengths with the FDTD-3D method using dielectric constants of constituents (Kunz and Luebbers, 1992; Palik, 1998). Dielectric constants of the dye were expressed in a damped harmonic oscillator model, $\varepsilon = 1 + S/(\omega_0^2 - \omega^2 - i\omega\Gamma)$, here an oscillation strength (S) of $5 \times 10^{31}\,\mathrm{s}^{-2}$, damping ($\Gamma$) of $2.5 \times 10^{14}\,\mathrm{s}^{-1}$ and resonance wavelength (ω_0) of $3.2 \times 10^{15}\,\mathrm{s}^{-1}$ ($\lambda_0 = 589\,\mathrm{nm}$) based on the experimental observations for rhodamine (Creighead and Glass, 1981). In order to discuss the coupling efficiency between the LSP and dye absorption, the resonance wavelength of dye was tuned between 400 and 700 nm. We adopted the various model structures such as isolated spheres, spheroid or adjacent spherical Ag particles with/without dye molecules. Yee cells are built from 160 × 120 × 120 meshes (2.3×10^6 meshes) with 2 nm/mesh along x, y and z directions using PML (perfectly matched layer) absorbing conditions (Kunz and Luebbers, 1992), as larger numbers of meshes or smaller sized meshes do not yield significant differences in scattering cross section. An incident light as a plane wave with linear polarization irradiates the cell at various wavelengths. Typically, one or two hours are necessary to yield a converged solution using a CPU (Pentium 4, 3.0 GHz) and Windows XP, which is roughly 10 times longer than in 2D calculation. Finer meshes at the junction do not give significant differences in far-field scattering, whereas providing much accurate near-field distribution compared to poorly resolved meshes. Therefore, the mesh sizes of 2 nm are adopted through the present evaluation.

§ 3. Results and discussion

3.1. Hot particles in SERS

Optical images in the bright-field, Raman and topographic (AFM) images for the same Ag particles were observed as shown in figs. 3(a)–3(d) using the facility in fig. 1. At surface coverage of ca. 300 R6G molecules/particle, only several Ag particles showed prominent SERS signal (bright spots in fig. 3(b)) among many particles in the sampled area (ca. 20 μm × 20 μm). All of the hot particles consist of Ag aggregates with a typical size of 1 μm

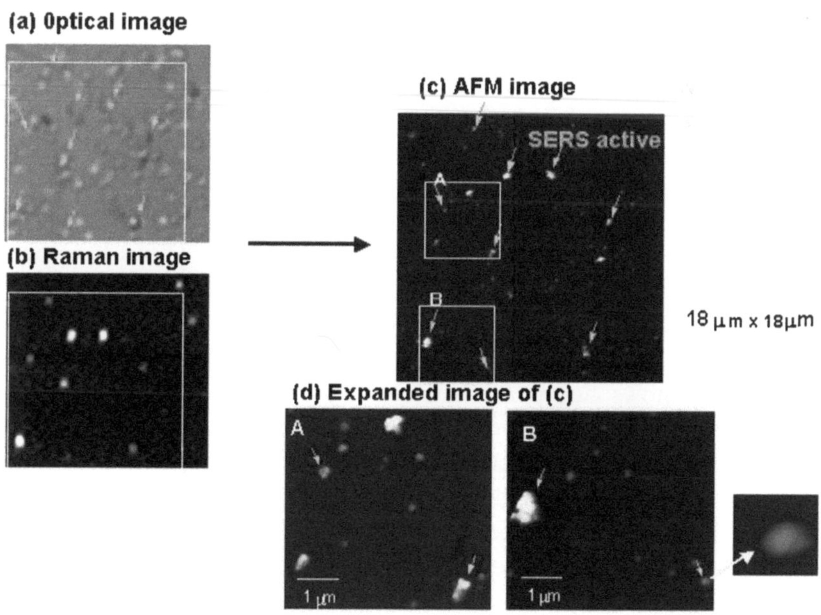

Fig. 3. Correlation of optical image (bright-field) (a), Raman image (b) and AFM images ((c) and (d)) of R6G/Ag particles. The same area in the same sample was measured in (a), (b) and (c) (18 μm × 18 μm).

or touching several particles as clearly visualized in the AFM images (figs. 3c–3d). At the surface coverage of ca. 300 molecules/particle, the enhancement factor was estimated to be about 2×10^7 by comparing the Raman scattering intensity from the hot particles with that for bulk solution. The LSP extinction spectra are apparently broadened and extended to longer wavelength upon coalescence of Ag particles (Futamata et al., 2002) due to the overlap and coupling of the LSP from different Ag particles. For example, one of the hot particles composed of two touching Ag spheroid with a similar size of ca. 210 nm × 160 nm × 120 nm (height) gives the main peak at 470 nm as well as shoulder peaks at ca. 490 and ca. 510 nm. Then, the Ar^+ laser at 488 nm can resonantly excite the LSP to enhance Raman signal from adsorbates on Ag aggregates.

This is in contrast to the observation that detectable SERS signal was not observed for isolated Ag particles that possess fairly sharp LSP bands at ca. 400 nm (Futamata et al., 2002). We also noted that slightly elongated Ag spheroid gives the LSP extinction at longer wavelength, which is in resonance with the same laser. Nevertheless, isolated particles are not hot. Accordingly, the junction of touching particles is crucial to give prominent enhancement in SERS. In addition, the Ag aggregates in figs. 3(c) and 3(d) are not always hot because of the polarization dependence of the local field under LSP resonance. Concerning this issue, high resolution transmission electron microscopy and electron energy loss spectroscopy have been applied by Kahn et al. to correlate SERS activity of Ag particles with their nanoscale structures and electronic state (Kahn et al., 2006). They found closely packed Ag particles with a gap of 0.5 nm showed pronounced SERS activity when parallel polarization to the touching axis was adopted. However, obviously no clear links have been observed between them, possibly due to essential experimental constraints, such as microscopically uncontrollable shape of particles or difficulty to adjust adsorbate amount as in other SMS-SERS experiments. Figure 4(a) shows hot particles consisting of two touching particles. Behind the bright spot, another particle is located, which is slightly shifted along the lateral direction. In this case, the polarization parallel to the touching axis (lateral direction in the figure) shows significantly stronger intensity, whereas perpendicular polarization gives only humble (see figs. 4(b) and 4(c)). The polarization dependence of the SERS enhancement was confirmed for various touching particles with different configurations. These discussions are supported by theoretical simulation for the local electric field on Ag particles using the FDTD method as follows.

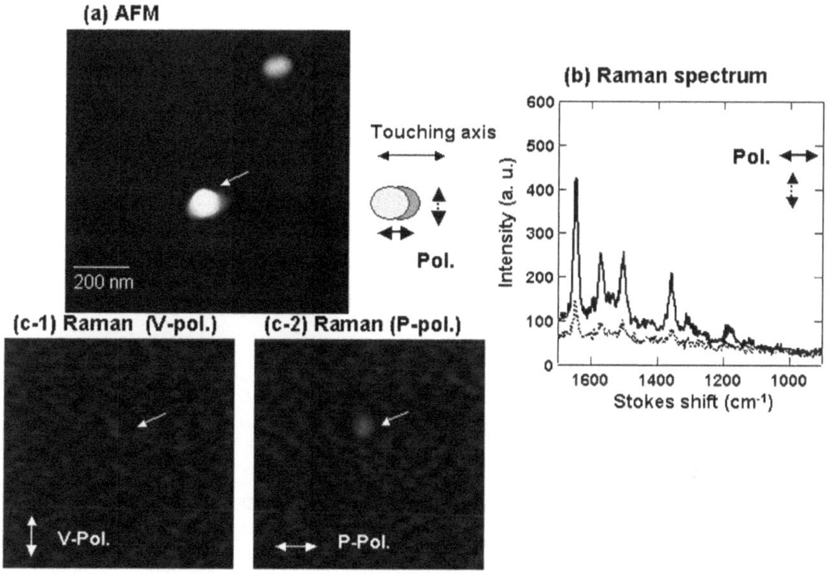

Fig. 4. Polarization dependence of SERS signal: (a) AFM image, (b) Raman spectra and (c) Raman images. Behind a bright spot in (a), another particle is located, which is slightly shifted to the lateral direction. Vertical and parallel polarization to the touching axis gave distinct enhancement in SERS.

3.2. Local field evaluation on the Ag nanoparticles

At first, the relevance of the FDTD method was confirmed for the electromagnetic field around the metal particles in the near- and far-field calculation. For this purpose, we adopt an Ag sphere with 10 nm in radius placed above the Ag flat substrate with a gap size of 0.5 nm (see fig. 5), as the analytical solution was obtained in the Maxwell equations for this case using bispherical coordinates (Aravind and Metiu, 1983). A plane wave is incident with the angle of 45° and p-polarization (see the inset of fig. 5). Consequently, quite similar local field enhancement of 4.2×10^4 at 450 nm and 1.29×10^4 at 400 nm was obtained by the FDTD method at the center between the Ag particle and substrate compared to the values of 4.2×10^4 at 450 nm and 1.3×10^4 at 410 nm obtained by analytical solutions (Aravind and Metiu, 1983). Essentially the identical results were obtained for various gap sizes in these different calculations. While these were obtained by the 2D calculation for p-polarization to the substrate, similar results were given by the 3D calculation as well as for s-polarization. Thus, we can safely use the FDTD method to evaluate the local electric field in the vicinity of the metal particles. Note that the vast enhancement of

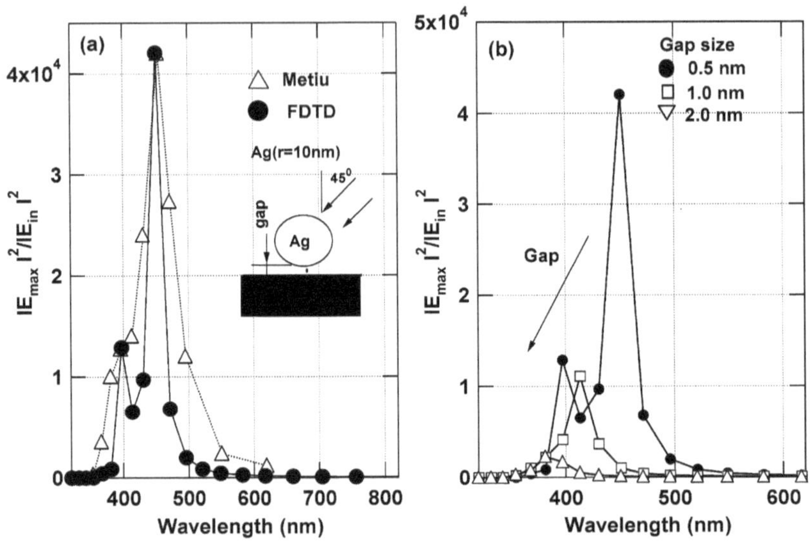

Fig. 5. Electric field intensity at the gap between the Ag sphere and substrate: (a) results by the analytical solution (triangles) and FDTD (filled circles) for the Ag sphere with $r = 10$ nm, and gap size of 0.5 nm, and (b) by FDTD for various gap size. Inserted figure is a schematic drawing of the sample configuration.

10^{10}–10^{11} for Raman scattering[2] was predicted for the gap size of 0 nm, which is comparable with those for touching Ag particles with various shape and sizes as described in the following section. Moreover, this result suggests tremendously large enhancement for Raman scattering from adsorbates at the gap between Ag particles and smooth Ag films, where the propagating SPP is excited using a prism.

Here we do not hold any substrates beneath the metal particles to evaluate the electric field in the near- and far-field for the isolated metal particles. In our real experiments, a glass substrate was used to immobilize the metal particles. However, dielectric substrate like a slide glass does not give significant changes in electric field intensity in contrast to the above result for metal substrate, while LSP extinction peak slightly

[2] As the first order approximation, the vast amplitude enhancement of >330 for the incident field obtained at the junction yields the Raman scattering enhancement of $>10^{10}$. As the scattering intensity from induced Raman dipole is resonantly enhanced by the LSP excitation of metal particles as well as the incident channel, the Raman scattering intensity is approximately proportional to fourth power of incident electric field, $I_{Raman} \propto |E_s \alpha E_i|^2 \approx |E_i|^4 \times |\alpha|^2$ (here, E_s, α, E_i are scattering field, Raman tensor and incident field intensity, respectively). The value 10^{10} corresponds to the SMS as it yields detectable signal of 5–10 counts s^{-1} with our facility.

shifts to longer wavelength by a few tens of nanometer. Accordingly, the dielectric substrate was extinguished to reduce size of the sample cell for saving computation time.

Scattering cross section for an isolated Ag circular cylinder shows the peak at ca. 370 nm irrespective of size between 10 and 80 nm. The peak is clearly arisen from the LSP excitation, of which wavelength accords with the analytical solutions for spherical particles with the same size (Bohren and Hoffman, 1983). In contrast, two distinct peaks were obtained for triangular structure at 430 and 500 nm for the right-angle (80 nm × 40 nm), or 400 nm (shoulder) and 450 nm for the equilateral right-angle (80 nm × 80 nm). The peak for ellipsoid slightly shifts to longer wavelength with increasing the aspect ratio, e.g. 370 nm (2:1) shifts to 400 (3:1) or 430 nm (4:1), which is much smaller than the observed values, i.e. 565 nm for the diameter (*D*) 95 and height (*H*) 48 nm (2:1) to 782 nm for D145 and H50 nm (3:1). For tetrahedral samples with a size of 80 nm (width, *W*) ×80 nm (*H*) at the cross section, the LSP peak was obtained at 380 nm irrespective of polarization direction quite similar to those for circular particles. In contrast to the results in 2D simulations, much larger red shift of the LSP peak was obtained by the 3D calculation. The scattering cross section (SCS) peak shifts from 380 (for sphere $r_1 = r_2 = 40$ nm) to 450 nm (2:1 spheroid, $r_a = 40$ nm, $r_b = r_c = 20$ nm) and 580 nm (4:1, $r_a = 40$ nm, $r_b = r_c = 10$ nm), which is consistent with the experimental data. In addition, these results fairly accord with analytical solutions for surface mode frequencies determined by geometrical factors (Bohren and Hoffman, 1983). For instance, the resonance condition for an ellipsoidal cylinder is given by $\varepsilon'/\varepsilon_m \approx -1$ at $\lambda_1 \approx \lambda_2 \approx 400$ nm, which splits into two branches for a prolate spheroid;$\varepsilon'/\varepsilon_m \approx -1$ at $\lambda_1 \approx$ 400 nm and $\varepsilon'/\varepsilon_m << -1$ at $\lambda_2 >> 400$ nm (here, ε', ε_m, λ_1 and λ_2 denote dielectric constant (real part) of particles, media, first and second resonance wavelength for the surface modes, respectively). It suggests that we should utilize 3D simulations to characterize the LSP extinction spectra and local electric field for real metal particles and then to explore the optimum nanostructure. Nevertheless, 2D calculation gives valuable insight into the vast enhancement at the junction as described here.

The local electric field on Ag circular cylinder surfaces shows the maximum intensity (*G*) of 10–15 at ca. 380 nm for different sizes, while three peaks are obtained for right-angle triangular cylinder (80 nm × 40 nm) at 380 nm ($G = 180$), 430 nm ($G = 370$), 500 nm (300) with different polarizations (see figs. 6(a) to (c), 7(a), 7(b), and 7(d)). The equilateral right-angle triangular shows the maximum ($G =$ ca. 500) at 430 nm with a shoulder at ca. 380 nm (figs. 6(c) and 7(c)). With increasing the aspect

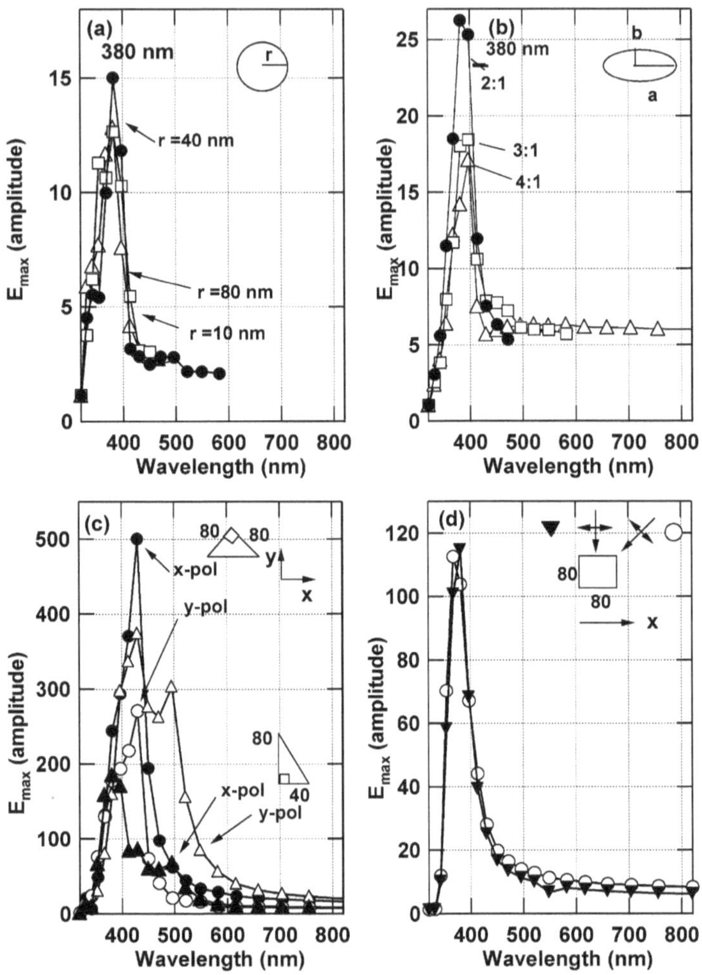

Fig. 6. Maximum electric field for isolated Ag particles: (a) circular tubes with radii of 10, 40, 80 nm, (b) ellipsoidal particles with different aspect ratio between 2:1, 3:1, 4:1 for fixed long axis (40 nm), (c) triangles with different shape and (d) tetragonal particles (80 nm (W) × 80 nm (H)) with different polarization. Polarization parallel to the long axis was used for ellipsoidal particles.

ratio of ellipsoidal particles, where the longer axis is fixed to 40 nm and shorter axis is changed from 40 nm to 10 nm, significant spectral changes are not observed (as drawn in fig. 6(b)). An isolated tetragonal particle (80 nm (W)×80 nm (H)) shows rather modest enhancement of ca. 110 at 380 nm. Thus, only triangular particles give vast enhancement at the sharp edge, while only modest enhancement was predicted for isolated circular, ellipsoidal and tetragonal particles. In addition, wavelength

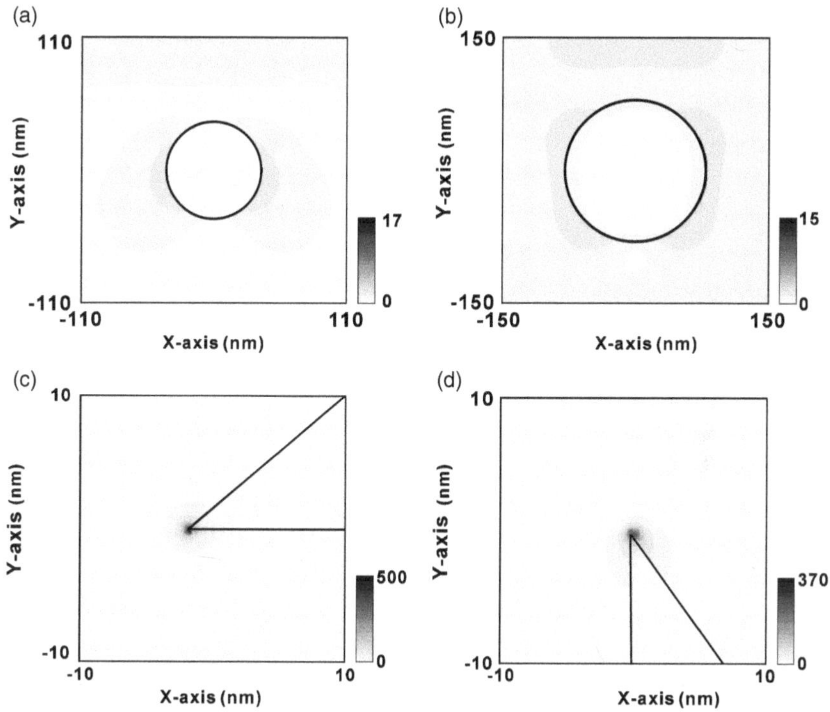

Fig. 7. Spatial distribution of the electric field on isolated Ag particles at the peak wavelength: (a) circular cylinder ($r = 40$ nm), (b) circular cylinder ($r = 80$ nm), (c) equilateral right-angle triangular (80 nm × 80 nm) and (d) right-angle triangular particles (80 nm × 40 nm). Polarized excitation light was used at 380 nm for (a), (b) and at 430 nm for (c, with x-pol.) and (*d*, with y-pol.). See also fig. 6. Electric field is shown as an amplitude enhancement relative to the incident field.

dependence of the scattering cross section is similar to that of the local field maximum for isolated particles. Slight differences of the peak wavelength and width, e.g. for triangular see fig. 6(c), can be explained based on the localization of the LSP field at the edge.

The scattering cross section is given by the shape-averaged far-field intensity contributed from the entire Ag surfaces. Accordingly, a particular site like a sharp edge of triangular may not be dominant. In contrast, the electric field maximum is determined by the local structure and the LSP resonance at the particular site, e.g. the enhancement is confined within a few nm from the edge, although these are not completely distinguished (see figs. 7(c) and 7(d)). Thus, the observed discrepancies between the resonant Rayleigh scattering and the SERS excitation (Michaels et al., 2000) are rationalized.

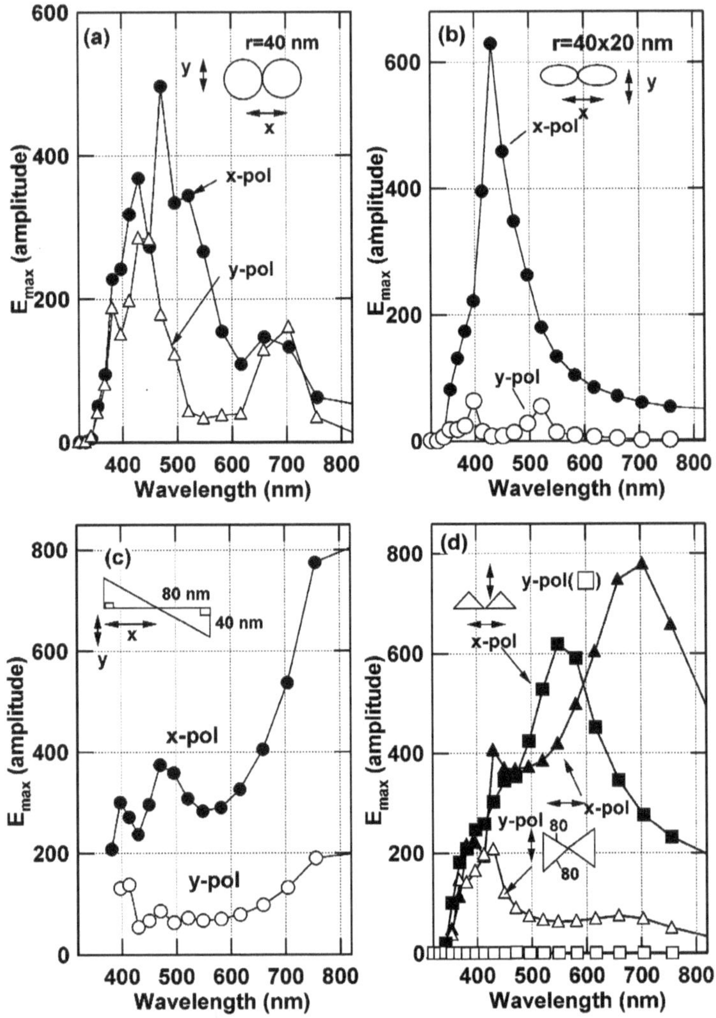

Fig. 8. Local electric field maximum for two touching Ag particles: (a) circular, (b) ellipsoidal tubes, (c) right-triangular and (d) equilateral right-angle triangular particles with different polarization.

For two touching particles, distinct spectral features were obtained for the local electric field maximum compared to the scattering cross section (Futamata et al., 2003) as summarized in figs. 8(a)–8(d). This is again because the local field maximum is determined by the LSP resonance and local nanostructure, whereas SCS is a shape-averaged far-field intensity contributed from the entire surface of Ag particles. Several peaks appear at 430, 480, 510 and 700 nm for touching circulars in addition to the original one at 370 nm for the isolated particle. Note that much larger electric field $G = 500$, which is in the SMS level, is formed at 480 nm for the polarization parallel to the touching direction in contrast to modest values for isolated circular cylinders. Even for a vertical polarization, prominent factor of ca. 270 is given at 440 nm. In addition, the maximum peak shifts to longer wavelength with the particle size, e.g. 410 nm at $r = 10$ nm, 520 nm at $r = 80$ nm.

Quite similar enhancement is obtained at 430 nm ($G = 600$) for touching ellipsoidal particles with the parallel polarization as well as the modest intensity ($G = 60$ at 400 and 520 nm) for vertical polarization (fig. 8(b)). Rather complicate features at longer wavelength are obtained for triangular particles as the prominent enhancement at the edge is overlapped. The tremendous enhancement factors are obtained for triangular particles with parallel polarization, while slightly modest values similar to isolated particles are predicted for vertical polarization. Right-angle triangular particles (80 nm × 40 nm) give vast enhancement $G = 400$ at 480 nm and $G > 800$ at > 800 nm, while equilateral right-angle triangular give $G > 600$ at 560 nm for side-by-side and 700 nm for edge-to-edge with parallel polarization (see figs. 8(c) and 8(d)).

These features do not essentially depend on the particle sizes, e.g. for spheres with a radii > 20 nm. As depicted in fig. 9, the enhancement rapidly decreases with increasing the gap size due to diminished LSP coupling and edge effect, i.e. $G \approx (G_0)/e$ at 1.5 nm gap size and $G = 20$ for 10 nm, here G_0 the enhancement value for the contacting particles. The local field maximum decays in a similar way irrespective of particle sizes between $r = 40 \sim 160$ nm (Futamata et al., 2003). As anticipated, the local field maximum for the triangular edge does not depend on the gap size at the resonance wavelength, i.e. 500 at ca. 400 nm for the equilateral right angle particle. However, the enhancement arose from the LSP coupling at the junction with the resonance wavelength of 660 nm rapidly decreases even for triangles at the gap size < 1 nm similar to the circular particles. The sharp decay of the local field maximum is clearly explained by the peak shift of the LSP resonance to shorter wavelength and by the decreased coupling with increasing the spacing.

These results indicate that the critical conditions to give the vast enhancement at SMS level are the LSP resonance and also the nanostructures such as edges or junctions, where the dense electric field is confined. This is rationalized by the following simulation that only modest electric field is obtained for two tetrahedral particles placed in a side-by-side configuration with sufficiently small spacing of $d < 0.5$ nm. In fact, the maximum enhancement (70 at 380 nm for a gap $d = 0$ nm) is quite similar

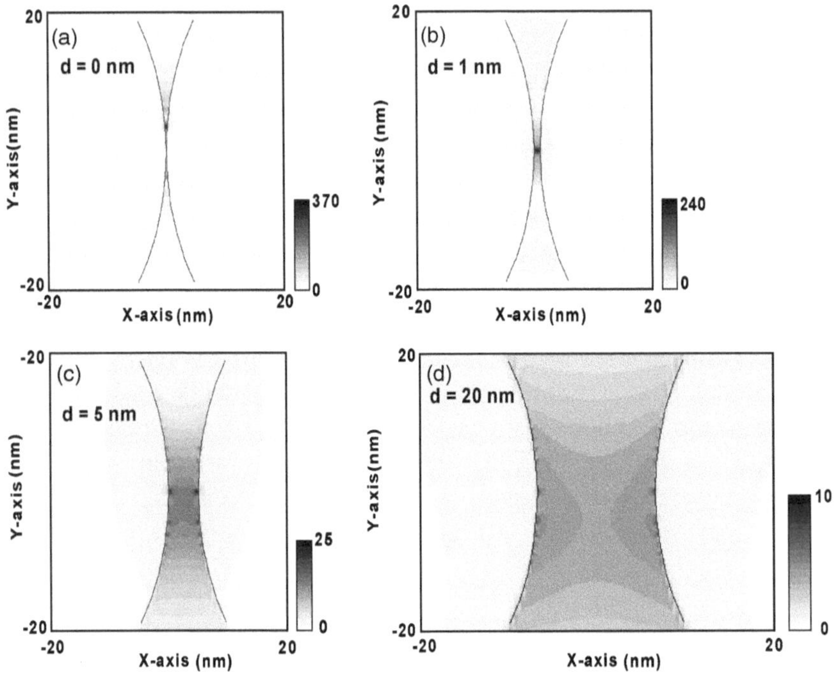

Fig. 9. Electric field distribution for two adjacent Ag circular tubes ($r = 40$ nm) as a function of the spacing at 480 nm (wavelength): (a) gap $d = 0$ nm, (b) $d = 1$ nm, (c) $d = 5$ nm and (d) $d = 20$ nm.

to that for isolated tetragonal particles (110 at 380 nm). In addition, the scattering cross section for this configuration is similar to isolated one, as no additional bands are observed at longer wavelength. In contrast, the edge-to-edge configuration of the same tetragonal particles give the vast enhancement at ca. 550 nm with a shoulder at 400 nm as well as an additional peak in scattering cross section at ca. 540 nm (not shown).

Conclusively, a sharp edge as well as the LSP coupling is indispensable to give the enormously large enhancement. Note that the junction between touching circular nanowires consists of moderate curvature at the metal and much sharper one at an air-side, which contributes electric field confinement enhanced by the LSP resonance. In addition, the local field maximum at a small protrusion with sufficiently sharp edge placed on a circular cylinder is quite similar to those for triangular edge or the junctions of particles, while the LSP spectra are similar to an isolated circular tube. This result suggests two plausible reasons for the discrepancies between the observed wavelength dependence of the optimum size and theoretical predictions (Krug et al., 1999): (1) the second particle is located behind the first one, or (2) an isolated particle with small protrusions, which could not be imaged due to insufficient spatial resolution of the AFM. Thus, the spatial distribution of the LSP resonance and local electric field should be further studied with the SNOM method in addition to topographic image of the metal particles. Moreover, as shown in Section 3.4.2, it was confirmed by 3D calculation that the similar enhancement of ca. 300 at 500 nm corresponding to SMS level in SERS is obtained at the junction with the comparable wavelength dependence for two touching Ag spheres. Slightly smaller values compared to 2D calculation is due to rather large mesh sizes adopted to save computing time similar to the SCS spectra. It indicates that the vast enhancement at the junction is already obtained by the 2D nanostructures, although distinct spectra for scattering cross section and the local field maximum are obtained for circular or ellipsoidal cylinder (2D) in comparison with spheres or spheroidal particles (3D).[3]

3.3. *Origin of the blinking*

3.3.1. *Blinking at room temperature*

As described above, Ag particles that yielded enormous enhancement are aggregates with a typical size <ca.1 μm or touching particles, which give higher enhancement for polarization parallel to the connection axis compared to the vertical direction. Only a few of these particles showed the blinking. In addition, the fluctuation of the peak frequency within ca. $10\,cm^{-1}$ and narrower bandwidth (ca. 1/2) than that for higher surface coverage were observed as shown in figs. 10–11. It is noteworthy that these spectra were sequentially measured with an accumulation time of 1 s to give sufficiently high signal to noise ratio with the CCD, while actual blinking frequency is several Hz, slightly faster than the accumulation time. Nevertheless, intensity and peak frequency fluctuations were clearly observed, while prominent intermittent features observed by Krug et al. (1999) using avalanche photodiode. These observations suggest the existence of various adsorption sites on Ag particles with different interactions and enhancement. Adsorbed molecules possibly diffuse between these sites.

[3] The accuracy of the obtained electric field distribution is not sufficient for closely adjacent metal nanoparticles with a tiny gap size, e.g. < 1 nm, according to spill-out of electron from the boundary. Much precise evaluation is necessary for quantitative discussions. Accordingly, rather crude calculation in this section is performed to demonstrate the critical importance of the junction of touching particles.

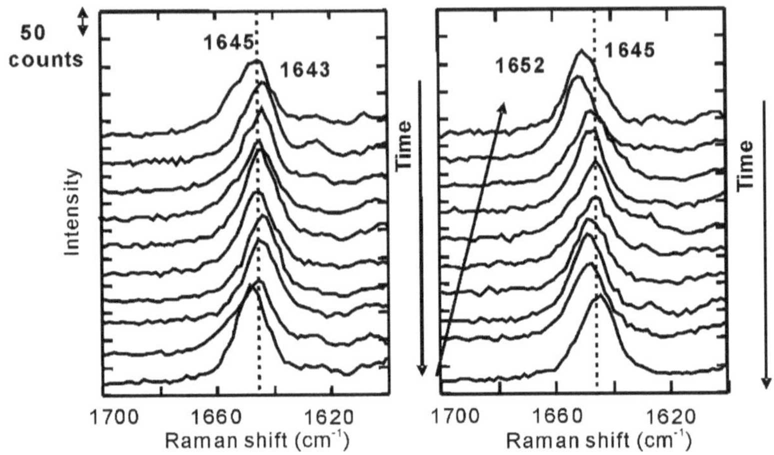

Fig. 10. Peak frequency fluctuation of the SERS spectra from R6G on Ag nanoparticles. Average surface coverage of R6G is 1 molecule per Ag particle. Each spectrum was obtained by sequential measurement of accumulation time of 1 s at $\lambda = 488\,\text{nm}$ and $70\,\mu\text{W}\,\mu\text{m}^{-2}$.

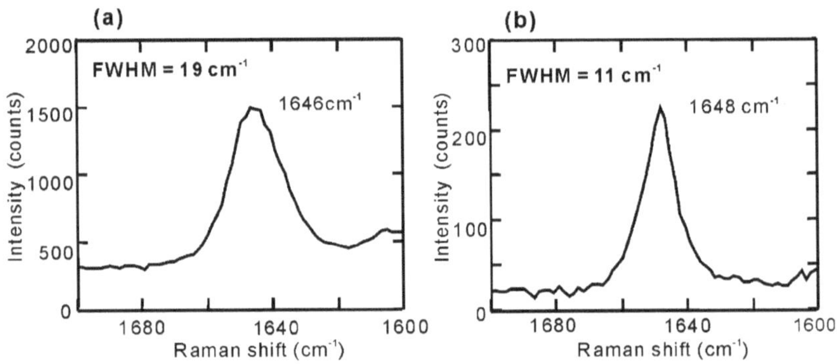

Fig. 11. SERS peak profile (a) from Ag particles with higher surface coverage and (b) from blinking particles. Average surface coverage of R6G in (a) and (b) are 300 molecules and 1 molecule per Ag particle, respectively. SERS spectra were obtained with the same condition as in fig. 10.

Before going into the temperature dependence, it is useful to discuss possible temperature increase of Ag nanoparticles by the excitation laser, which may cause significant difference in the LSP resonance and enhancement in SERS. At first, in our experiments with quite weak laser power of $70\,\mu\text{W}\,\mu\text{m}^{-2}$ at wavelength of 488 nm, the observed Raman bands are safely assigned to vibrational modes from original species. Accumulated spectra for a long duration of measurement, e.g. for 100 s or longer, do not contain any pronounced Raman bands from plausible contaminants such as amorphous carbon (Kudelski and Pettinger, 2000). Therefore, the blinking of SERS signal is attributed to R6G and adenine adsorbates. On the other hand, if the temperature of Ag particles increases even up to ca. 315 K, the morphology of the Ag particles could be irreversibly changed as reported by Semin et al. (1996). However, AFM measurement before and after the laser irradiation did not give any distinct change in nanoscale morphology of our Ag particles (not shown). The laser power ($\lambda = 488$ nm, $100\,\mu W\,\mu m^{-2}$ for 10 min) adopted here is the maximal intensity in our experiments to measure SERS spectra, while only modest intensity of $1\,\mu W\,\mu m^{-2}$ was used for the SERS image measurements here. In contrast, Ag nanoparticles prepared with the same procedure significantly changed their morphology by annealing at 350 K or at 400 K for 1 h in accordance with the observation for Ag island films (Semin et al., 1996).

Thus, invariant morphology of Ag particles under the laser irradiation indicates only negligible temperature increase of the samples during our SERS experiments. Namely the LSP resonance of Ag nanoparticles and thereby the enhancement are not modulated by the excitation laser. In accordance with these results, the temperature increase of Ag particles is estimated to be negligible (<4.5 K) for our laser power used for SERS spectral measurements ($\leqslant 100\,\mu W\,\mu m^{-2}$) using eq. (3) in Goulet et al. (2005) based on stationary heat diffusion from Ag nanoparticles to hydrating water layer. In addition, similar blinking features were observed for SERS spectra of adenine on the Ag nanoparticles at RT (Maruyama et al., 2001), which has no electronic transitions in visible wavelength region in a bulk solid or solution state. From these observations, it seems that photochemical bleaching, relaxation via triplet electronic state or morphology changes by laser irradiation are not concerned with the blinking. The origin of the blinking is possibly due to thermal process such as thermal diffusion of individual molecule on the Ag particle presuming physisorbed molecules, of which binding energy is comparable with thermal energy. This is rationalized, if (1) the Ag surface has different sites with distinct enhancement, and (2) sufficiently large thermal energy to overcome an activation barrier for diffusion.

Concerning the first point, it was demonstrated by numerical simulations using the FDTD method (see Section 3.2, Futamata et al., 2002, 2003) that vast enhancement of $>10^{11}$ in Raman scattering is obtained at a junction between two touching Ag particles with various shapes and sizes in addition to an edge of isolated trigonal prism under the LSP resonance (see figs. 7 and 9). The enhanced field sharply decays with increasing the gap size (Futamata et al., 2003). Other sites apart from the junctions or edges of the touching particles and of isolated triangular prisms give only modest enhancement of $\leqslant 30$. These results by the numerical simulations agree with the experimental observations, since only Ag aggregates show the vast enhancement with parallel polarization to touching axis. Thus, we may attribute the blinking to thermal diffusion of adsorbed molecules on the Ag surface between the junctions with vast enhancement and other ordinary sites with modest enhancement. If the blinking arises from thermal diffusion of adsorbates on the Ag particles with respect to the second point, the fluctuation frequency should be decreased or blinking is completely suppressed with decreasing the temperature according to simple consideration of hopping. Therefore, we measured the temperature dependence of the blinking in SERS signal from R6G at a surface coverage of ca. 3 molecules/Ag particle.

3.3.2. Blinking at low temperature

At first, the blinking particles were found at RT, and then cooled to 77 K. As clearly depicted in figs. 12(a) and 12(b), a bright spot became completely darkened and never turned to a bright image at 77 K in contrast to repeated intensity changes at RT, e.g. bright spot with intensity fluctuation between 1.6–3.4 s, and 3.9–4.4 s. We also observed alternative cases

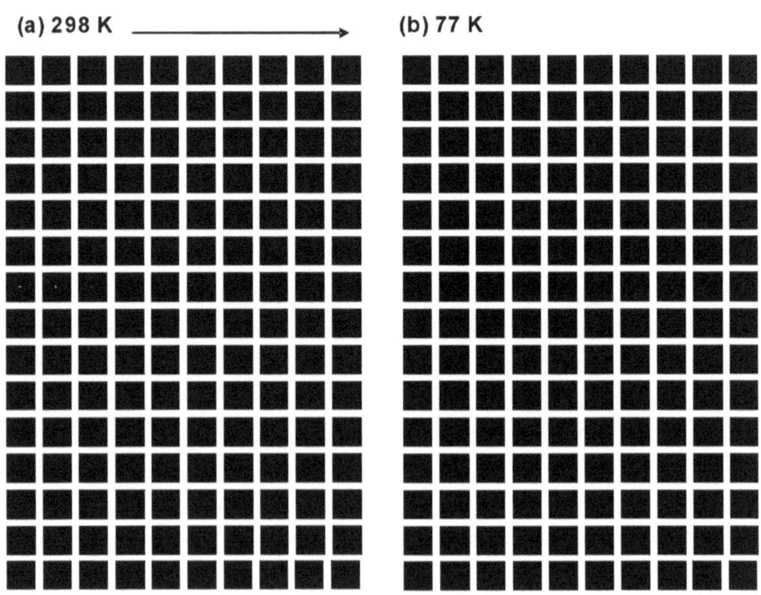

Fig. 12. Temperature dependence (I) of blinking: (a) at RT, (b) at 77 K. Darkened spot was observed at 77 K through the experiments (for ca. 10 min), indicating blinking is suppressed at inactive sites.

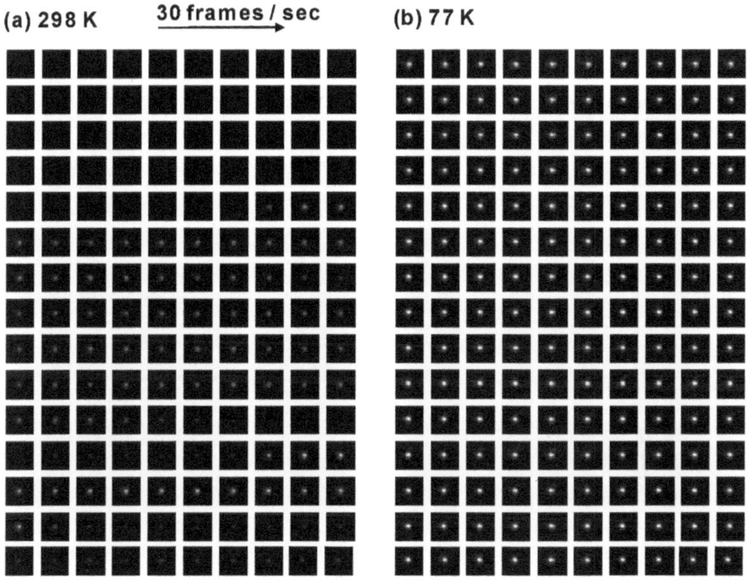

Fig. 13. Temperature dependence (II) of blinking: (a) at RT, (b) at 77 K. Bright spot was observed at 77 K through the experiments (for ca. 10 min), indicating blinking was suppressed at active sites.

that the blinking spot at RT, e.g. bright spot between 1.3 and 2.1 s did not change its intensity at 77 K through the measurement, > 10 s in reality, as shown in figs. 13(a) and 13(b). Both these observations enable us to conclude that the blinking in SERS signal at RT is suppressed at 77 K, indicating the blinking is a thermally activated phenomenon: when individual dye molecule is immobilized at the sites with vast enhancement, bright invariant spot was observed at 77 K, whereas at modest enhancement sites, dark images were given. It should also be noted that roughly 1/3–1/4 of blinking Ag particles were frozen, suggesting most of the blinking particles have much smaller activation energy for the process compared to thermal energy at liq. N_2 temperature. This is not surprising, since each adsorbed molecule can possess different bound energy on polycrystalline Ag particles according to locally different surface electronic state. Moreover, Raman spectra from R6G on Ag at blinking (RT) and at frozen (77 K) conditions, were safely assigned to intramolecular vibrations of R6G, e.g. 1,653 (ϕ C–C str.), 1,582 (ϕ C–C str.), 1,539, 1,510 (ϕ C–C str.), 1, 358 cm^{-1} (ϕ C–C str.) in good agreement with the former report (Hildebrandt and Stockburger, 1984). Rather poor signal to noise ratio of the SERS spectra compared to the previous one (Hildebrandt and Stockburger, 1984) is due to lower optical throughput and/or collection

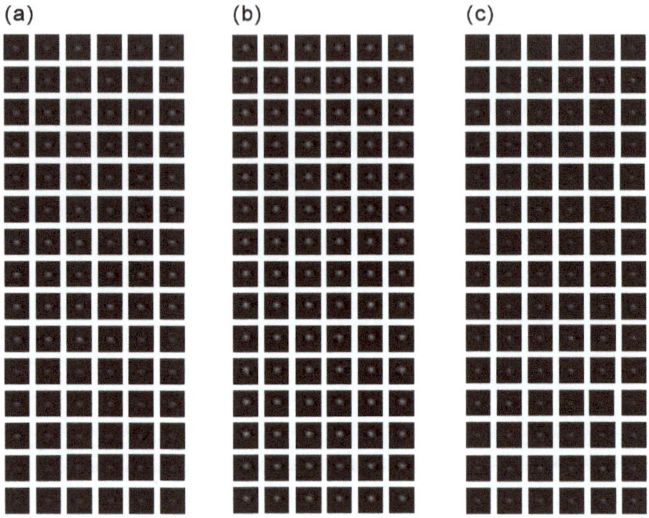

Fig. 14. Temperature dependence (III) of blinking: (a) at RT, (b) at 77 K and (c) after warmed up to RT. Blinking was suppressed at 77 K and recovered at RT after warmed up, indicating it is thermally activated.

efficiency for the sample in a liq. N_2 cryostat. Occasional intensity difference of these SERS spectra is due to rather long accumulation time of 1 s with respect to the blinking frequency of a few Hz.

Interestingly, the frozen particles at 77 K recovered the blinking, when they were warmed to RT as shown in fig. 14. Thus, the suppression of the blinking is intrinsic and reversible with the temperature variation between RT and 77 K. It clearly suggests that the temperature dependence observed here is not an experimental artifact, such as irreversible photochemical reaction of adsorbates by excitation light. Consequently, the blinking is thermally activated, most probably due to thermal diffusion of adsorbed molecules between the particular sites with vast enhancement and with modest enhancement on Ag surfaces. These sites are attributed to the junction (ca. 2–3 nm, Futamata et al., 2002, 2003) of touching particles and other ordinary sites far from the junctions based on the theoretical simulation. Relative intensity changes of SERS bands during the blinking (Michaels et al., 2000; Bosnick et al., 2002), can also be explained by orientation changes of molecules during diffusion with respect to the anisotropic electric field at the junctions (Garcia-Vidal and Pendry, 1996; Futamata et al., 2003).

Weiss and Haran (2001) reported for R6G on Ag particles using 532 nm excitation that (1) background intensity as well as SERS signal shows the intensity fluctuation, (2) fluctuation rate is proportional to the laser power, while thermal effect is negligible in their experiments (excited at 532 nm with 1 μW μm^{-2}). Then, they concluded the origin of the increased fluctuation is not a thermal but a photochemical process, possibly due to molecular diffusion that are mediated by desorption triggered by electron tunneling between the metal surface and molecules. To avoid a possible confusion, it should be noted that the laser power dependence was studied by Weiss and Haran at a constant temperature, while we explicitly changed the sample temperature at a fixed laser power. As described in Section 3.1, in our experiments temperature of the Ag particles was not significantly raised by the excitation laser (at 488 nm with <100 μW μm^{-2}). Clearly, photochemical activation and thermal activation suggested here are not exclusive, but compatible for particular molecules such as R6G or other dyes which have an appropriate electronic resonance with the excitation light. Thermal diffusion can also work for other molecules which have no electronic transitions in visible wavelength at adsorbed states. Haran and coworkers further reported that the preferential fluctuation was observed for the SERS signal at 614 and 773 cm^{-1} with the 532 nm excitation, which proves the fluctuation is not due to re-orientation of molecules. Instead, it suggests the charge transfer (CT) resonance for these bands in accordance with their distinct excitation profiles from others (Hildebrandt and Stockburger, 1984). They pointed out that: (1) the fluctuation in SERS could be originated by that in local work function induced or sampled by molecular diffusion or Ag atoms motion in surface relaxation dynamics. It causes the fluctuation in the CT resonance and in SERS enhancement (Haran, 2004). Furthermore, (2) the polarization dependence of these bands was definitely deviated from other bands, which provides the information on the orientation of molecules and charge transfer dipole (Shegai and Haran, 2006). These approaches are essentially important to evaluate the charge transfer interaction, in particular, at proximity of nanoscale roughness or junctions.

With respect to vast electric field gradient at the junction, an optical confinement of molecules is prospected analogous to the optically biased diffusion of R6G molecule in solution (Osborne et al., 1998). However, as predicted by theoretical simulation for surface plasmon enhanced optical forces (Xu and Käll, 2002), even if the enormously large electric field is formed at the junction of touching Ag particles, only modest trapping force is induced for R6G adsorbates under our experimental condition ($\leqslant 10$ μW μm^{-2} at 488 nm, see also figs. 1 and 3 in Xu and Käll, 2002). In fact, they obtained the trapping energy of ca. 0.03 $k_BT/(100$ μW μm$^{-2})$ at the junction of two touching Ag particles with a size of 40 nm in diameter at $\lambda = 540$ nm, roughly similar to the above estimation based on the trapping potential in solution (Osborne et al., 1998). Conclusively, the optical trapping force works to immobilize adsorbates onto the junction, but the value is much smaller than thermal energy. Thus, thermal diffusion is not frozen by the optical confinement effect.

At this stage, it is quite important to get definite evidences that adsorbed molecules are sitting at the junction when vast enhancement in SERS is observed. On this issue, it was convinced here based on the correlation between elastic scattering spectra and SERS activity together with the FDTD-3D simulation.

3.4. Critical importance of the junction for SMS-SERS

3.4.1. Elastic scattering experiments

As reported in our recent paper (Futamata et al., 2004a), the elastic scattering spectra from the touching particles provide marked differences according to the on/off feature of SERS activity. For example, the Ag particles consisting of a few ellipsoids as in the AFM image (fig. 15(c)) give the elastic scattering peaks at 530 and 730 nm according to the LSP excitation. In addition, another peak appears at 625 nm when tiny amounts of dye molecules, e.g. 30 molecules/Ag particle, adsorb to give pronounced SERS signal as in figs. 15(a) and 15(b). Actually, the enhancement factor observed for the Raman bands of R6G at 1,654, 1,574, 1,510 or $1,365\,cm^{-1}$ is estimated to be 10^8–10^9 compared to the signal intensity at bulk solution state. Very interestingly, the scattering peak at 625 nm disappears when the SERS activity is lost after duration of measurement for typically 30 min. This is probably due to diffusion to marginal sites or desorption of the dye molecules as discussed in the next section. In addition, intermediate scattering spectrum was observed for partially active state. These distinct elastic scattering spectra are inherently related to the SERS activity, since identical features were observed for every hot particle. Moreover, these features were confirmed for the blinking particles as indicated in fig. 16. Although two distinct LSP peaks were observed at ca. 475 nm and at ca. 630 nm according to rather complicate morphology of individual particles, appreciable spectral changes were observed at ca. 500–600 nm. Namely, definite scattering peak emerged at ca. 590 nm by the addition of R6G is extinguished when the SERS activity is lost. Experimental errors such as fluctuation of the signal intensity due to instable alignment of the optics are negligibly small, as confirmed for Ag particles at entirely inactive state (not shown). The

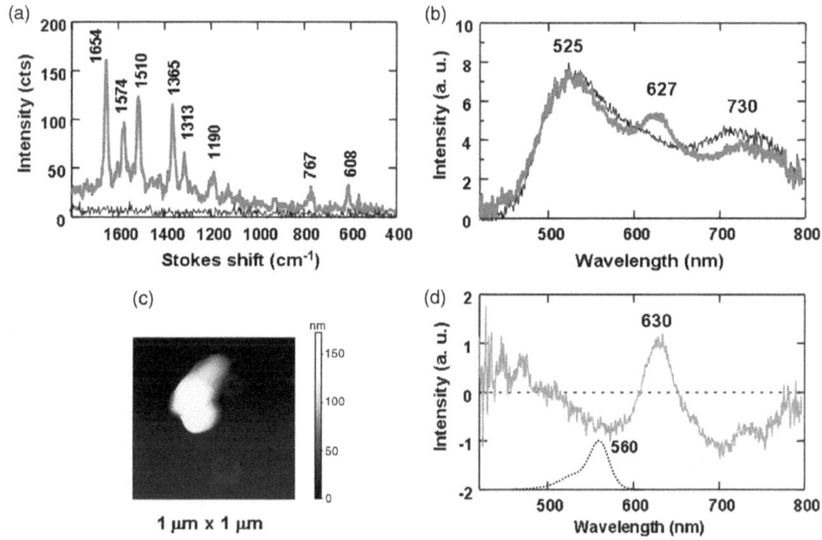

Fig. 15. Topography and scattering spectra from the same hot Ag particles (30 molecules/Ag particle): (a) SERS spectra, (b) elastic scattering spectra, (c) AFM images of the hot particles and (d) differential scattering spectra in b between SERS-active and SERS-inactive particles. Dotted line in (d) denotes the absorption of dye at bulk state.

additional peak in elastic scattering spectra observed at SERS-active state probably arises from the electronic absorption of the dye molecules, located at slightly shorter wavelength, i.e. ca. 560 nm as discussed in the next section. The LSP does not give such additional peaks unless quite formidable coalescence or morphology variations are induced by the dye adsorption. This is not the case, since only a few molecules are attached and desorbed from each Ag particle immobilized on the glass substrate, of which morphology and aggregation state are invariant through the measurement as confirmed by AFM images.

Concerning the observed spectral changes, it was reported independently by Creighead and Glass (1981) and Garoff et al. (1981) in the early stage of SERS study that the extinction spectra of Ag evaporated films varies with deposition of thin dye film (thickness <1 nm). They could qualitatively reproduce the observed spectra based on Garnet or Mie theory: (1) obvious splitting of the LSP peak, in other words new peak appeared at slightly longer wavelength than bulk absorption peak of dye molecules, and (2) faint changes in the LSP peak. Thus, together with analytical solution, the effective medium theory that describes averaged dielectric properties based on the volume fraction of the constituents is conveniently utilized to yield macroscopic optical response from composite or stratified

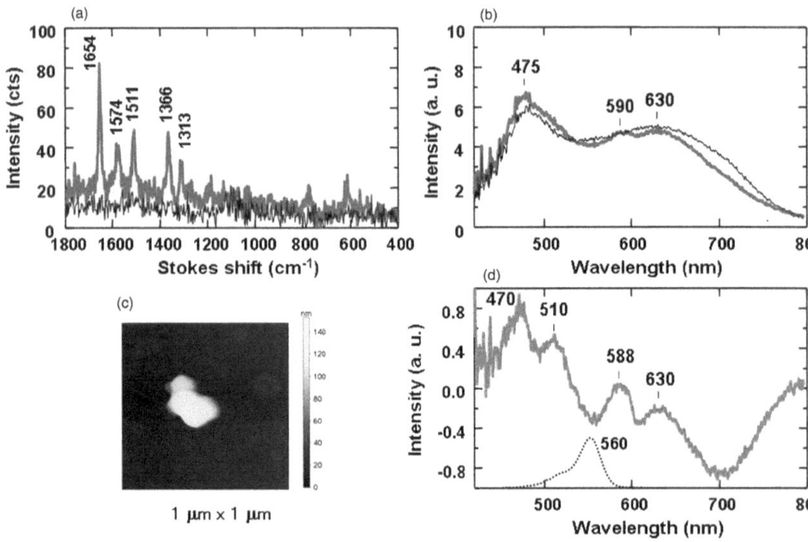

Fig. 16. Topography and scattering spectra from the same blinking Ag particles with tiny amount of dye (3 molecules/Ag particle): (a) SERS spectra, (b) elastic scattering spectra, (c) AFM images of the hot particles and (d) differential scattering spectra in b between SERS-active and SERS-inactive particles. The dotted line in d denotes the absorption of dye at bulk state.

samples. These efforts encourage us to evaluate electromagnetic field in near- and far-field for various nanostructures and locations of tiny amount of molecules with using numerical simulations.

3.4.2. Numerical simulations of elastic scattering spectra

We adopted three-dimensional FDTD (FDTD-3D) method to evaluate the scattering cross section for adjacent Ag nanoparticles at various wavelengths between 300 and 800 nm. Small amounts of the dye molecules are placed onto different sites on the isolated or touching Ag particles to investigate the effect of adsorbed positions. Actual SERS active Ag particles consist of 3–4 particles with different shape and sizes as shown in figs. 15(c) and 16(c). However, the model structure is relevant to discuss the effect of dye adsorption onto the junction, since we are studying single molecule phenomena plausibly arose from an individual junction. Two adjacent bare Ag spheres ($r = 40$ nm) with various gap sizes (g, nm) provide distinct LSP peaks attributed to isolated LSP resonance at 370 nm and coupled LSP at 430–540 nm as shown in fig. 17(a). With increasing the gap sizes, the coupled LSP peak shifts to shorter

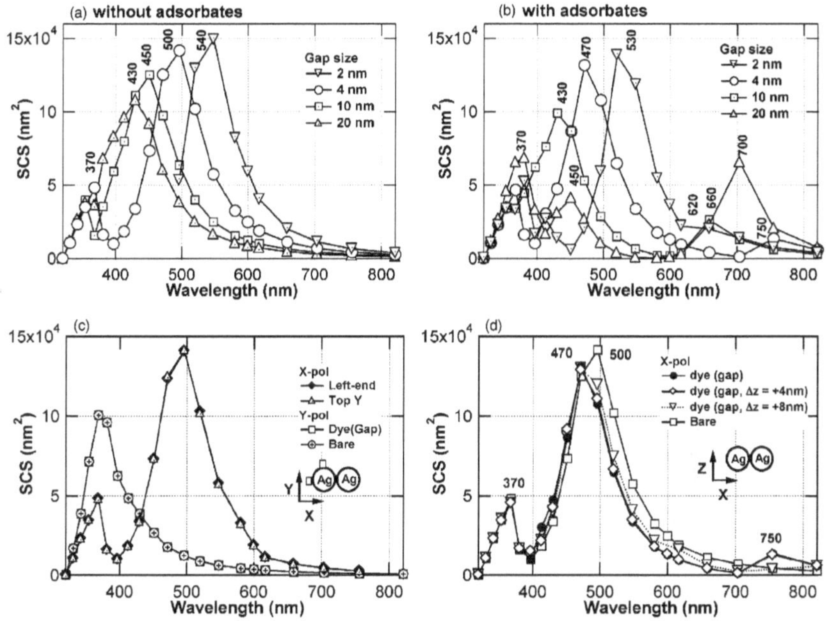

Fig. 17. Elastic scattering spectra calculated for the Ag neighboring spheres (diameter of 80 nm) with different gap sizes ($g = 2, 4, 10$ and 20 nm) using the FDTD-3D method. (a) Without dye, (b) with dye filling the gap, (c) and (d) with dye ($4 \times 4 \times 4\,\text{nm}^3$) for various positions and polarizations. Parallel polarization (x-pol.) to the touching axis was used unless otherwise noted.

wavelength according to diminished coupling, i.e. 540, 500, 450 and 430 nm for $g = 2, 4, 10$ and 20 nm, respectively.

Significant spectral changes were observed by filling the gap with dye. For instance, an additional peak was observed at 750 nm for the gap (4 nm) when illuminated with the polarized light parallel to the touching axis, here the dye molecules with volume of $4 \times 4 \times 4\,\text{nm}^3$ were used (see fig. 17(b), smaller volume of dye such as $2 \times 2 \times 2\,\text{nm}^3$ provides significant spectral changes as well). The LSP peak at 370 and 500 nm showed only little spectral changes. It should also be noted that: (1) the additional peak was not observed for other location of dye molecules, e.g. left end or top position of the left particle (see fig. 17(c)). (2) Incident light with y-polarization (vertical to the touching axis) does not give such spectral changes because of faint LSP coupling of neighboring Ag particles (fig. 17(c)) (Xu et al., 2000; Kottmann et al., 2001; Futamata et al., 2003). (3) The additional peak position and intensity are quite sensitive to the detailed location of dye, e.g. slight migration (by 8 nm) along z (or y)-axis within the gap causes a weak shoulder at ca. 600 nm, while minor diffusion (4 nm) does not modify the 750 nm peak (fig. 17(d)). These results are feasibly understood by considering that marginal positions even in the gap or y-(or z-) polarized light do not increase the LSP coupling of neighboring Ag particles. At larger gap sizes, which are filled by the dye molecules, the LSP peak manifests a blue shift with diminished intensity, e.g. 530 nm ($g = 2$ nm), 470 nm ($g = 4$ nm), 430 nm ($g = 10$ nm) or 370 nm ($g = 20$ nm) as in fig. 17(a). These peak deviations are significantly larger compared to the bare particles due to increased LSP coupling through conductive dye molecules. Monotonous peak shift of the additional peak was not observed according to distinct coupling efficiency, as the LSP peak positions for the bare Ag particles are already dissimilar at different gap sizes. In addition, only faint spectral changes without the additional peak were obtained for small amount of dye ($4 \times 4 \times 4\,nm^3$) located at the center of larger gap ($g = 10$ or 20 nm), indicating negligible coupling of the LSP at each Ag particle. The dye filling the gap with prominently small sizes increases the LSP coupling to give definite red-shift of the LSP peak together with the additional peak. If quite large amount of molecules such as $20 \times 20 \times 20\,nm^3$ or more adsorb and continuously detach from the junction of Ag particles presumably due to photochemical decomposition or evaporation under laser irradiation, one can expect drastic red-shift of the LSP peak and simultaneous blue-shift of the additional peak with diminished intensity. Much smaller amount of dye like $4 \times 4 \times 4\,nm^3$ ($= 64\,nm^3 < 1/100$ ($20 \times 20 \times 20\,nm^3$)), in contrast, would provide only slight spectral changes that are elimination of the additional peak at 750 nm and concomitant LSP peak shift from 500 to 470 nm. These are in good agreement with our observations for extremely low surface coverage.

It is noteworthy that the absorption of dye is crucial to yield drastic spectral variations, since transparent materials with fairly high refractive index do not give similar changes (Futamata and Maruyama, 2005). For instance, water molecules with the dielectric constant of $\varepsilon = 1.82$, or SiO_2 of $\varepsilon = 2.10$ with a size of $4 \times 4 \times 4\,nm^3$ filling the gap ($g = 4$ nm) does not give additional peaks at longer wavelength. Furthermore, the additional peak position shifts monotonously from 580 to 750 nm by tuning the absorption wavelength (λ_0) from 400 to 700 nm, in contrast to the invariant LSP peaks at 370 and 470 nm as depicted in figs. 18(a) and 18(b). This is clearly explained by variation of the implicit coupling between the LSP and dye.

Pronounced coupling was obtained for the dye absorption (λ_0) between 400 and 500 nm with the LSP resonance at 470 nm, whereas gradual decrease with increasing their separation, at $\lambda_0 > 500$ nm, as evidenced by weakened peak intensity. Rather similar spectral changes were obtained

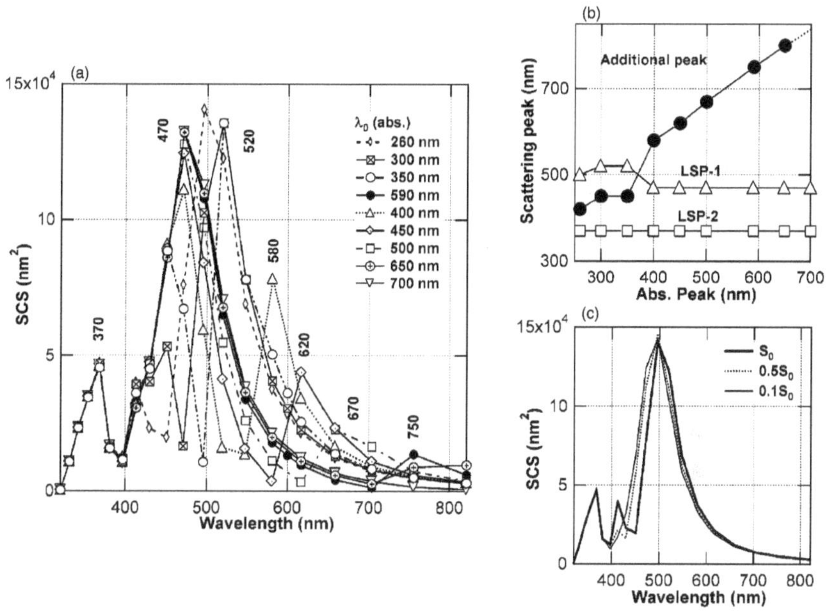

Fig. 18. Elastic scattering spectra adjacent Ag nanospheres with a gap size of 4 nm calculated with FDTD-3D: (a) R6G filling at the gap with different absorption peak wavelength (λ_0) between 260 and 700 nm, (b) plot of the peak position as a function of absorption peak wavelengths, (c) the same as (a) but with different strength S (see the Experimental section). Parallel polarization was used to the touching axis.

for the absorptions at 350, 300 and 260 nm that provide peak shift for the LSP band from 470 to 500–520 nm as well as an additional peak at 450–420 nm. Again the new peak originates mainly from the absorption of adsorbates sitting at the junction, since significant spectral changes, viz. weaker peak intensity, were obtained only for this peak at diminished oscillation strength (fig. 18(c)). The additional band appears at shorter wavelength side of the in-phase LSP peak (470 nm) for absorption between 260 and 350 nm, whereas longer wavelength for absorption between 400 and 700 nm. Clearly, the absorption does not couple with the other LSP peak at 370 nm similar to the case for isolated particles. Increased coupling by adsorbates for parallel polarization is afforded by favorable electric conductivity (σ), e.g. ca. 3.3×10^4 at $\lambda = 450$ nm or 1.8×10^6 at $\lambda = 589$ nm (at absorption peak) as estimated from the equation, $\sigma = \omega \varepsilon_0 \varepsilon''$, where ω, ε_0 and ε'' are frequency of the excitation light, free space dielectric constant and imaginary part of dielectric constant of dye (Futamata et al., 2004a). Correspondingly, enhanced absorption of dye was obtained at the same wavelength (750 nm) (Futamata and Maruyama, 2005). Moreover, we found the increased coupling of the

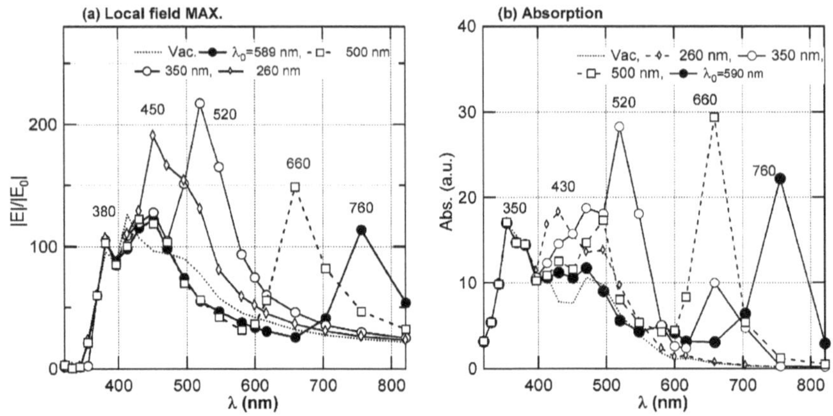

Fig. 19. Enhanced local field (a) and absorption (b) induced by adsorbates at the gap for various absorption wavelengths (λ_0) of molecules for adjacent Ag spheres ($r = 40\,\text{nm}, g = 4\,\text{nm}$). The gap is filled by adsorbates ($v = 4 \times 4 \times 4\,\text{nm}^3$) with different absorption peak (λ_0) (see the text).

LSP via adsorbates provides additional enhancement in the electric field at the gap, and thereby that in SERS as depicted in fig. 19.

As shown in fig. 19(a), the local field maximum without dye molecules ('Vac.' in the figure) shows distinct peaks at ca. 400 nm attributed to isolated LSP on each particle or out-of-phase LSP of adjacent particles, and at 450–500 nm originated from the coupled LSP. Moreover, the Ag nanoparticles with dye at the junction provide the additional field maximum, i.e. ~ 200 as an amplitude enhancement $|E|/|E_0|$ corresponding to 10^9 as a Raman enhancement, comparable with that from the LSP resonances. Since the peak wavelengths accord with those in elastic scattering and enhanced absorption (figs. 18(a) and 18(b) and 19(b)), it is straightforwardly attributed to EM coupling of the LSPs and molecular electronic states. Absorptions at 350 or 260 nm that are quite apart from the LSP resonance at 470 nm surprisingly yield prominent enhancement of about 200 at 450–520 nm. Namely, four distinct DNA bases can provide equivalent coupling with the LSP, since they have similar absorption at 250–300 nm. Therefore, it is not always necessary for adenine or guanine to have particular charge transfer state with Ag surfaces for prominent SERS activity. More detailed analysis in elastic scattering, electric field and emission spectra is awaited to conclude this point.[4] The

[4] Local electric field at the gap was evaluated with a homogeneous grid of 2 nm. More precise and quantitative analysis is being progressed with finer meshes at the gap region, although essential features were already obtained here.

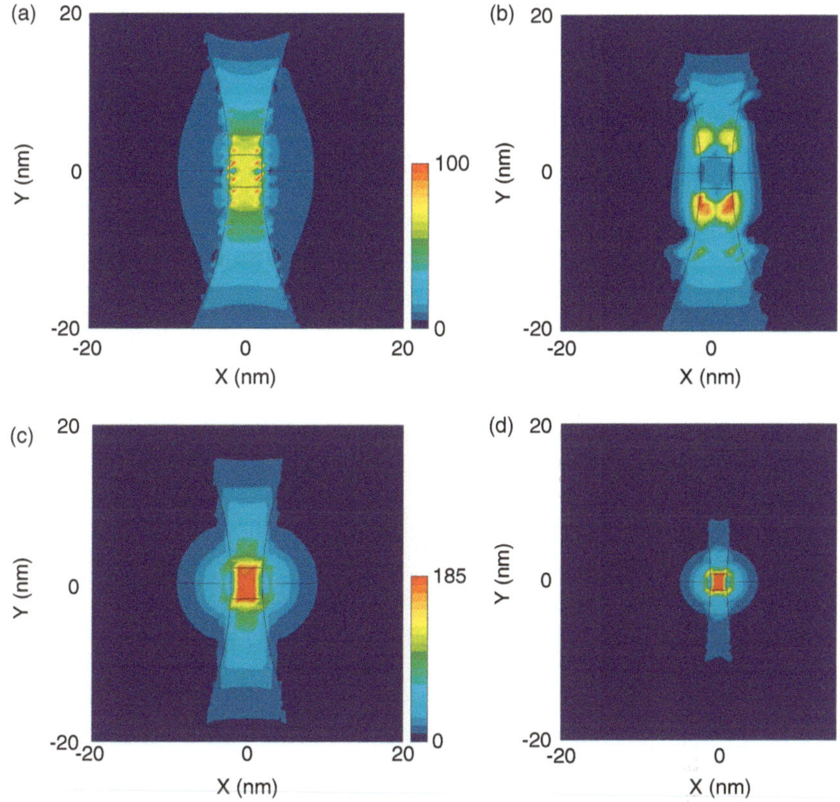

Fig. 20. Electric field distribution (|E|) for two Ag spheres ($r = 40$ nm) with/without adsorbates at the gap: (a) without adsorbates at $\lambda_{ex} = 410$ nm, (b) with adsorbates ($\lambda_0 = 350$ nm) at $\lambda_{ex} = 400$ nm, (c) $\lambda_0 = 350$ nm at $\lambda_{ex} = 500$ nm, (d) $\lambda_0 = 350$ nm at $\lambda_{ex} = 470$ nm. The gap size (g) and adsorbates volume (v) are 4 nm and $4 \times 4 \times 4$ nm^3 for (a)–(c), while $g = 2$ nm and $2 \times 2 \times 2$ nm^3 for (d).

enhancement factor depends on the gap size, and thus on the amount of adsorbate. For instance, more than 400 of amplitude enhancement that corresponds to 2.6×10^{10} of SMS level was obtained for the gap of 2 nm at the additional peak $\lambda_{es} = 470$ nm for $\lambda_0 = 350$ nm ($2 \times 2 \times 2$ nm of adsorbates, see fig. 20(d)). These discussions are based on the electric field maximum at the gap region containing Ag adjacent side and adsorbates, i.e. inside or outside of adsorbates (see figs. 20(a)–20(d)). It is invaluable to demonstrate the actual electric field distribution at the gap, because inherent damping of enhanced incident field is anticipated in the adsorbed layer. The absorption and thermal dissipation in adsorbates may reduce the enhancement by several orders of magnitudes. In this aspect, the local field enhancement reported for bare Ag particles as described above and in literatures (Xu et al., 2000; Kottmann et al., 2001; Futamata et al., 2003) can be overestimated. However, the electric field distribution inside the adsorbed layer is prominently enhanced as depicted in figs. 20(a)–20(d) at the additional peak $\lambda_{es} = 450\,\text{nm}$ (for $\lambda_0 = 350\,\text{nm}$, vide supra), whereas only outside is enhanced at the shorter LSP resonance ($\lambda_{ex} = 400\,\text{nm}$ in fig. 20(b)). This is again due to increased coupling of LSPs at the longer peak through electronic state of molecules at the junction. At shorter wavelength peak, electric field is intrinsically diminished inside of adsorbate due to marked absorption. Much different features were obtained for the same systems without adsorbate as exemplified in fig. 20(a), where enhanced field is induced even in the junction at most wavelengths between 320 and 820 nm. At this stage, it is also useful to discuss the effect of static dielectric constant (ε_s) in a damped oscillator model for adsorbed molecules, e.g. 5.9, 4.5, 3.3, 2.3 for $\lambda_0 = 590, 500, 400$ and 300 nm, respectively. The higher refractive index of the materials located at the junction contributes to reduce the effective gap size d (nm) to $d/\sqrt{\varepsilon_s}$ as in high refractive index media. For instance, the values of 5.9 (ε_s), and 2.3 apparently reduce the gap size (d) to 1.6 and 2.7 nm from the nominal value of 4 nm, respectively. It results in significant but conservative shift of the coupled LSP resonances (fig. 17(a)) and in enhanced field maximum (fig. 5(b)), although particular gap sizes induce different LSP resonances as known in the STM tip-induced emission (Johansson, 1998).

Accordingly, enormous electric field is formed even inside the adsorbates at particular wavelength close to the additional peak in elastic scattering and enhanced absorption. This is attributed to enhanced EM coupling between neighboring LSPs on each Ag nanoparticle with electronic interaction with molecules located at the junction.

In conclusion, it should be noted that most of the observations for SERS active particles, e.g. enormous SERS intensity obtained only for closely adjacent particles, the origin of blinking, the additional peak in elastic scattering, enhanced absorption and additionally enhanced field in the adsorbed molecules are rationalized by this junction model. Furthermore, the inelastic scattering of surface electron simultaneously observed with vast SERS signal can be explained with the similar way, as it was observed only for SERS active state with dye at the junction (as seen in the next section). Thus, it was suggested that the junction benefited by EM coupling between LSP and molecules plays the equivalent role with the atomic scale roughness on the cold-deposited metal films. It was confirmed by Otto et al. (1992) that atomic scale roughness on metal surfaces has specific electronic state relevant for dynamic charge transfer to excited state of molecules, yielding additional enhancement. Adsorbates at the junction could facilitate charge transfer with Ag particles which accompanies inelastic scattering. In addition, new type of resonance effect was suggested that absorption at 260 nm in UV region can couple with the LSP at 500 nm to provide the additional field enhancement at ca. 400 nm, which is much wider compared to ordinary resonance Raman effect for bulk materials. Possibly, it concerned with huge SERS signal or spectral changes in elastic scattering for adenine or guanine on Ag nanoparticles (Futamata et al., 2004b).

3.5. Emission spectra

Bare Ag nanoparticles do not give detectable emission or inelastic scattering spectra in accordance with the recent report for Ag nanoparticles (Jiang et al., 2003), whereas cold deposited clean Cu or Ag surface showed inelastic scattering (Otto, 2001; Otto et al., 2003). This is possibly because smooth facetted Ag surface prepared here by annealing at around 370 K for 1 h possess negligible electron–photon coupling in contrast to those films evaporated at low temperature with marked roughness or defect. On the contrary, hot or SERS-active Ag particles adsorbed with dye molecules provided notable emission or inelastic scattering peaks at $\lambda = 550$–710 nm. Figures 21–23 demonstrate these spectra together with the SERS signal from R6G, here surface coverage was adjusted to be 3–300 molecules/Ag particle (Maruyama and Futamata, 2005ab). These spectra were accumulated for 15–30 min, in which each spectrum was obtained in 90 s and thus 10–20 spectra were coadded. Albeit fluctuations of the peak positions and intensity were evident, they were statistically averaged for sufficiently long time to give reproducible and stable spectra (Maruyama and Futamata, 2005bb). Two distinct group of emission or inelastic scattering peaks were observed at 650–710 nm and 550–600 nm for R6G, CV and MG. These are originated from different mechanisms, since the former peak intensity grows with the surface coverage, whereas the latter shows almost invariant intensity as shown in figs. 21(a)–21(d). For instance, CV with the coverage of 30 molecules/Ag particle gives notable emission at 555 and 645 nm (fig. 21(c)), while MG with 300 molecules/Ag particle shows stronger fluorescence at 710 nm in addition to weaker peak at 570 and 620 nm (see fig. 21(d)). These peak positions could vary for different particles with various sizes, shapes and touching conditions according to distinct LSP resonance wavelengths or inelastic scattering spectra of excited electrons by different surface roughness (Otto, 2001; Otto et al., 2003; Maruyama and Futamata, 2005ab). More clearly,

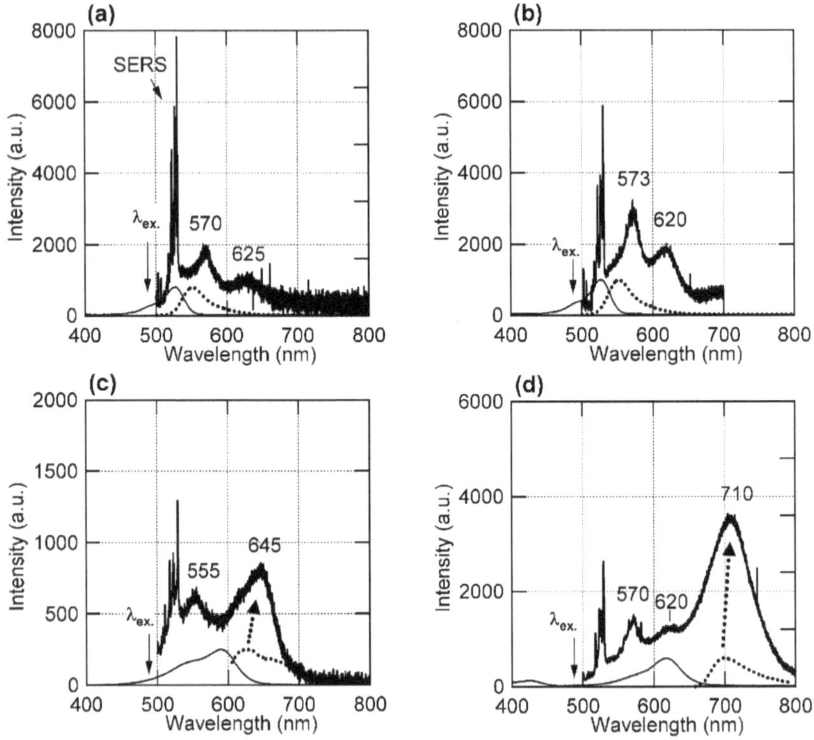

Fig. 21. SERS, emission and inelastic scattering spectra for various dye molecules adsorbed on Ag nanoparticles: (a) R6G (3 molecule/particle), (b) R6G (30 molecule/particle), (c) CV (30 molecule/particle) and (d) MG (300 molecule/particle). Marked differences were observed for the emission bands at 600–750 nm in contrast to similar features at 550–580 nm. Each spectrum was obtained by the accumulation of 20 sequential measurements. The full and dotted line denote absorption and fluorescence spectrum of each dye molecule in solution, respectively.

the fluorescence intensity of dye (at 600–700 nm) is comparable with the shorter wavelength peak (at 550–570 nm) at lower concentration, which prominently increased with the coverage. The former result is partially due to extremely low surface coverage of dye adsorbed on the Ag particles, where fluorescence can be markedly quenched by the energy transfer to metals. Being in accordance with the luminescence of dye on glass substrate as seen in figs. 21(a)–21(d), the emission peak at 650–710 nm is attributed to enhanced fluorescence of dye molecules. The other peak at shorter wavelengths (550–570 nm) shows almost identical peak positions for different adsorbates, although monolayer R6G, CV or MG on the cover slip gives distinct absorption centered at about 535, 605 and 640 nm,

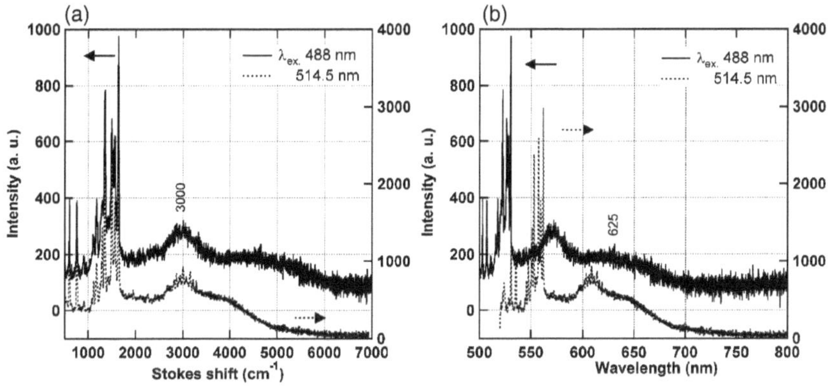

Fig. 22. SERS, emission and inelastic scattering spectra for R6G molecules (30 molecule/particle) adsorbed on Ag nanoparticles using different excitation wavelengths: (a) in Stokes shift and (b) in wavelength scales. Even for different particles, they gave the same Stokes shift ($3,000\,cm^{-1}$) for the shorter wavelength peak.

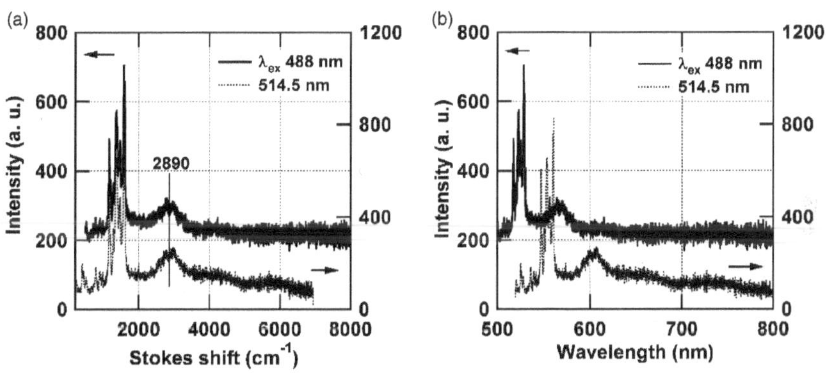

Fig. 23. SERS, emission and inelastic scattering spectra for MG molecules (300 molecule/particle) adsorbed on Ag nanoparticles using different excitation wavelengths: (a) in Stokes shift and (b) in wavelength scales. This figure was obtained for the same Ag particles but with different excitation wavelengths. Two distinct peaks were obtained at the same Stokes shift ($2,890\,cm^{-1}$).

respectively. Based on the Stokes law that luminescence appears at lower energy region of the absorption peak, it seems the emission band at 550–600 nm does not arise from fluorescence of dye. Also this peak was not observed for the bare or SERS-inactive Ag particles. From these observations, it is possibly attributed to the metal, e.g. radiative relaxation of the LSP or inelastic scattering of the excited electrons on metal surfaces by surface roughness. Both these mechanisms could be induced or enhanced by adsorbed dye that works as a defect or specific roughness. Rather similar spectra were reported using STM-induced photo emission measurement for metal substrate with or without adsorbed molecules (Fujita et al., 2001; Meguro et al., 2002). In these cases, radiative relaxation of the excited LSP was assured as the same peak was observed at 1.7 eV (ca. 730 nm) on the bare metal surfaces, whereas fluorescence peaks at 2.0 eV only with adsorbates. In our former paper, we presumably attributed the shorter wavelength peak at 550–600 nm for 488 nm excitation to the same origin (Maruyama and Futamata, 2005ab).

However, it turned out that the shorter wavelength peak gives exactly the same Stokes shift for the excitation at 488 and 514.5 nm similar to ordinary Raman bands for molecular vibrations. As shown in fig. 22, we noted that the same Ag particles with 30 R6G molecules/particle provides exactly the same Stokes shifted peak at $3,000\,\mathrm{cm}^{-1}$ for different excitation wavelengths. On the contrary, the discrepancy of these peaks in the wavelength scale was evident. Surprisingly, the same Stokes shifted peak at $3,000\,\mathrm{cm}^{-1}$ was commonly observed for different hot Ag nanoparticles. It should be noted again that each spectrum in fig. 22 was observed for the same Ag particles only with different excitation wavelengths as well as in fig. 23. This is not arisen from an emissive relaxation of the LSP that should have invariant wavelength, such as 390 nm (calculated value) for an Ag sphere with a radius of 50 nm, instead of the same Stokes shift at $3,000\,\mathrm{cm}^{-1}$. Consequently, this peak is attributed to much smaller energy transition, most probably to inelastic scattering of excited electron by surface roughness (Otto, 2001; Otto et al., 2003) or by adsorbed dye. In the former case, photon annihilation occurs by electronic excitation from occupied state, Shockley surface state (S_2), to image states (S_1), followed by photon creation with electronic transition from S_1 to unfilled S_2.

Here the electron is scattered in S_1 by surface roughness and in S_2 by phonons. In a similar fashion, the electron of our Ag nanoparticles can be excited to an energy level with $\Delta \approx 3,000\,\mathrm{cm}^{-1}$ higher than the original state resulting in an electron–hole pair formation, which is an electronic Raman process. Moreover, the same Stokes shift of $3,000\,\mathrm{cm}^{-1}$ was obtained for different Ag particles. If this peak originates from inelastic scattering of the excited electron by surface roughness, it indicates the Ag nanoparticles possess essentially identical nanoscale morphology. This is not plausible, as the citrate method used here with prolonged annealing provides various shapes of Ag nanoparticles with smooth facetted surfaces (Jiang et al., 2003; Futamata et al., 2004b; Maruyama and Futamata, 2005ab). Instead, adsorbed molecules could inelastically scatter the excited electron, since this was observed only for hot Ag nanoparticles with adsorbates.[5]

In contrast to the shorter wavelength peak, the other peak at 650–710 nm shows the same wavelength instead of the same Stokes shift for different excitation wavelengths. For instance, the Ag particles in fig. 22 have the peaks at 625 and 630 nm, respectively. This is clearly consistent with their assignment to the enhanced fluorescence of adsorbed dye molecules. Essentially identical results were observed for CV and MG. As shown in fig. 23, distinct emission spectra were observed for MG at ca. $2,900\,cm^{-1}$ for 488 and 514.5 nm excitation. Thus, it was demonstrated that inelastic scattering of surface electron was observed with invariant Stokes shift, i.e. ca. $3,000\,cm^{-1}$, irrespective of adsorption species or excitation wavelength only for hot Ag nanoparticles. This was induced by the adsorption of dye probably due to electron transfer interaction between the adsorbed dye and Ag particles.[6]

§ 4. Summary

In the first part of this chapter, recent progress in SMS in SERS was briefly reviewed. SMS in SERS cannot be directly proved in contrast to fluorescence spectroscopy. However, there are various supporting evidences such as complete intermittent features of the blinking signal, or Poisson distribution observed in mixed dye experiments adsorbed on Ag nanoparticles. Also vibrational pumping as well as heating by LSP resonance was ensured by detailed analysis of the temperature dependences of Stokes/anti-Stokes intensity. Various advanced fabrication methods such as NSL or electron beam lithography have been progressed with respect to utilize the LSP most efficiently, and to apply SERS active substrate to bio-medical or environmental science. In order to fabricate metal nanostructures with SMS in SERS, much more precise control of morphology, sizes or spacing is necessary as well as to characterize them at a nanometer scale.

---

[5] Next issue to conclude this is to fabricate Ag nanoparticles with various morphologies in addition to adopt various adsorbates with distinct molecular vibration or electronic states. Preliminary, we confirmed that the citrate method with much shorter annealing time at 373 K, e.g. 5–10 min, compared to the standard time of 45 min, which was used in most of the experiments here, did not change the inelastic peak position.

[6] Further effort is requisite to reach the definitive conclusion, since each dye molecule provides significantly different fluorescence spectra associated with distinct electronic state, and then often overlaps with the short wavelength peak.

The following part was devoted to demonstrate the critical importance of the junction of closely adjacent metal nanoparticles. In good accordance between experiments and theoretical evaluation, only touching metal nanoparticles provide enormous electric field to yield SMS, whereas isolated particles, except nanostructures with sharp edges, gave only modest enhancement. This is attributed to the increased coupling of the LSP of individual particles at the junction. Blinking is raised from thermal activation, most probably thermal diffusion of adsorbed molecules between the junction and marginal region. An elastic scattering peak at ca. 600 nm was extinguished during inactivation process of enormous enhancement in SERS of Ag touching particles with adsorbates. Numerical simulations using the FDTD-3D method proved that this peak originates from implicit electromagnetic coupling between the LSP of the Ag particles and absorption of dye located at their junction. Furthermore, it was found that pronouncedly enhanced electric field is formed at the additional peak even inside the adsorbed molecules due to increased LSP coupling. Thus, the junction model could rationalize most of observations in SMS-SERS, such as enormous SERS signal only for touching particles, the origin of the blinking, additional peak in elastic scattering for SERS active state and inelastic scattering of surface electron. Further extensive study would provide conclusive evidences on the contribution of electronic excited state of adsorbates, in particular, at the junction. In addition, the charge transfer interaction was evidenced by distinct fluctuation and polarization properties observed for particular modes. Detailed evaluation using various molecules with distinct absorption wavelengths is necessary to get deeper insight into both of these mechanisms.

Acknowledgment

The authors appreciate Prof. Andreas Otto (Henrich-Heine-Universität Düsseldorf) for his instructive discussion on various aspects of SERS, especially on inelastic scattering of metal electron. We also thank Dr. Mitsuru Ishikawa (AIST) for useful collaboration. This research was financially supported in part by Grant-in-Aid for Scientific Research (B) 17350013 by Japan Society for the Promotion of Science (JSPS).

References

Aravind, P.K., Metiu, H., 1983, Surf. Sci. **124**, 506.

Bohren, F., Hoffman, D.R., 1983, Adsorption and Scattering of Light by Small Particles, Wiley, New York, p. 344.

Bosnick, K.A., Jiang, J., Brus, L.E., 2002, J. Phys. Chem. **B106**, 8096.

Chew, H., Kerker, M., 1985, J. Opt. Soc. Am. B **2**, 1025.

Constantino, C.J.L., Lemma, T., Antunes, P.A., Aroca, R., 2002, Spectrochim. Acta **A58**, 403.

Creighead, H.G., Glass, A.M., 1981, Opt. Lett. **6**, 248.

Dick, L.A., Hayes, A.J., Van Duyne, R.P., 2000, J. Phys. Chem. B **104**, 11752.

Dick, L.A., McFarland, A.D., Haynes, L.C., Van Duyne, R.P., 2002, J. Phys. Chem. B **106**, 853.

Dieringer, J.A., McFarland, A.D., Shah, N.C., Stuart, D.A., Whitney, A.V., Yonzon, C.R., Young, M.A., Zhang, X., Van Duyne, R.P., 2006, Faraday Discussions, **132**, 9.

Doering, W.E., Nie, S., 2003, Anal. Chem. **75**, 6171.

Emory, S.R., Nie, S.M., 1998, J. Phys. Chem. B **102**, 493.

Félidj, N., Truong, S.L., Aubard, J., Lévi, G., Krenn, J.R., Hohenau, A., Leitner, A., Aussenegg, F.R., 2004, J. Chem. Phys. **120**, 7141.

Fujita, D., Ohgi, T., Deng, W.-L., Ishige, K., Okamoto, T., Yokoyama, S., Kamikado, T., Mashiko, S., 2001, Surf. Sci. **493**, 702.

Futamata, M., Maruyama, Y., 2005, J. Mol. Struct. **735–736**, 75.

Futamata, M., Maruyama, Y., Ishikawa, M., 2002, Vibrational Spectrosc. **30**, 11965.

Futamata, M., Maruyama, Y., Ishikawa, M., 2003, J. Phys. Chem. **B107**, 7607.

Futamata, M., Maruyama, Y., Ishikawa, M., 2004a, J. Phys. Chem. **B108**, 13119.

Futamata, M., Maruyama, Y., Ishikawa, M., 2004b, Vibrational Spectrosc. **35**, 121.

Garcia-Vidal, F.J., Pendry, J.B., 1996, Phys. Rev. Lett. **77**, 1163.

Garoff, S., Weitz, D.A., Gramila, T.J., Hanson, C.D., 1981, Opt. Lett. **6**, 245.

Goulet, P.J.G., Pieczonka, N.P.W., Aroca, R.F., 2005, J. Raman Spectrosc. **36**, 574.

Hao, E., Schatz, G.C., 2004, J. Chem. Phys. **120**, 357.

Haran, G., 2004, Israel J. Chem. **44**, 385.

Haynes, L., Van Duyne, R.P., 2001, J. Phys. Chem. B **105**, 5599.

Haynes, L., Van Duyne, R.P., 2003a, J. Phys. Chem. B **107**, 7426.

Haynes, L., Van Duyne, R.P., 2003b, Nano Lett. **3**, 939.

Hildebrandt, P., Stockburger, M., 1984, J. Phys. Chem. **88**, 5935.

Itoh, T., Hashimoto, K., Ikehata, A., Ozaki, Y., 2003, Appl. Phys. Lett. **83**, 5557.

Jensen, T.R., Duval, M.L., Kelly, K.L., Lazarides, A.A., Schatz, G.C., Van Duyne, 1999, J. Phys. Chem. B **103**, 9846.

Jiang, J., Bosnick, K., Maillard, M., Brus, L., 2003, J. Phys. Chem. **B107**, 9964.

Johansson, P., 1998, Phys. Rev. B **58**, 10823.

Kahn, I., Cunnigham, D., Lazar, S., Graham, D., Smith, W.E., McComb, D.W., 2006, Faraday Discussions **132**, 171.

Kerker, M., (Ed.) 1990, Surface Enhanced Raman Scattering, SPIE, Vol. **MS10**.

Kneipp, K., Kneipp, H., Itzkan, I., Dasari, R.R., Feld, M.S., 1999, Chem. Rev. **99**, 2957.

Kneipp, K., Wang, Y., Kneipp, H., Perelman, L.T., Itzkan, I., Dassari, R.P., Feld, M.S., 1997, Phys. Rev. Lett. **78**, 1667.

Kottmann, J.P., Martin, O.J.F., Smith, D.R., Schultz, S., 2001, Chem. Phys. Lett. **341**, 1.

Krug, J.T., Wang, G.D., Emory, S.R., Nie, S., 1999, J. Am. Chem. Soc. **121**, 9208.

Kudelski, A., Pettinger, B., 2000, Chem. Phys. Lett. **321**, 356.

Kunz, K.S., Luebbers, R.J., 1992, The Finite Difference Time Domain Method for Electromagnetics, CRC Press, Boca Raton.

Le Ru, E.C., Meyer, M., Etchegoin, P.G., J. 2006, Phys. Chem. **B110**, 1944.

Lee, P.C., Meisel, D.P., 1982, J. Phys. Chem. **86**, 3391.

Liver, N., Nitzan, A., Gersten, J.I., 1984, Chem. Phys. Lett. **111**, 449.

Maher, R.C., Etchegoin, P.G., Le Ru, E.C., Cohen, L.F., 2006a, J. Phys. Chem. **B110**, 11757.

Maher, R.C., Cohen, L.F., Gallop, J.C., Le Ru, E.C., Etchegoin, P.G., 2006b, J. Phys. Chem. **B110**, 6797.

Maruyama, Y., Ishikawa, M., Futamata, M., 2001, Chem. Lett. **30**, 834.
Maruyama, Y., Ishikawa, M., Futamata, M., 2004, J. Phys. Chem. **B108**, 673.
Maruyama, Y., Futamata, M., 2005a, Chem. Phys. Lett. **412**, 65.
Maruyama, Y., Futamata, M., 2005b, J. Raman Spectrosc. **36**, 581.
Meguro, K., Sakamoto, K., Arafune, R., Satoh, M., Ushioda, S., 2002, Phys. Rev. B **65**, 165405.
Michaels, M., Nirmal, M., Brus, L.E., 2000, J. Phys. Chem. B **104**, 11965.
Nie, S., Emory, S.R., 1997, Science **275**, 1102.
Osborne, M.A., Balasubramanian, S., Furey, W.S., Klenerman, D., 1998, J. Phys. Chem. **102**, 3160.
Otto, A., 2001, Phys. Stat. Sol. (a) **188**, 1455.
Otto, A., 2006, J. Raman Spectrosc. **37**, 937.
Otto, A., Bruckbauer, A., Chen, Y.X., 2003, J. Mol. Struct. **661–662**, 501.
Otto, A., Mrozek, I., Grabhorn, H., Akemann, W., 1992, J. Phys.: Condens. Matter **4**, 1143.
Palik, D., 1998, Handbook of Optical Constants of Solids, Academic Press, London, p. 351.
Peyser-Capadona, L., Zheng, J., González, J.I., Lee, T.-H., Patel, S.A., Dickson, R.M., 2005, Phys. Rev. Lett. **94**, 058301.
Semin, D.J., Lo, A., Roark, S.E., Skodje, R.T., Rowlen, K.L., 1996, J. Chem. Phys. **105**, 5542.
Sharaabi, Y., Shegai, T., Haran, G., 2005, Chem. Phys. **318**, 44.
Shegai, T.O., Haran, G., 2006, J. Phys. Chem. B **110**, 2459.
Svedberg, F., Käll, M., 2006, Faraday Discussions **132**, 35.
Taflove A. (Ed.), 2000, Computational Electrodynamics: The Finite-Difference Time-Domain Method (2nd ed.), Artech House, Norwood.
Weiss, A., Haran, G., 2001, J. Phys. Chem. B **105**, 12348.
Xu, H., Aizpurua, J., Käll, M., Apell, P., 2000, Phys. Rev. E **62**, 4318.
Xu, H., Käll, K., 2002, Phys. Rev. Lett. **89**, 246802.
Yamaguchi, Y., Maruyama, Y., Ishikawa, M., Futamata, M., 2005, J. Kor. Phys. Soc. **47**, S1.
Yee, K.S., 1966, IEEE Trans. Antennas Propag. **14**, 302.
Zheng, J., Dickson, R.M., 2002, J. Am. Chem. Soc. **124**, 13982.

Chapter 3

Near-field effects in tip-enhanced Raman scattering

by

Yasushi Inouye[a,c], Prabhat Verma[b], Taro Ichimura[a], Satoshi Kawata[b,c]

[a]*Graduate School of Frontier Biosciences, Osaka University, Osaka, Japan*

[b]*Graduate School of Engineering, Osaka University, Osaka, Japan*

[c]*RIKEN, Japan*

Tip Enhancement
Advances in Nano-Optics and Nano-Photonics
ISSN: 1871-0018

S. Kawata and V.M. Shalaev (Editors)
© *2007 Elsevier B.V. All rights reserved.*
DOI: 10.1016/S1871-0018(06)01003-X

Contents

§ 1. Introduction

A metallic needle having a nanotip generates surface plasmon polaritons (SPPs) locally in the vicinity of the tip apex. As the SPPs couple with photons resonantly, photons are densely confined in close vicinity of the nanotip. These confined photons illuminate specimen or interact with matters at nanometric scale. Such near-field scanning optical microscopy (NSOM) (Kawata, 2001) was invented by two groups independently, including our group, in 1994 (Inouye and Kawata, 1994; Gleyzez et al., 1995). Then, we proposed and developed nano-Raman spectroscopy using the tip-enhanced NSOM in 1999 for chemical nanoanalysis of molecules and nanomaterials.

Tip-enhanced Raman scattering (TERS) is based on the same phenomenon as surface-enhanced Raman scattering (SERS), which occurs on island structures consisting of metallic nanoparticles. Chemical enhancement effect, which occurs due to the formation of a complex of sample molecules and metal atoms, as well as physical enhancement effect (e.g. SPPs) can be expected in TERS. The differences between TERS and SERS are as follows:

(1) TERS occurs on a single metallic nanotip which specifies position for observation and analysis with nanometric accuracy, while SERS is induced on plenty of metallic nanostructures.
(2) A nanotip is able to perturb molecules dynamically, electrically, or magnetically as the tip is controlled by another scanning probe microscopy, e.g. atomic force microscopy (AFM) or scanning tunneling microscopy. Hence, TERS provides especial circumstances for Raman spectroscopy like high pressure, intense electric field, and so on.

In this chapter, we will describe such tip effects on TERS including local induction of nonlinear Raman scattering, coherent anti-Stokes Raman scattering for novel vibrational spectroscopy. In Section 2, the mechanism of electromagnetic field enhancement in TERS will be described. In Section 3, TERS spectra of Rhodamine 6G will be shown and its unique spectral shape due to the chemical interaction between the tip and the molecules will be discussed. In Section 4, TERS spectra of C_{60} molecules will be shown and the spectral shift due to the tip force will be discussed. In Section 5, tip-enhanced CARS imaging will be introduced and the enhancement mechanism due to the nonlinearity will be discussed.

§ 2. Tip enhancement of Raman scattering

2.1. Metallic probe as a nanolight source

When a metallic probe which has a nanometric tip is illuminated with an optical field, conductive free electrons collectively oscillate at the surface of the metal (fig. 1). The quantum of the induced oscillation is referred to as SPP. The electrons (and the positive charge) are concentrated at the tip apex and strongly generate external electric field. Photon energy is confined in the local vicinity of the tip. Therefore, the metallic tip works as a photon reservoir.

The local electric field is used as a nanolight source to excite photon–matter interaction, including fluorescence (Hayazawa et al., 1999; Hamman et al., 2000), second harmonic generation (Zayats and Sandoghdar, 2000) and Raman scattering (Inouye et al., 1999; Hayazawa et al., 2000; Stöckle et al., 2000) under the tip. In particular, the enhancement of Raman scattering by metallic nanostructures has been widely investigated as SERS since 1970s (Chang and Furtak, 1981; Moskovits, 1985). For the SERS effect, nanoparticles and their aggregates are extensively used in various models to understand the physical behavior of the local electric field and the magnitude of the enhancement (Aravind and Metiu, 1983; Bohren and Huffman, 1983). Similarly, in case of the tip enhancement, the local electric field can be calculated with numerical analysis. Figure 2(a) shows the local field distribution near a metallic probe tip calculated via the finite-difference time-domain (FDTD) method, which gives a numerical solution of

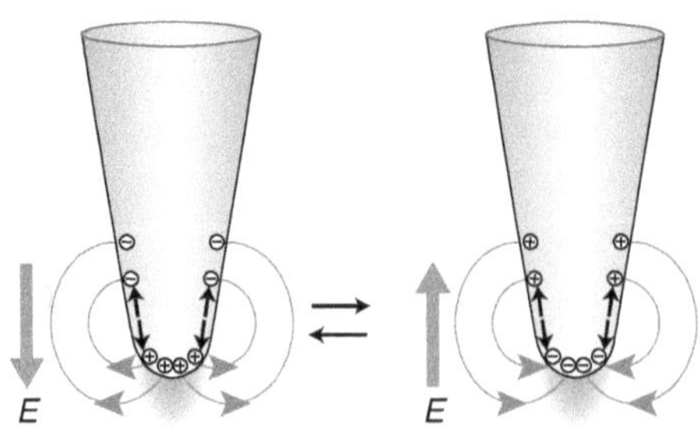

Fig. 1. Oscillation of electrons in a metallic tip structure.

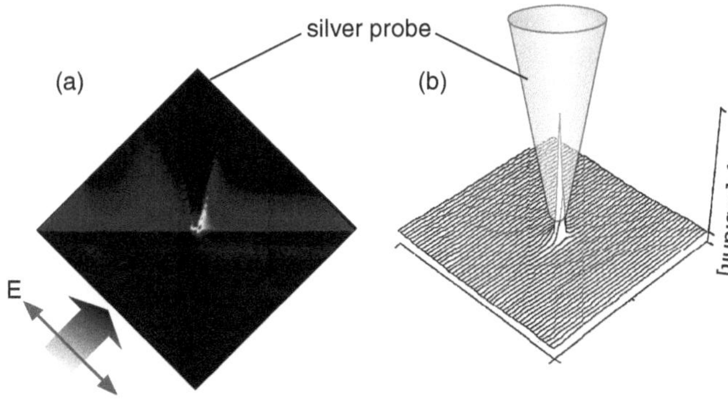

Fig. 2. Intensity distribution of enhanced electric field localized at a silver probe tip (a) in a plane containing the tip axis and (b) on a dielectric substrate, calculated with the FDTD method.

Maxwell's equations (Furukawa and Kawata, 1998). It is seen that the optical field is highly confined into a tiny volume near the tip end. In addition to the spatial confinement, the optical field is strongly amplified by a factor of 50 at the maximum. Figure 2(b) shows a light intensity distribution on the dielectric substrate. Figure 2 shows the three-dimensional confinement of the optical field and amplification of the light intensity in the confined spot. The size of the spot is comparable to that of the probe tip.

It has been shown that a metallic probe highly localizes and strongly amplifies optical field through the resonance effect of the plasmon polaritons at the probe tip. This concept has made it possible to optically observe the nanometric samples with a nanometric spatial resolution. Many groups have applied it to spectroscopic imaging which provide optical and spectroscopic information, differently from other types of scanning probe microscopy. There have been lots of studies aiming at optimization of the probe shape to improve both the localization and enhancement. Several reports suggested that a tip truncated to a finite size (< wavelength) gives a strong enhancement (Kottmann et al., 2001a; Martin et al., 2001; Krug et al., 2002). It is possible to fabricate a designed structure by using a focused ion beam (FIB).

§ 3. Enhancement mechanism for Rhodamine 6G

In this section, we compare resonant Raman spectra (RRS), SERS, surface enhanced resonant Raman spectra (SERRS) and TERS of Rhodamine 6G (R6G), and discuss the enhancement mechanism for R6G in the near-field scattering. Raman bands of R6G in experimental results have been assigned on the basis of density-functional theory (DFT) calculations.

For the SERRS and the TERS measurements, R6G molecules ($2.6\,\mathrm{mmol\,L^{-1}}$) dissolved in ethanol were cast and dried on a silver island film (thickness: 8 nm). For the measurement of the off-resonance Raman scattering as well as the RRS, R6G molecules ($50\,\mathrm{mmol\,L^{-1}}$ and $50\,\mu\mathrm{mol\,L^{-1}}$, respectively) dissolved in ethanol were used, with careful subtraction of solvent peaks.

Quantum chemical calculations to analyze and identify the vibrational frequencies, normal modes, and Raman intensities for R6G molecule were carried out using the B3LYP functional (Backe, 1993; Lee et al., 1988) with the basis set of 6-311++G(d,p). The calculations for electronic excited states of the R6G molecule were performed by the time-dependent DFT (TD-DFT) method (Bauernschmitt and Ahlrichs, 1996) using the B3LYP functional with the same basis set. These calculations were performed using the GAUSSIAN98 Revision A.9 program package (Frisch et al., 1998).

Raman bands observed in the experiments were assigned (Watanabe et al., 2005) on the basis of the DFT calculations. The band at $1389\,\mathrm{cm^{-1}}$ is assigned to degenerate vibrational modes of ν_{122}, ν_{123}, ν_{124}, and ν_{125}, which correspond to the symmetric bending motions of individual methyl groups in R6G molecules. Three bands are assigned to alternative vibrational modes; that is, the band at $705\,\mathrm{cm^{-1}}$ is assigned to the calculated mode of either ν_{58} or ν_{59}, the band at $1061\,\mathrm{cm^{-1}}$ is due to the mode of either ν_{90} or ν_{91}, and $1490\,\mathrm{cm^{-1}}$ is either of ν_{140} or ν_{141}, respectively. Nine Raman bands with strong scattering intensities at 613, 775, 1130, 1278, 1364, 1389, 1512, 1577, and $1651\,\mathrm{cm^{-1}}$ are assigned to the corresponding vibrational modes ν_{53}, ν_{65}, ν_{96}, ν_{111}, ν_{117}, ν_{122}, ν_{146}, ν_{151}, and ν_{154}, respectively.

In general, enhancement mechanisms of SERS are classified into two mechanisms (Moskovits, 1985; Campion and Kambhampati, 1998). One is an electromagnetic (EM) enhancement mechanism that enhances whole molecular Raman scatterings due to local SPP (Efrima and Metiu, 1979; Otto et al., 1980). Another is a chemical or charge transfer (CT) mechanism that is caused by the changes in the electronic structure of molecules adsorbed on the metal surfaces (Billmann et al., 1980; Campion et al., 1995), where the increase of specific SERRS signal and the changes in Raman frequencies against the RRS spectra take place. Therefore, the EM mechanism would be predominant, as for the SERRS measurement of R6G in the present experimental conditions.

3.1. *RRS and SERRS spectra of R6G*

Figure 3 shows RRS and SERRS spectra together with the off-resonance Raman spectrum of R6G. The RRS spectrum of R6G in ethanol solution

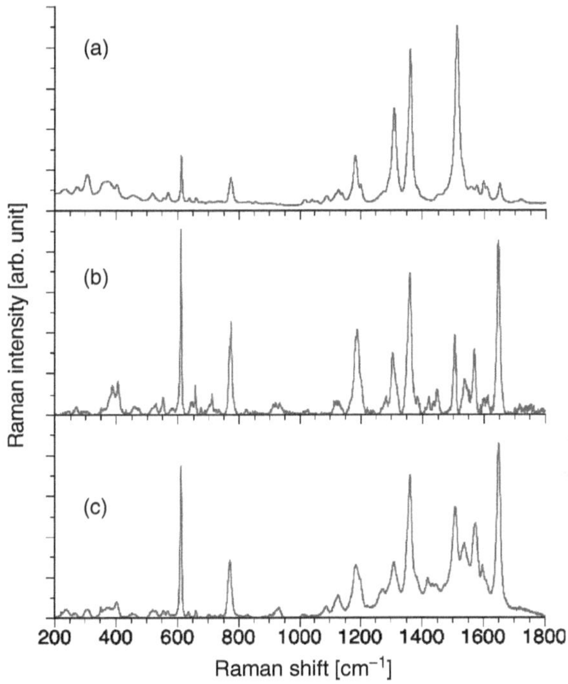

Fig. 3. Comparison of the Raman spectra of R6G. (a) The offresonance Raman spectrum (Ex: 1064 nm) of R6G in ethanol solution, (b) RRS spectrum (Ex: 488 nm) of R6G crystals, and (c) SERRS spectrum (Ex: 488 nm) of R6G adsorbed on colloidal silver are plotted against the corresponding frequencies (cm^{-1}). The peak assignment of some important peaks (ν_{55}, ν_{147}, ν_{151}, and ν_{154}) is shown.

(excitation wavelength = 488 nm) and the SERRS of R6G adsorbed on a silver island film (excitation wavelength = 488 nm) as well as the off-resonance Raman spectrum of R6G in ethanol solution (excitation wavelength = 1064 nm) are plotted versus corresponding frequencies in the region of 200–1800 cm^{-1}. The RRS spectral pattern (fig. 3(b)) differs from the off-resonance Raman spectrum (fig. 3(a)). In particular, RRS bands at 657 cm^{-1} (ν_{55}), at 1577 cm^{-1} (ν_{151}), and at 1649 cm^{-1} (ν_{154}) exhibit specific intensity enhancement against the corresponding off-resonance Raman bands. Among these specific RRS bands, the band ν_{154} is mainly due to a C =C totally symmetric stretch of the xanthene ring. The vibrational modes of both bands ν_{55} and ν_{151} are localized in the phenyl ring, whereas the band ν_{55} is due to a symmetric deformation and the band ν_{151} is due to a C =C symmetric stretching motion, respectively. In principle, the intensities of the RRS bands depend on the nature of the resonant excited state; that is, only a vibrational mode which is active to vibrational transitions with electronic excitations between the ground and resonant excited states demonstrates RRS intensity enhancement. In particular, totally symmetric vibrations, which coincide with the change in molecular geometry under electronic excitations (the Albrecht A term of RRS intensity (Albrecht, 1961)), are strongly enhanced. According to the TD-DFT calculation of an electronic transition of an R6G molecule that is excited at 488 nm of incident light for Raman excitation, the calculated ($S_0 \rightarrow S_1$) excitation energy (2.913 eV) corresponds to the main electronic absorption ($\lambda_{max} \approx 520$ nm) of R6G in ethanol solution. This $S_0 \rightarrow S_1$ excitation is assigned to the summation of two different transitions. The major transition corresponds to the promotion of an electron from the highest occupied orbital (ϕ_{118}) to the lowest unoccupied xanthene π orbital (ϕ_{119}), and the minor corresponds to a CT transition from an occupied xanthene π orbital (ϕ_{117}) to an unoccupied phenyl π orbital (ϕ_{121}). Hence, the three RRS signals ν_{55}, ν_{151}, and ν_{154} are strongly enhanced because their vibrational modes would mimic the changes in their molecular geometries between the ground and the resonant excited states. Remarkably, benzene ring vibrational modes (ν_{55} and ν_{151}), which would seem to have no contributions superficially to the main electronic absorption, are largely enhanced in the RRS spectra.

The SERRS spectrum of R6G adsorbed on silver thin film surfaces (fig. 3(c)) is quite analogous to the RRS spectral pattern of R6G in ethanol solution (fig. 3(b)). Only a SERRS band at 235 cm^{-1} has no corresponding peak in the RRS spectrum. This band is assigned to the Ag–N stretching mode in accordance with the previous assignments (Hildebrandt and Stockburger, 1984). An enhancement factor estimated from the SERRS/RRS signal ratio at 1650 cm^{-1} (ν_{154}) is approximately 1×10^4 under the same experimental condition, except for the samples with or without silver film. SERRS signals at 1538 cm^{-1} (ν_{147}) and at 1575 cm^{-1} (ν_{151}) relatively increase to the RRS spectral pattern; the factor is, at most, <1.5 times normalized by the peak intensity at 1650 cm^{-1} (ν_{154}). No significant changes in Raman frequencies are observed in comparison with the corresponding SERRS and RRS bands. Since the EM mechanism for enhancement is predominant, the adsorption of R6G on the silver surfaces through the Ag–N bonding would be expected to be weak.

3.2. *TERS spectra of R6G*

Figure 4 shows a spectral mapping of the TERS spectra of R6G molecules dispersed on a silver island film using the silver-coated probe tip at 30 nm intervals (Hayazawa et al., 2001). Among these TERS spectra, spectra-a, -b,

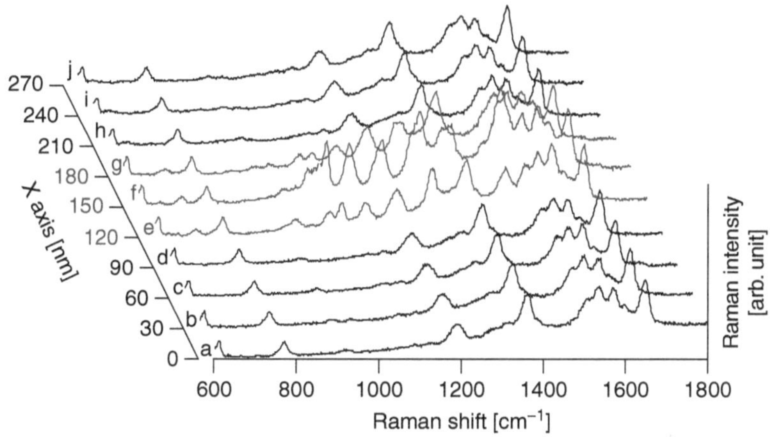

Fig. 4. TERS spectral mapping of R6G adsorbed on silver film at several positions. Alphabets (a–j) correspond to the tip position where the TERS spectra were observed.

-c, -d, -h, and -i together with spectrum-j are quite similar to the far-field SERRS spectrum (fig. 3(c)). On the other hand, some of the spectra, such as the spectra-e, -f, and -g exhibit anomalous spectral patterns that are different from those in the far-field SERRS (fig. 3(c)). Such anomalous spectral patterns, in detail, seem to have an overlap of several new peaks, in addition to the TERS spectral pattern analogous to the far-field SERRS. These anomalies suggest two possible phenomena: one is a case of coexistence of contaminant materials in the anomalous regions of spectra-e, -f, and -g, and another is a case that Raman peaks of R6G molecule itself change in intensities and/or in frequencies. The differential spectrum of spectrum-f (the anomalous region) against spectrum-a (the normal region) was analyzed from the former viewpoint by using a commercial library search (Know It All Information System) of Raman peaks. However, no significant results could be obtained from the library search. The anomalous TERS spectrum-f was then compared with the vibrational modes from the latter point of view. The frequencies of most TERS bands were found to be consistent with those of the SERRS peaks, although some minor SERRS peaks were missing in the TERS spectrum. In contrast, the intensities of some TERS bands such as bands at 702 cm^{-1} (either ν_{58} or ν_{59}), at 1027 cm^{-1} (ν_{86}), at 1061 cm^{-1} (either ν_{90} or ν_{91}), at 1120 cm^{-1} (ν_{96}), at 1269 cm^{-1} (ν_{111}), and at 1457 cm^{-1} (ν_{131}) are quite inconsistent with those of the SERRS being strongly enhanced.

In the TERS spectra, such as spectra-a, -b, -c, -d, etc., in fig. 4, an averaged enhancement factor by the metallic tip is estimated from the TERS/SERRS signal ratio at 1647 cm^{-1} (ν_{154}). The averaged enhancement factor by the tip reaches approximately 1×10^2 which is extrapolated for the spot area (400 nmϕ) from the metallic tip diameter (40 nmϕ). Meanwhile, a specific enhancement factor is estimated from the signal ratio of the most strongly enhanced peak at 1054 cm^{-1} (ν_{90} or ν_{91}) in spectrum-f. The specific enhancement factor reaches approximately 2×10^4. In TERS spectroscopy, enhancement mechanisms caused by the metallic tip are thought to be the same as the SERS, the EM, and the CT. A ratio of the specific/averaged enhancement factors at least correspond to the contribution of the CT mechanism toward the enhancement. The ratio obtained in the TERS experiment is less than or equal to 10^2; this magnitude is as much as a generally accepted degree of the enhancement effect caused by the CT mechanism (Moskovits, 1985; Campion and Kambhampati, 1998).

Among the seven peaks of the anomalous TERS signals, two modes (ν_{86} and ν_{131}) are mainly localized in a pair of methyl groups adjacent to the xanthene ring, and four modes (ν_{58}, ν_{59}, ν_{90}, and ν_{111}) are mainly localized in the phenyl region. In short, the vibrating regions of all these vibrational modes are located on the opposite side of the amino group that is bonded to the silver surface. The CT mechanism can be explained by the resonant Raman mechanism in which charge-transfer excitations (either from the metal to the adsorbed molecule, or vice versa) occur at the energy of incident laser frequency (Campion and Kambhampati, 1998). In the case of the TERS, the electronic wave function overlap between the silver atoms of metallic tip apex and molecular orbitals of the adsorbed species is a requirement for the charge-transfer excitations. With an excitation frequency at 488 nm (2.54 eV), adsorbates such as R6G having a narrow HOMO-LUMO band gap are electronically excited from the lower occupied orbitals to a Fermi level of silver surfaces, or from the Fermi level to higher virtual orbitals. In the R6G molecule, occupied orbitals (ϕ_{116} and ϕ_{117}) and a virtual orbital (ϕ_{120}) are the candidates for the CT electronic excitations in comparison with their energy differences from the Fermi level and the excitation frequency. An electron density distribution of the occupied orbitals (ϕ_{116} and ϕ_{117}) is localized in the xanthene ring, and that of the virtual orbital (ϕ_{120}) is centralized in the phenyl ring. In our experimental setup, the silver metallic tip approaches from the opposite direction toward the adsorbed R6G molecules on the silver island film and touches the molecules by the contact mode of the AFM standard operation. As the silver atoms of the tip at the apex interact with the virtual orbital (ϕ_{120}) at the phenyl ring of the R6G molecules, the bands (ν_{58}, ν_{59}, and ν_{90}) due to symmetric deforming motions of the phenyl ring could be very highly enhanced in their TERS intensities. Hence, the specific intensity enhancement of some of these TERS bands could be understood with the CT mechanism caused by the silver tip together with the approaching direction of the tip toward the molecule. From the viewpoint of the molecular orientations, it is strongly suggested that R6G molecules in the anomalous regions turn the opposite part of the molecules against the adsorption site to the direction of the metallic tip. In other words, the R6G molecules are in a state of higher order such as assembly or aggregation in the anomalous regions.

The TERS band ν_{90} due to a CH_2 twisting motion of the ethyl ester part of the phenyl region is away from the part where the electron density of the virtual orbital ϕ_{120} is significantly high. Accordingly, a large intensity enhancement effect would hardly be expected to the band by means of the CT mechanism. The TERS bands ν_{86} and ν_{131} due to a rocking motion of the methyl group adjacent to the xanthene ring also give the same result. Comparing both the enhancement factors of SERRS/RRS and TERS/SERRS, the averaged enhancement factor of TERS/RRS and the factor at the specific band are estimated to be approximately 1×10^6 and 2×10^8, respectively. In particular, the enhancement factor at the specific band is beyond the values that are usually obtained in the conventional SERS measurement. Recent progress in a single molecule detection of SERS reveals that a gigantic local plasmonic field is generated at a junction between two connecting metallic nanoparticles (Futamata et al., 2003). In our experiments, R6G molecules are tightly sandwiched between the silver island film (8 nm thickness) and the silver tip (40 nm ϕ) by the contact mode of the AFM. These contacting areas could be the junctions (hot spots). Supposing that the hot spot is an area where all particles (40 nm ϕ) are contacting within the gap of 1 nm, the diameter of the hot spot area would be less than 1 nm on the basis of the FDTD method (Futamata et al., 2003). Accordingly, the enhancement factor at the specific band would reach the $10^{11} - 10^{12}$ level which is estimated for the conventional TERS observable area by the area of the hot spot. This means a metallic tip-surface doubly enhanced Raman scattering has a possibility to detect the molecular vibration of sub-nanometer size at the level of single molecule. Under such a strong electric field, two enhancement mechanisms would be tentatively considered. One is an extended CT mechanism where the bands of SERRS inactive vibrational modes could be enhanced either with a vibronic coupling (the Albrecht B term 49) or with electronic excitations that are not excited by the classical, linear CT mechanism. Another is a polarized EM mechanism where Raman bands could be enhanced by a highly polarized electric field coupled with their Raman polarizability tensors.

§ 4. Near-field Raman scattering from Carbon-60

In this section, we will show some TERS results from the aggregates of carbon-60 (C_{60}) molecules (Verma et al., 2006). Apart from the near-field enhancement of various Raman modes, we would also show the gap-mode enhancement using a combination of SERS and TERS, and pressure effects on Raman spectra, when C_{60} molecules are uniaxially pressurized by the tip. Samples for these experiments were prepared by dispersing the commercially available C_{60} molecules on glass cover slip. This was done by dropping and evaporating C_{60}/toluene diluted solution (100 nM) on a cover slip, and then choosing an appropriate area from the dried sample. TERS spectra were measured from different parts of the sample by approaching a silver-coated silicon tip to the aggregates. The spectrum (a) in fig. 5 shows one of such TERS spectra from a 150-nm-large and 20-nm-thick aggregate of C_{60} molecules. A corresponding far-field spectrum, where the tip was taken off the sample, is presented by (b) in fig. 5. The far-field spectrum shows just a trace of the two Ag-modes at 496 and 1470 cm^{-1}. However, as seen in the near-field spectrum, all Raman modes, including two Ag-modes and seven Hg-modes, are enhanced. The degrees of enhancement are different for different modes.

4.1. The gap-mode enhancement

The confined size of the non-propagating photons is typically about 30 nm (same as the size of the tip apex). Therefore, for the best results in TERS experiments, it is better to choose the sample with a thickness of about

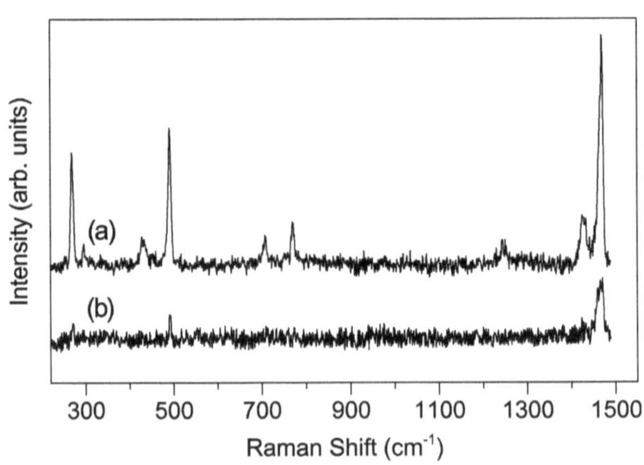

Fig. 5. (a) The near-field and (b) the far-field spectra of C_{60} aggregates.

20–25 nm. A thinner sample results in weaker spectra. In order to investigate thinner sample with further improved enhancement, a combination of SERS and TERS techniques was employed. It has been theoretically evaluated that if two metal nanoparticles, which are quite close to each other, are irradiated, then there is a huge field enhancement within the "gap" between them (Kottmann et al., 2001b; Futamata et al., 2003). The strength of this enhancement depends on the particle size, the gap size and the polarization direction. Hence it can be predicted that if there is a sample in this gap, the Raman scattering from that sample would be strongly enhanced. The Raman spectra thus obtained can be called the "gap-mode" Raman spectra. When a tip, which is used in TERS experiments, is approached to the metal-island-film, which is used as the substrate in SERS experiments, then a similar condition is obtained, and a "gap" is created between the tip and the metal film. In order to investigate the gap-mode enhancement, a sample was prepared by dispersing a diluted (100 nM) C_{60}/toluene solution on a 4-nm-thick silver layer, which was deposited on the glass cover slip by evaporation method. This sample serves for the SERS measurements. The sample was then investigated in the close proximity of a metallic tip under the TERS configuration. An schematics of the sample-tip combination can be seen in the inset of fig. 6. A 30-nm-silver-coated tip was approached to one of the C_{60} aggregates on 4-nm-thick silver film. It was estimated from the AFM image that there was only one C_{60} aggregate in the focal spot, and the aggregate was estimated to have about 10,000 molecules of C_{60}. However, the number of molecules near the tip apex (immersed in the evanescent field) was estimated to be only 3,000. Thus, the gap-mode Raman spectra originated

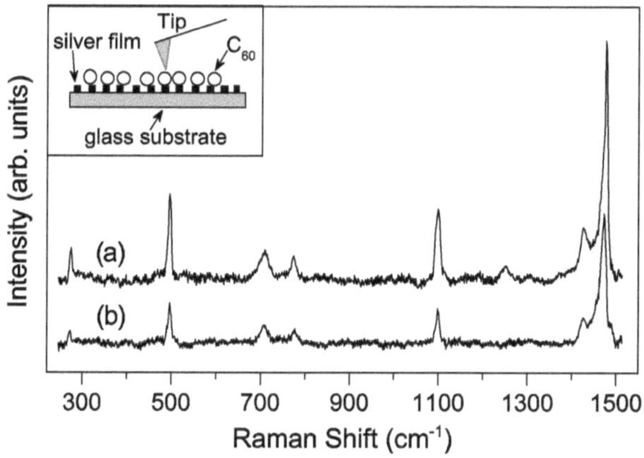

Fig. 6. (a) The "gap-mode" TERS and (b) the SERS spectra of C_{60} sample. The inset shows an schematics of the tip-sample combination for the gap-mode experiments.

only from these 3,000 molecules. The obtained gap-mode Raman spectrum is shown by (a) in fig. 6. For comparison, a SERS spectrum from the same aggregate, when the tip was removed, is also presented by (b) in fig. 6. Even though the number of molecules under the tip were much smaller than the number of molecules scattering in the SERS experiment, fig. 6 shows that spectrum (a) is stronger compared to spectrum (b), which confirms a definite enhancement in the gap-mode configuration.

4.2. *Tip-force effect on C_{60}*

With a cage-like near-spherical hollow structure, C_{60} molecules are suitable for studying the effect of uniaxial pressure. The C_{60} molecules were pressurized by the AFM tip in controlled manner, and the mechanical interaction between the tip and the C_{60} molecules was studied quantitatively under uniaxial pressure. Far-field Raman scattering from C_{60} under isotropic hydraulic pressure has been reported in the past (Tolbert et al., 1992), where a uniform pressure was applied in all directions. The shape of the molecule remains unchanged under a hydraulic pressure. However, when C_{60} molecules are squashed with the tip apex uniaxially, the shape of the molecules deform, the bond-lengths change anisotropically and the molecules lose their degree of symmetry. Therefore, the spectral changes in pressure-dependent Raman spectra in the two cases differ from each other. As the symmetry of C_{60} molecule collapses with increasing uniaxial force, it is expected that the Hg-modes get broadened. This is because the degeneracies of the five-fold degenerate Hg-modes start loosing, and the five modes start appearing individually at slightly different frequencies. This effectively broadens the whole Hg-band. On the other hand, it is expected that the molecular vibrations for the non-degenerate radial-breathing Ag-mode would be restricted by the external pressure, and hence the vibrational energy would be increased.

In order to justify our experimental results, the spectral changes in Raman modes of C_{60} molecules under uniaxial force were also estimated theoretically using the DFT calculations, derived from the molecular deformation caused by the tip-force. The C_{60} molecules are composed of 60 carbon atoms, arranged in 12 pentagons and 20 hexagons, forming the shape of a soccer ball with a diameter of 0.71 nm. Each pentagon is surrounded by five hexagons. With this shape, there would be two independent configurations in which C_{60} molecule could be deformed by the tip-force, when the molecule is squashed between the tip and the substrate. One is the so called D3D spheroidal deformation, where both apexes of the spheroid are formed by the hexagons, as shown in fig. 7(a). The other is the D5D deformation, where the apexes are formed by the pentagons, as shown in fig. 7(b). The geometrical structure of C_{60} molecule was fully optimized at the B3LYP functional (Backe, 1993; Lee et al., 1988) using 6-31G(d) basis set (Frisch et al., 1998), which was then followed by calculations of the vibrational properties at the same level of the theory using the same basis set. The forces are calculated from a harmonic oscillation of the binding energy differences. The calculated geometry was a sphere shape (IH symmetry) which consisted of both the pentagons and the hexagons. Figure 7(c) shows the calculated relation between the tip-force (which is same as the van der Waals repulsive force accompanying the D3D and the D5D molecular deformation) and the change in the molecule dimension in the direction of the force, which we call the compression depth. As can be seen from fig. 7(c), the elastic properties of C_{60} molecules differ slightly in the D3D and the D5D configurations. Since our experiments involve several C_{60} molecules, both D3D and D5D configurations are expected. Therefore, it is expected that the experimental results would compare with the average of the two theoretical calculation. The calculated vibrational frequencies had a constant factor difference with the experimental results. Multiplying the calculated vibrational frequencies by a constant scale factor of 0.985 provided a good fit between the calculated and the observed frequencies (Bethune et al., 1991). The curves in figs. 8(a), 8(b), 8(c) shows the calculated relations between the compression depth and the shift in the Ag-mode at 496 cm^{-1}, the full width at half maxima (FWHM)

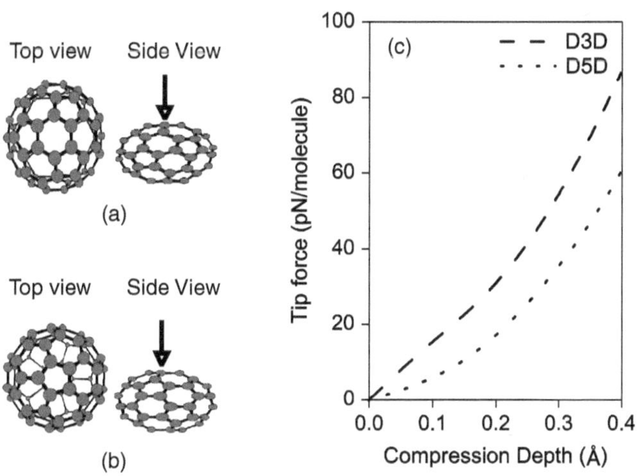

Fig. 7. Schematics of C_{60} molecule under uniaxial tip-force in (a) D3D configuration and (b) in D5D configuration. (c) Theoretical plots showing the tip-force-dependency on the compression depth of C_{60} molecule.

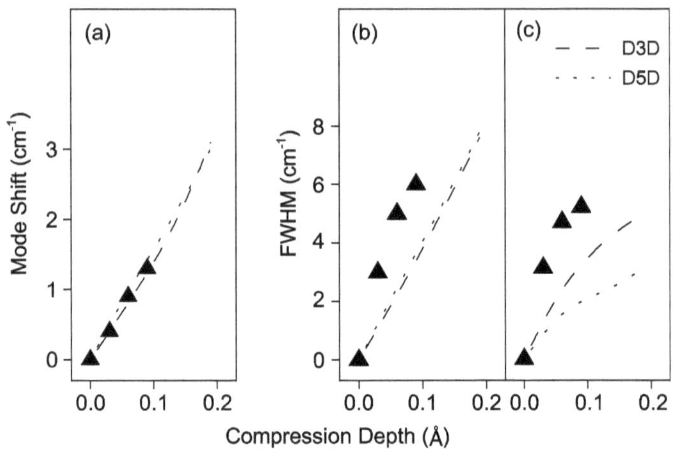

Fig. 8. Curves show the theoretical plots and the triangles show the experimental points for the relation between the compression depth and (a) shift in the frequency position of the Ag-mode at 496 cm^{-1}, (b) FWHM for the Hg-mode at 272 cm^{-1}, and (c) FWHM for the Hg-mode at 710 cm^{-1}. The intrinsic width has been subtracted from the experimental points in (b) and (c).

for the Hg-mode at 272 cm^{-1}, and the FWHM for the Hg-modes at 710 cm^{-1}, respectively. The dashed and the dotted lines in fig. 8 correspond to the D3D and the D5D configurations, respectively.

The tip-force effect was experimentally investigated in two steps. In the first step, the SERS spectra were measured when the C_{60} molecules were pressurized by an un-coated silicon tip, and in the second step, the gap-mode spectra were measured when the C_{60} molecules were pressurized by a silver-coated silicon tip. In the first step, the tip was not used for the near-field enhancement, rather it was simply used for applying a uniaxial force on the molecules. Therefore, it was neither necessary to have metal coating on the tip, nor was it necessary to have a tip with very sharp apex. In fact, if the size of the molecule aggregate under the focal spot is larger than the tip apex, then only a small part of aggregate is pressurized by the tip, however, Raman scattering originates from the whole aggregate in the focal spot. Raman scattering from the pressurized molecules in such a situation is eclipsed by that from the unpressurized molecules. Therefore, in the SERS experiment, it is rather preferable to have a tip with apex larger than the size of aggregate. For this reason, a special tip was prepared by chopping off the tip apex of an AFM silicon cantilever, using the FIB. The tip was prepared with a flat tip apex of 300 nm, which coincides with the focal spot size. The flatness of the FIB-modified tip was confirmed by observing the SEM image of the tip apex. The sample used was the same as that used in the previous gap-mode experiment. A

100-nm-large C_{60} aggregate was placed at the center of the focal spot, and the FIB-modified tip was approached to the aggregate in order to apply a controlled uniaxial pressure on the aggregate. The force applied by the tip to the aggregate was sequentially increased from 0.7 nN to 8.7 nN (the corresponding force-per-molecule was estimated to range from about 0.8 pN to about 12 pN), and SERS spectra were measured under different forces, especially for the five-fold degenerated Hg-modes at 273 and 710 cm^{-1}, and for the radial breathing Ag-mode at 496 cm^{-1}.

Figure 9 shows the SERS spectra for C_{60} molecules pressurized uniaxially under the varying tip-force. The two Hg-modes at 273 and 710 cm^{-1} are shown in figs. 9(a), 9(c), respectively, and the Ag-mode at 496 cm^{-1} is shown in fig. 9(b). The intensity scales in fig. 9 are arbitrarily adjusted for better viewing. The peak position of the Ag-mode at 496 cm^{-1} shifts to higher frequency by about 1.5 cm^{-1} with increasing uniaxial pressure. The DFT calculation predicts a shift of 1.5 cm^{-1}, when the C_{60} molecule is uniaxially deformed to the diameter 7.00 Å from its original diameter of 7.10 Å. Figure 8(a) compared the theoretical plots with the experimental data (solid triangles in the figure) for the spectral shift in the Ag-mode at 496 cm^{-1}, and the results are in excellent agreement.

Even though the Hg-modes are weak in our experiments, a careful observation indicated a gradual increase in the FWHM of these modes, as expected. No frequency shifts were observed for these Hg-modes. The experimental results for the two Hg-modes were compared with the

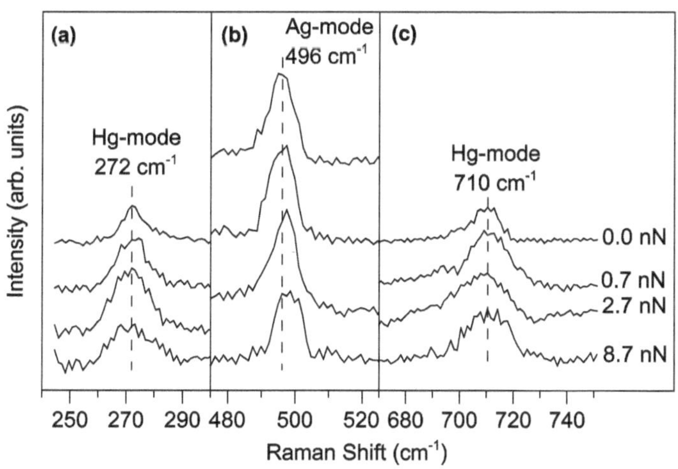

Fig. 9. SERS spectra under indicated uniaxial force applied by the tip. The Hg-modes at 272 and 710 cm^{-1} in (a) and (c), respectively, show a broadening, while the Ag-mode at 496 cm^{-1} in (b) shows a shift.

theoretical calculations, as shown by the solid triangles in figs. 8(b), 8(c), respectively. Since these modes are quite weak, there is a possibility of overestimation of the FWHM in experiments. Keeping that in mind, figs. 8(b), 8(c) show good agreement between theory and experiments.

In the next step, we investigated the tip-force effect by measuring the tip-enhanced gap-mode Raman spectra from uniaxially pressurized C_{60} molecules. The tip used here was the same as the one used in the TERS experiments (sharp tip, coated with 30-nm-silver layer). In order to reduce the background signal from the unpressurized molecules, the sample was prepared by dispersing a low concentrated C_{60} solution (0.1 μM) on a 4-nm-silver film. An AFM-image indicated that the sample had C_{60} aggregates with a width of 40 nm and height of 5 nm. This aggregation contained about 1,500 molecules. The tip was approached to one such aggregate and the gap-mode Raman spectra were measured at different tip-forces. Since the scattering was collected from only few molecules (about 1,000 or less) directly under the tip, the Raman modes were very weak. Even after gap-mode enhancement, out of the three modes in fig. 9, only the radial-breathing Ag-mode at 496 cm^{-1} was clearly observed. Obtained Raman spectra for this mode with increasing applied tip-force are presented in fig. 10. Since the number of C_{60} molecules under the tip was only about 1,000, the force-per-molecule for a given tip-force was much higher for this experiment, compared to the previous experiment. The molecules experienced much higher pressure for the same tip-force. When the applied force was increased to 0.7 nN, the intensity of the mode drastically reduced and it shifted to the higher frequency side by 3 cm^{-1}. This behavior is expected from the DFT calculations. However, when the applied force was further increased to 2.7 nN and 4.7 nN, the mode shifted in lower frequency direction by 1 and 3 cm^{-1}, respectively. This behavior is opposite of what is expected from the DFT calculations. In order to understand this low frequency shift with increasing pressure, one needs to consider a different pressure-induced effect, independently responsible for a low-frequency shift. One possible reason is the polymerization of C_{60} molecules. It has been reported (Davydov et al., 2000) that if a high pressure (about 1 GPa) is applied to C_{60} aggregates, the molecules tend to polymerize, and this phenomenon gives a low frequency shift to the radial breathing mode. Considering the tip apex to be 30 nm, assuming that there is no inter-molecular gap for the molecules under the tip, and assuming a uniform pressure on all the molecules, we estimate that the uniaxial pressure on every molecule under the tip is about 7 MPa for the tip-force of 4.7 nN. It looks difficult that the C_{60} molecules get polymerized at this pressure. However, if we consider the photo

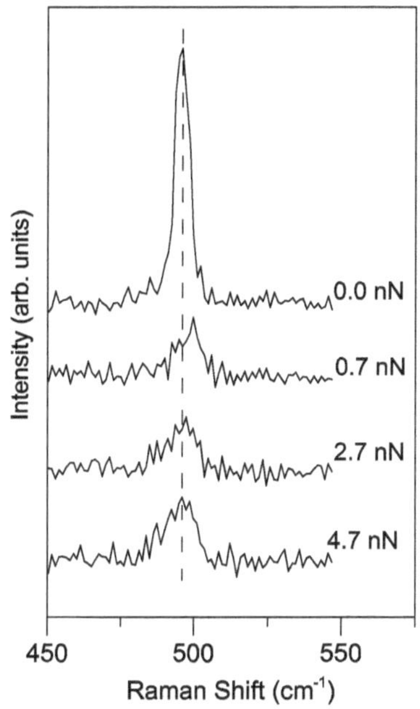

Fig. 10. The Ag-mode at 496 cm^{-1} measured in the gap mode configuration under indicated uniaxial force applied by the tip. As the tip force increase, the mode first shifts to the higher frequency direction, and then towards the lower frequency direction.

absorption at the probing wavelength ($\lambda = 532$ nm), and consider photo-polymerization at high pressure, then it is understandable to have polymerization of C_{60} molecules under a combination of probing light irradiation and a tip-applied pressure of 7 MPa. Photo-polymerization of C_{60} molecules can be easily observed (Rao et al., 1993) when the probing laser power-density is increased. Also, our assumptions in calculating the pressure due to tip-force always under-estimated the actual pressure, which means the actual pressure on the C_{60} molecules could be much larger. The possible reasons could be that the actual tip-end which touches the C_{60} molecules could be much smaller than 30 nm due to local curvatures, which is not observed in the SEM image; or that there are some inter-molecular gaps between the C_{60} molecules under the tip.

§ 5. Tip-enhanced nonlinear optical spectroscopy

5.1. Photon confinement due to nonlinear optical effect

Nonlinear optical effects contribute to spatial confinement of photons in a smaller volume than the size of the diffraction-limited focal spot. The

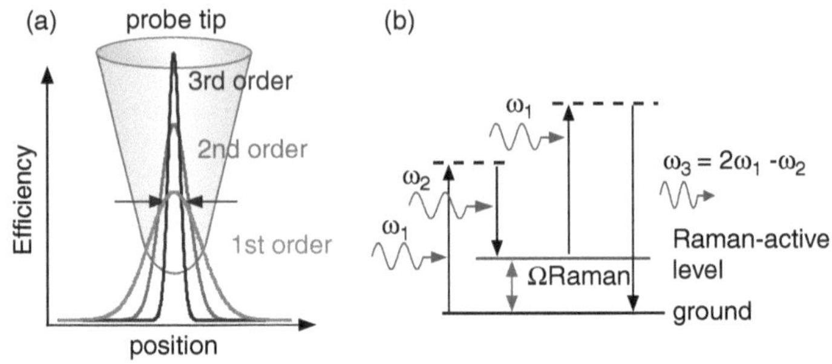

Fig. 11. (a) Spatial confinement of the excitation efficiency of high-order nonlinear effects. (b) Energy diagram of the CARS process.

intensity of the nonlinear effects is proportional to the high-order powers (square, cube, etc.) of the excitation light intensity. The spatial distribution of the signal emission (harmonic signal, two photon excited fluorescence, etc.) becomes narrower than the intensity distribution of the excitation field. This leads to the reduction of the effective volume of light–matter interaction beyond the diffraction limit. Higher-order optical effects give much finer spatial response (fig. 11(a)). So far, many types of microscopy and fabrication methods using nonlinear optical effects have been reported (Denk et al., 1990; Kawata et al., 2001).

The tip-enhanced near-field microscopy also gets benefited from the nonlinear effects. The volume of the light–matter interaction can be further confined to a tiny volume at the tip end due to the nonlinearity. Because of the nonlinear responses, even a small enhancement of the excitation field could lead to a huge enhancement of the emitted signal, allowing a reduction of the far-field background.

5.2. Tip-enhanced coherent anti-Stokes Raman scattering

In order to realize tip-enhanced nonlinear molecular-vibration spectroscopy, coherent anti-Stokes Raman scattering (CARS) spectroscopy is employed, which is one of the most widely used nonlinear Raman spectroscopies (Shen, 1984). The CARS spectroscopy uses three incident fields including a pump field (ω_1), a Stokes field (ω_2; $\omega_2 < \omega_1$), and a probe field ($\omega_1' = \omega_1$), and induces a nonlinear polarization at the frequency of $\omega_3 = 2\omega_1 - \omega_2$, which is given by

$$P^{(3)}_{\mathrm{CARS}}(\omega_3 = 2\omega_1 - \omega_2) = \chi^{(3)} E_1(\omega_1) E_2^*(\omega_2) E_1(\omega_1), \qquad (5.2.1)$$

where $\chi^{(3)}$ represents the third-order nonlinear susceptibility, $E_1(\omega_1)$ and $E_2(\omega_2)$ are the electric fields for excitation light. $E_2^*(\omega_2)$ denotes the complex conjugate of $E_2(\omega_2)$. Here, the electric fields are regarded as scalars for simplicity, while the quantities are basically vectors. The nonlinear susceptibility is expressed by vibration-resonant term ($\chi_{\rm R}^{(3)}$) and nonresonant term ($\chi_{\rm NR}^{(3)}$).

$$\chi^{(3)} = \chi_R^{(3)} + \chi_{\rm NR}^{(3)} = \frac{A}{\Omega - (\omega_1 - \omega_2) - i\Gamma} + \chi_{\rm NR}^{(3)} \qquad (5.2.2)$$

The coefficient of the fraction, A, denotes a constant related to the strength of the vibration, Ω denotes one of the specific molecular vibrational frequencies of a given sample, and Γ corresponds to the spectral band width of the same vibration mode. When the frequency difference of ω_1 and ω_2 ($\omega_1 - \omega_2$) coincides with Ω, the anti-Stokes Raman signal is resonantly generated. Figure 11(b) shows an energy diagram for the CARS process. $\chi_{\rm NR}^{(3)}$ is a contribution from transition process which does not undergo the vibrational state. In particular, the process which undergoes the $2\omega_1$ state may be resonant or pre-resonant to an electronic state, resulting in strong contribution to the susceptibility given by eq. (5.2.2).

The nonlinear polarization given by eq. (5.2.1) provides a light radiation corresponding to CARS signal. One can obtain a CARS spectrum by plotting the CARS signal intensity with sweeping ω_2. The CARS spectrum gives essentially identical information with spontaneous Raman spectra (Levenson and Kano, 1988). In ordinary CARS spectroscopy, the propagation angles of incident electric fields have to fulfill the phase matching condition, $\boldsymbol{k}_{\rm CARS} = 2\boldsymbol{k}_1 - \boldsymbol{k}_2$, to induce CARS polarization (Shen, 1984). However, when the CARS polarization is induced in a volume smaller than the wavelength of CARS field, the phase matching condition is automatically satisfied. In the small volume, the induced polarization can oscillate in phase, and the wavevector of CARS field loses the relation with the incident excitation field. This concept has been commonly noticed in laser scanning CARS microscopy (Zumbusch et al., 1999; Hashimoto et al., 2000; Hashimoto and Araki, 2001; Cheng et al., 2002), in which excitation beams are focused into a volume smaller than wavelengths by a high NA objective lens. Furthermore, surface enhanced CARS was also reported, in which CARS was amplified by isolated gold nanoparticles (Liang et al., 1994; Ichimura et al., 2003). These reports verified the possibility of the local enhancement of CARS by a metallic nanostructure.

On the basis of the concept mentioned above, one can observe CARS signals generated by the enhanced electric field at a metallic tip end of nanometric scale. Owing to the nonlinearity of the CARS process, the CARS polarization can be further confined to the very end of the tip, in addition to the field confinement due to the near-field effect. If one assumes the spatial distribution of the excitation fields ($E_1(\boldsymbol{r}, \omega_1)$ and $E_2(\boldsymbol{r}, \omega_2)$) as that of a Bessel or a Gaussian function, the spatial distribution of the CARS field ($P^{(3)}_{\mathrm{CARS}}(\boldsymbol{r}, \omega_1)$) turns out to be much narrower due to the nonlinearities (fig. 11(a)). Here a system consisting of a nanosphere and a single molecule, where the sizes of both are much smaller compared to the wavelength of light, is considered. in this case, the near field generated around the nanosphere can be regarded as a field generated by an oscillating dipole. The intensity of the field is inversely proportional to the third power of the distance from the dipole ($1/R^3$) (Okamoto, 2001). Then, the distribution of CARS polarization is proportional to the third power of $1/R^3$, i.e. $1/R^9$. In contrast, spatial distribution of spontaneous Raman scattering polarization is simply proportional to $1/R^3$. Actually, the metallic tip scatters the CARS signal with another enhancement of signal. In consequence, spatial distribution of CARS and spontaneous Raman are inversely proportional to the twelfth power ($1/R^{12}$) and the sixth power ($1/R^6$) of the distance, R, respectively. This is the major difference between the CARS and spontaneous Raman in terms of spatial resolution.

Nonlinearity also contributes to the enhancement factor. Field enhancement is effective to each of the fields including the CARS field. The nonlinear polarization of tip enhanced CARS is expressed by

$$P^{(3)}_{\mathrm{CARS}}(\omega_3) = L(\omega_3)\{\chi^{(3)}[L(\omega_1)E_1(\omega_1)][L(\omega_2)E_2(\omega_2)]^*[L(\omega_1)E_1(\omega_1)]\} \qquad (5.2.3)$$

$$= L(\omega_3)L(\omega_1)L(\omega_2)^*L(\omega_1)[\chi^{(3)}E_1(\omega_1)E_2^*(\omega_2)E_1(\omega_1)], \qquad (5.2.4)$$

where $L(\omega_i)$\$, $(i = 1, 2, 3)$, is referred to as "local field factor" which represents the factor of enhancement at a given wavelength at a particular position, e.g., at the tip apex. The polarization of spontaneous Raman scattering, however, can be expressed by

$$P^{(3)}_{RS}(\omega_2) = L(\omega_2)\{\chi^{(1)}[L(\omega_1)E_1(\omega_1)]\} \qquad (5.2.5)$$

$$= L(\omega_2)L(\omega_1) \times [\chi^{(1)}E_1(\omega_1)], \qquad (5.2.6)$$

where $\chi^{(1)}$ is linear susceptibility for Raman scattering. Eventually, the enhancement factor for CARS signal has higher order dependence. It results in not only increase of the signal intensity but also relative reduction of far-field background, which is the CARS/Raman signal generated at an area far from the enhanced spot. At times, far-field background makes near-field optical images difficult to interpret, and was previously discussed in tip enhanced Raman imaging (Hayazawa et al., 2001; Mehtani et al., 2005). In tip-enhanced CARS microscopy, the near-field contribution becomes dominant to the far-field contribution, allowing one to interpret the obtained images in a simpler way.

Consequently, one can expect three advantages in the use of nonlinear optical effect: (1) improvement of spatial resolution, (2) enhancement of signal intensity, and (3) reduction of far-field background.

5.3. Experimental system

Figure 12 shows the experimental system of tip-enhanced CARS microscopy (Ichimura et al., 2004a,b). As similar to the tip-enhanced normal Raman spectroscopy (Hayazawa et al., 2000), the system mainly consists of an excitation laser, an inverted microscope, an AFM using a silver-coated probe, and a monochromator. Two mode-locked Ti:sapphire lasers (pulse duration: 5 ps, spectral band width: $4\,\mathrm{cm}^{-1}$, repetition rate: 80 MHz) are used for the excitation of CARS. The ω_1 and ω_2 beams are collinearly combined in time and space, and introduced into the microscope with an oil-immersion objective lens (NA = 1.4) focused onto the sample surface. The repetition rate of the excitation lasers is controlled by an electro-optically modulated pulse picker. The backscattered CARS emission enhanced by the probe tip is collected with the objective lens and detected with an avalanche-photodiode based photon-counting module through an excitation-cut filter and the monochromator. The observing spectral width through the detection system is about $12\,\mathrm{cm}^{-1}$. The pulse signal from the APD are counted by a time-gated photon counter synchronously triggered with the pulse picker, which effectively reduces the dark counts down to almost 0 counts sec^{-1}.

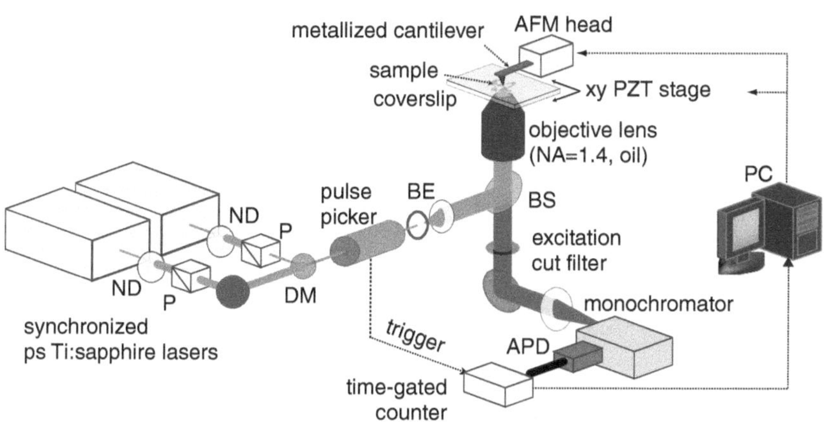

Fig. 12. An experimental system of tip-enhanced CARS microscopy. See the text for detail.

5.4. *Tip-enhanced CARS images of DNA clusters*

The DNA molecules of poly(dA-dT) aggregated into clusters were used as a model sample for CARS imaging. The poly(dA-dT) solution in water ($250\,\mu g\,ml^{-1}$) was cast and dried on a glass substrate at the room temperature with the fixation time of ~ 24 h. The dimensions of the clusters are typically 20 nm in height and 100 nm in width. The frequency difference of the two excitation lasers for CARS imaging was set to $1337\,cm^{-1}$, corresponding to a Raman mode of adenine (ring-stretching mode of diazole) (Watanabe et al., 2004), by tuning the excitation frequencies ω_1 and ω_2 to be $12710\,cm^{-1}$ (λ_1: 786.77 nm) and $11373\,cm^{-1}$ (λ_2: 879.25 nm), respectively. After the imaging in this resonant condition, the frequency of ω_2 was changed such that the frequency difference corresponds to none of the Raman-active vibrations. Figure 13 shows a spontaneous Stokes Raman spectrum of a DNA sample in a part of the fingerprint region. The solid arrows on the spectrum denote the frequencies adopted for the "on-resonant" and "off-resonant" conditions in TE-CARS imaging.

Figure 14 shows the CARS images of the DNA clusters obtained by our experimental system. Figures 14(a), 14(b) are the tip-enhanced CARS image in the on-resonant condition ($\omega_1 - \omega_2 = 1337\,cm^{-1}$) and the simultaneously acquired topographic AFM image, respectively. The DNA clusters of about 100 nm diameter are visualized in fig. 14(a). The two DNA clusters with separation of about 160 nm are clearly distinguished by the tip-enhanced CARS imaging. This indicates that the CARS imaging can successfully achieve the super-resolving capability beyond the diffraction limit of light. In the off-resonant condition ($\omega_1 - \omega_2 = 1278\,cm^{-1}$), the CARS signals mostly

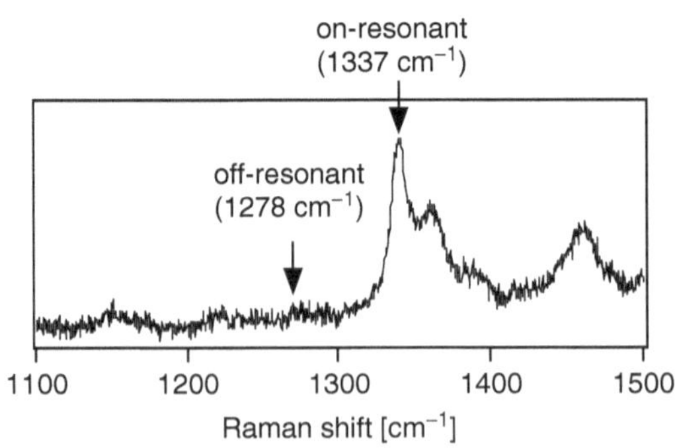

Fig. 13. A spontaneous Raman spectrum of the DNA of poly(dA-dT)-poly(dA-dT). The two frequencies adopted for our TE-CARS imaging are indicated by the downward arrows. The on-resonant frequency at $1337\,cm^{-1}$ can be assigned to the ring-breathing mode of diazole adenine molecule in the DNA.

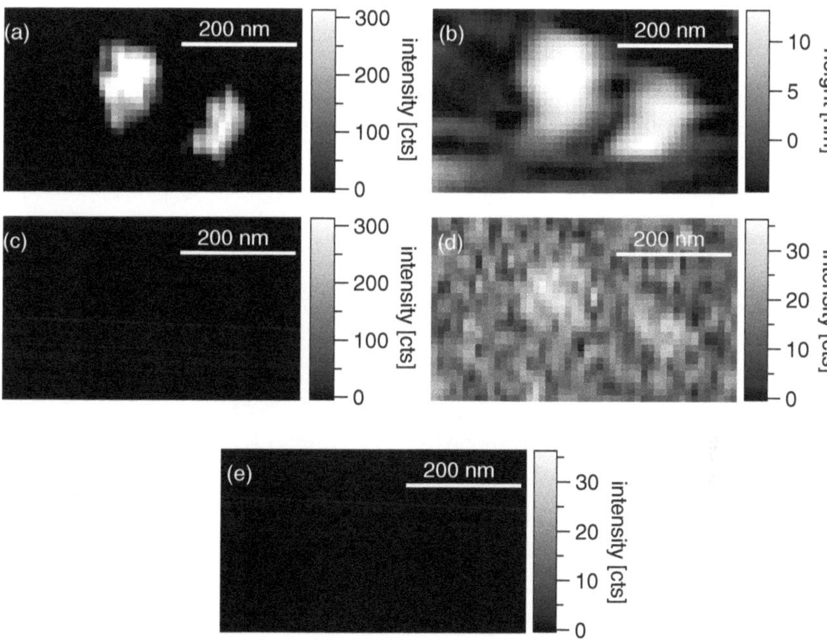

Fig. 14. CARS images of the DNA clusters. (a) Tip-enhanced CARS image at on-resonant frequency (1337 cm^{-1}), (b) the simultaneously obtained topographic image, (c) Tip-enhanced CARS image at the off-resonant frequency (1278 cm^{-1}), (d) The same image as (c) shown with a different gray scale, and (e) CARS image of the corresponding area obtained without the silver tip. The scanned area is 500 nm by 300 nm. The number of photons counted in 100 ms was recorded for one pixel. The acquisition time was ~ 3 min for the image. The average powers of the ω_1 and ω_2 beams were 30 μW and 15 μW at the 800 kHz repetition rate.

vanish in fig. 14(c). Figures 14(a), 14(c) verify that vibrationally resonant CARS is emitted from the DNA molecules at a specific frequency. However, a slight trace of the clusters could be seen in the off-resonant condition in fig. 14(d), which is the same as fig. 14(c) but is shown with a different gray scale. This can be caused by both the frequency-invariant (non-resonant) component of the nonlinear susceptibility of DNA (Shen, 1984) and the topographic artifact (Hecht et al., 1997). Figure 14(e) is a CARS image at the on-resonant frequency which was obtained after retracting the tip from the sample. The CARS signal was not detected in the CARS image without the silver tip, which confirms that the CARS polarization in fig. 14(a) is effectively induced by the tip-enhanced field. The background signal seen in the fig. 14(d) is attributed to both local four-wave-mixing emission of silver probe due to its nonlinear susceptibility and white light continuum of silver due to the electron-hole recombination (Boyd et al., 1986; Wilcoxon and Martin, 1998). These two components compete with the CARS process and degrade the image contrast and signal-to-noise ratio. Further, they subsequently limit the smallest number of molecules that can be observed. In this experiment, however, the tip-enhanced CARS signal intensity largely surpasses the background because the number of molecules in the excited volume is enough to induce the signal.

§ 6. Conclusion

Tip effects significantly improves Raman scattering spectroscopy to a novel analytical tool that provides important information in especial circumstances. For example, Raman spectroscopy under high pressure (~ GPa) is easily achieved by giving atomic force to molecules via the nanotip. Molecular orientation can be determined by detecting peak shift of Raman bands when molecules are chemically adsorbed on metallic atoms of the nanotip. Spatial resolution of molecular imaging reaches up to 15 nm by inducing nonlinear Raman scattering at the tip. If such tip effects work on a single molecule, there is possibility that atomic resolution for chemical analysis and molecule imaging is realized by detecting Raman scattering of a perturbed molecule.

References

Albrecht, A.C., 1961, J. Chem. Phys. **34**, 1476.
Aravind, P.K., Metiu, H., 1983, Surf. Sci. **124**, 506.
Backe, A.D., 1993, J. Chem. Phys. **98**, 5648.
Bauernschmitt, R., Ahlrichs, R., 1996, Chem. Phys. Lett. **256**, 454.
Bethune, D.S., Meiger, G., Tang, W.C., et al., 1991, Chem. Phys. Lett. **179**, 181.
Billmann, J., Kovacs, G., Otto, A., 1980, Surf. Sci. **92**, 153.
Bohren, C.F., Huffman, D.R., 1983, Absorption and Scattering of Light by Small Particles, Wiley, New York.
Boyd, G.T., Yu, Z.H., Shen, Y., 1986, Phys. Rev. B **33**, 7923.
Campion, A., Ivanecky, J.E., Child, C.M., Foster, M., 1995, J. Am. Chem. Soc. **117**, 11807.
Campion, A., Kambhampati, P., 1998, Chem. Soc. Rev. **27**, 241.
Chang, R.K., Furtak, T.E. (Eds.), 1981. Surface Enhanced Raman Scattering, Plenum Press, New York.
Cheng, J.-X., Volkmer, A., Xie, X.S., 2002, J. Opt. Soc. Am. B **19**, 1363.
Davydov, V.A., Kashevarova, L.S., Rakhmanina, A.V., et al., 2000, Phys. Rev. B **61**, 11936.
Denk, W., Strickler, J.H., Webb, W.W., 1990, Science **248**, 73.
Efrima, S., Meitu, H., 1979, J. Chem. Phys. **70**, 1939.
Frisch, M.J., Trucks, G.W., Schlegel, H.B., et al., 1998, Computer Code GAUSSIAN98 Revision A.9, Gaussian Inc., Pittsburgh, PA.
Furukawa, H., Kawata, S., 1998, Opt. Commun. **148**, 221.
Futamata, M., Maruyama, Y., Ishikawa, M., 2003, J. Phys. Chem. B **107**, 7607.
Gleyzez, P., Boccara, A.C., Bachelot, R., 1995, Ultramicrosc. **57**, 318.
Hamman, H.F., Gallagher, A., Nesbitt, D.J., 2000, Appl. Phys. Lett. **76**, 1953.
Hashimoto, M., Araki, T., Kawata, S., 2000, Opt. Lett. **25**, 1768.

Hashimoto, M., Araki, T., 2001, J. Opt. Soc. Am. A **18**, 771.
Hayazawa, N., Inouye, Y., Kawata, S., 1999, J. Microsc. **194**, 472.
Hayazawa, N., Inouye, Y., Sekkat, Z., Kawata, S., 2000, Opt. Commun. **183**, 333.
Hayazawa, N., Inouye, Y., Sekkat, Z., Kawata, S., 2001, Chem. Phys. Lett. **335**, 369.
Hecht, B., Bielefeldt, H., Inouye, Y., Pohl, D.W., Novotny, L., 1997, J. Appl. Phys. **81**, 2492.
Hildebrandt, P., Stockburger, M., 1984, J. Phys. Chem. **88**, 5935.
Ichimura, T., Hayazawa, N., Hashimoto, M., Inouye, Y., Kawata, S., 2003, J. Raman Spectrosc. **34**, 651.
Ichimura, T., Hayazawa, N., Hashimoto, M., Inouye, Y., Kawata, S., 2004a, Appl. Phys. Lett. **84**, 1768.
Ichimura, T., Hayazawa, N., Hashimoto, M., Inouye, Y., Kawata, S., 2004b, Phys. Rev. Lett. **92**, 220801.
Inouye, Y., Hayazawa, N., Hayashi, K., Sekkat, Z., Kawata, S., 1999, Proc. SPIE **3791**, 40.
Inouye, Y., Kawata, S., 1994, Opt. Lett. **19**, 159.
Kawata, S. (Ed.), 2001. Near-Field Optics and Surface Plasmon Polaritons, Springer, Berlin.
Kawata, S., Sun, H.-B., Tanaka, T., Takada, T., 2001, Nature **412**, 667.
Kottmann, J.P., Martin, O.J.F., Smith, D.R., Schultz, S., 2001a, Chem. Phys. Lett. **341**, 1.
Kottmann, J.P., Martin, O.J.F., Smith, D.R., Schultz, S., 2001b, J. Microsc. **202**, 60.
Krug J.T., II, Sânchez, E.J., Xie, X.S., 2002, J. Chem. Phys. **116**, 10895.
Lee, C., Yang, W., Parr, R.G., 1988, Phys. Rev. B **37**, 785.
Levenson, M.D., Kano, S.S., 1988, Introduction to Nonlinear Optical Spectroscopy, Academic Press, Orlando.
Liang, E.J., Weippert, A., Funk, J.M., Materny, A., Kiefer, W., 1994, Chem. Phys. Lett. **227**, 115.
Martin, Y.C., Hamman, H.F., Wickramasinghe, H.K., 2001, J. Appl. Phys. **89**, 5774.
Mehtani, D., Lee, N., Hartschuh, R.D., Kisliuk, A., Foster, M.D., Sokolov, A.P., Maguire, J.F., 2005, J. Raman Spectrosc. **36**, 1068.
Moskovits, M., 1985, Rev. Mod. Phys. **57**, 783.
Okamoto, T., 2001. in Ref. (Kawata, 2001), p. 97.
Otto, A., Timper, J., Billmann, J., Kovacs, G., Pockrand, I., 1980, Surf. Sci. **92**, L55.
Rao, A.M., Zhou, P., Wang, K.A., et al., 1993, Science **259**, 955.
Shen, Y.R., 1984, The Principles of Nonlinear Optics, Wiley, New York.
Stöckle, R.M., Suh, Y.D., Deckert, V., Zenobi, R., 2000, Chem. Phys. Lett. **318**, 131.
Tolbert, S.H., Alivisatos, A.P., Lorenzana, H.E., Kruger, M.B., Jeanloz, R., 1992, Chem. Phys. Lett. **188**, 163.
Verma, P., Yamada, K., Watanabe, H., Inouye, Y., Kawata, S., 2006, Phys. Rev. B **73**, 045416.
Watanabe, H., Hayazawa, N., Inouye, Y., Kawata, S., 2005, J. Phys. Chem. B **109**, 5012.
Watanabe, H., Ishida, Y., Hayazawa, N., Inouye, Y., Kawata, S., 2004, Phys. Rev. B **69**, 155418.
Wilcoxon, J.P., Martin, J.E., 1998, J. Chem. Phys. **108**, 9137.
Zayats, A.V., Sandoghdar, V., 2000, Opt. Commun. **178**, 245.
Zumbusch, A., Holton, G.R., Xie, X.S., 1999, Phys. Rev. Lett. **82**, 4142.

Chapter 4

Use of tip-enhanced vibrational spectroscopy for analytical applications in chemistry, biology, and materials science

by

T. Schmid, B.-S. Yeo, W. Zhang, R. Zenobi

Department of Chemistry and Applied Biosciences, ETH Zurich, 8093 Zurich, Switzerland;
e-mail: zenobi@org.chem.ethz.ch

Tip Enhancement
Advances in Nano-Optics and Nano-Photonics
ISSN: 1871-0018

S. Kawata and V.M. Shalaev (Editors)
© *2007 Elsevier B.V. All rights reserved.*
DOI: 10.1016/S1871-0018(06)01004-1

Contents

§ 1. Introduction

Proposed as early as 1985 (Wessel, 1985), the combination of optical field confinement by a metallic tip with vibrational spectroscopy has opened a way for molecular analysis and identification with true nanoscale spatial resolution. Three different principles to achieve this have appeared in the literature.

- Raman spectroscopy: by dispersing inelastically scattered photons, usually from Stokes scattering, in a spectrometer. Owing to its similarity to surface-enhanced Raman spectroscopy (SERS), the apertureless near-field version is commonly referred to as tip-enhanced Raman spectroscopy (TERS). It can be performed at irradiation wavelengths in the visible or the near IR.
- Coherent anti-Stokes Raman scattering (CARS): CARS is a nonlinear optical spectroscopic method based on four-wave mixing. It requires two synchronized tunable lasers and has advantages because it eliminates much of the background inherent in the other two methods. However, it is technically more difficult to implement, which is evident from the relative paucity of literature employing a CARS setup.
- IR (scattering) spectroscopy: by detecting a contrast in the scattering efficiency as a function of wavelength. This arrangement is usually called scattering scanning near-field optical microscopy (s-SNOM). It requires infrared (IR) light, preferably in the form of a tunable IR laser, for excitation of the sample.

In this chapter we will discuss the experimental setup for implementing TERS, tip-enhanced CARS, and s-SNOM. We will then focus on the heart of such experiments, the tip, describe tip fabrication methods, and discuss enhancement factors, contrast, and spatial resolution in detail. This is then followed by a thorough review of the literature on applications in the fields of organic molecules, catalysis, biology, nanotubes, and semiconductors.

§ 2. Setups for tip-enhanced vibrational spectroscopy

2.1. *Tip-enhanced Raman spectroscopy (TERS)*

The key point of TERS is to create a strong and localized light source using a sharp metal tip. Classical electromagnetic theory predicts that the incident field will be enhanced greatly at the end of the tip at resonance because of the excitation of localized plasma polaritons (Zayats, 1999). There are two different types of localized plasma polaritons which are used in TERS. One is referred to as the "lightning rod" effect. It suggests that when an isolated sharp metal tip is illuminated, the *p* component of the field will be greatly enhanced at resonance (Novotny et al., 1997). Another method uses the "gap mode" of the electromagnetic field (Aravind and Metiu, 1983). In theory, the *p* component of the electromagnetic field at a tip–substrate junction can be enhanced more than 100 times under the resonant condition of the gap mode. With this method, a 10^6 times enhancement of Raman signal intensity was achieved recently (Pettinger et al., 2004). It should be mentioned that besides these two methods, isolated noble metal nanoparticles can also enhance the electric field efficiently. For example, a single gold nanoparticle has been attached to the apex of a fiber tip for a similar purpose (Kalkbrenner et al., 2001), but to date, there is no reports utilizing such probes for TERS.

For TERS based on the "lightning rod" effect, epi-illumination (transmission mode) is popular (fig. 1), because a large numerical aperture (NA)

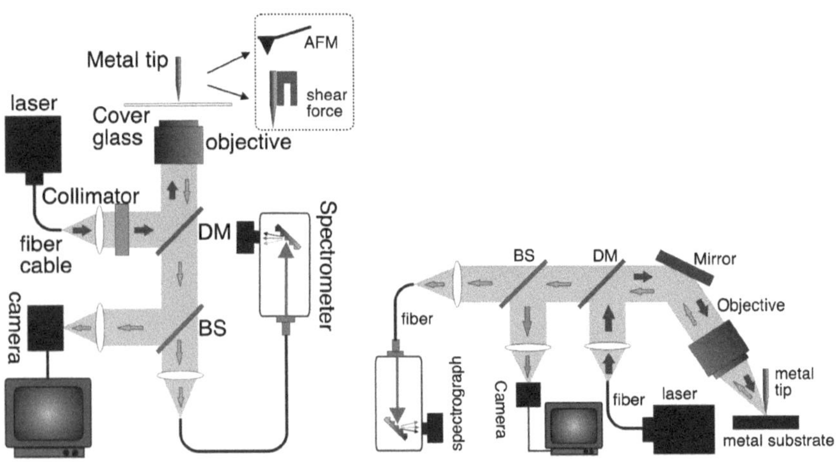

Fig. 1. Illumination and detection schemes in tip-enhanced vibrational spectroscopy. The left figure is the transmission mode. The right figure is the side illumination mode. DM: dichroic mirror, BS: beam splitter.

microscope objective can be used to afford a tight focus. This is crucial for decreasing the background noise induced by far-field illumination. There are two disadvantages of this type of transmission–illumination setup. First, a transparent sample is necessary. Second, this setup is not efficient for creating a p-polarized component from a linearly polarized illumination beam. To overcome this problem, different solutions have been adopted, annular illumination (Hayazawa et al., 2000) or modifying the wave front using a wave plate (Novotny et al., 1998; Saito et al., 2005). It should be mentioned that no matter which method is chosen to create a *p* component for the illumination, a high NA objective is required.

For gap mode TERS, side illumination scheme is normally employed, because an opaque metal substrate is used (fig. 1). This also leads to an inherent disadvantage, that a long-working distance objective must be used. Consequently, the NA of the objective is low and the area of far-field illumination is large. Considering that the enhanced electromagnetic field at the tip–sample junction is localized in an area of tens of nanometers, the background induced from the far-field illumination is relatively larger than in the epi-illumination method discussed above. Similar to the lightning rod effect, the existence of a p-polarized component of the illumination is crucial to excite the gap mode and only this component can be greatly enhanced (Aravind and Metiu, 1983). Table 1 summarizes the experimental configurations used for tip-enhanced Raman scattering.

2.2. Tip-enhanced coherent anti-Stokes Raman scattering (TE-CARS)

Field enhancement by a sharp tip is also used in nonlinear Raman spectroscopy. The most interesting method in this area is CARS. It has great potential for bio-sample imaging, because of its low noise and high penetration depth ability in biological samples. CARS is a four-wave mixing spectroscopic method. It uses three laser beams with two frequencies – twice ω_1 and once ω_2 ($\omega_1 > \omega_2$) – to generate an anti-Stokes field with a frequency of $\omega_{as} = 2\omega_1 - \omega_2$ as shown in fig. 2. The frequency of the signal is higher than that of the excitation lasers making it a fluorescence-free method. Both far-field and near-field CARS have been demonstrated recently (Zumbusch et al., 1999; Ichimura et al., 2004a). For near-field CARS measurements, the group of Kawata employed an inverted microscope for epi-illumination of an Ag-coated AFM tip (Ichimura et al., 2004a, 2004b). For excitation, two Ti:sapphire lasers were used, which generate near-IR radiation with frequencies of ω_1 and ω_2, respectively. By keeping ω_1 (pump laser) constant and tuning ω_2 (Stokes laser) over a certain range, bands at Raman shifts of $\omega_{Raman} = \omega_1 - \omega_2$ can be investigated.

Table 1. Setups for tip-enhanced Raman spectroscopy (TERS).

Illumination/detection		Angle (beam-tip axis)	Feedback mechanism	Refs.
Epi-illumination/transmission	Inverted microscope (NA 1.4)	180°	AFM (contact mode)	(Stöckle et al., 2000; Wang et al., 2003)
		180°	AFM (tapping mode)	(Bulgarevich and Futamata, 2004)
		180°	Shear-force tip feedback (tuning fork)	(Stöckle et al., 2000; Hartschuh et al., 2003a, 2003b)
		180°	STM	(Pettinger et al., 2003)
	Illumination under total internal reflection (TIR) using an inverted microscope (NA 1.4)	TIR illumination, 180° detection	AFM (contact mode)	(Hayazawa et al., 2000, 2001, 2002b)
	Conventional microscope	180°	AFM (contact mode)	(Anderson, 2000b)
	Laser beam focused using one single lens	180°	Ag-coated silica fiber tip in x–y–z micropositioning stage	(Stokes et al., 2004)
Side illumination/reflection	Conventional microscope (NA 0.3–0.6)	~ 90°	AFM (contact mode)	(Anderson and Pike, 2002)
		65°	AFM (contact mode)	(Mehtani et al., 2005a, 2005b)
		60°	STM	(Pettinger et al., 2003)
		45°	Shear-force tip feedback (tuning fork)	(Atesang and Geer, 2005)
		45°	AFM (contact mode)	(Wu et al., 2005)
		~ 0°	AFM (contact mode)	(Sun and Shen, 2001)
Side illumination/transmission	Conventional microscope (NA 0.35–0.55), detection using an inverted microscope (NA 0.6–1.4)	80° illumination, 180° detection	AFM (contact mode)	(Hayazawa et al., 2002b)
		45° illumination, 180° detection	AFM (tapping mode)	(Nieman et al., 2001)

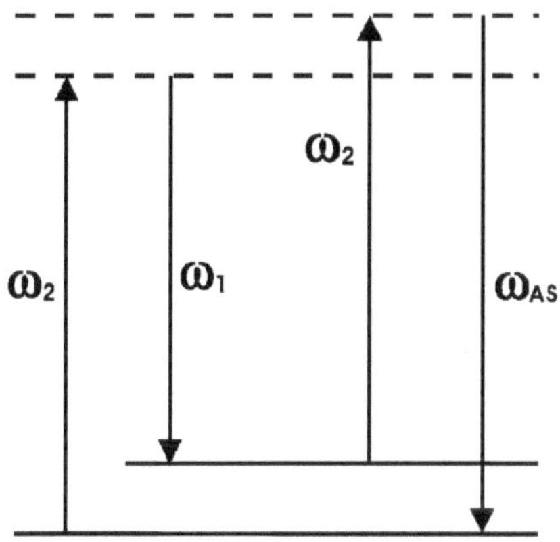

Fig. 2. Illustration of CARS.

2.3. *Scattering scanning near-field optical microscopy (s-SNOM)*

Near-field IR microscopy has also been developed in various groups during the past 10 years. One of most successful schemes is the s-SNOM setup developed in the group of Keilmann (Knoll and Keilmann, 1999; Hillenbrand et al., 2002). Different from TERS, it is not based on the giant enhancement of EM field at the tip end, but the localized perturbation of the electromagnetic field induced by the tip–sample interaction (Taubner et al., 2004). Modeling the tip as a sphere with a radius of a, at a distance d from a substrate, this interaction can be described by an effective polarizability using the dipole approximation:

$$\alpha_{\perp}^{\text{eff}} = \frac{\alpha(1+\beta)}{1-(\alpha\beta/16\pi(a+d)^3)},$$

where $\beta = (\varepsilon_s - 1)/(\varepsilon_s + 1)$, $\alpha = 4\pi a^3(\varepsilon_t - 1)/(\varepsilon_t + 2)$, ε_t and ε_s denote the dielectric constants of the tip and the sample, respectively. In this case, the scattering response can be described by $\sigma = E_s/E_i = \alpha_{\perp}^{\text{eff}}$ where the E_s and E_i denote the intensities of the scattering field and the incident field, respectively. In practice, the value of $\alpha_{\perp}^{\text{eff}}$ can be obtained by an interferometric s-SNOM (Taubner et al., 2003). Consequently, the value of the optical constants of the sample can be determined with high spatial resolution. This is an elegant and successful approach. It has been proven experimentally with special samples of flat polymer films. On the other hand, this mode of operation is not yet suitable for investigating complex, real-life samples, because it requires a very flat surface. Also, a spherical tip is assumed to derive the polarizability, and in our opinion the methodology still needs further validation.

Narita and Kimura presented a combination of FTIR with scanning probe microscopy (SPM), where an Au-coated etched glass fiber probe was side-illuminated by the output of a FTIR spectrometer (Narita and Kimura, 2001). Detection was performed in reflection mode using a InSb detector whose signal was sent back to the FTIR spectrometer. Since no enhancement could be achieved for white light illumination, the contrast between near-field signal and background was comparatively low. The estimated lateral resolution of $<2.5\,\mu m$ is also poor compared to s-SNOM experiments with monochromatic excitation but demonstrates sub-wavelength resolution with respect to the mid-infrared wavelength used for illumination. Table 2 summarizes the experimental s-SNOM configurations used by various research groups.

An alternate way to measure optical absorption at the nanometer scale is to combine SPM techniques with the photothermal effect, which describes the heating of a sample due to absorption of electromagnetic radiation. Chopped irradiation of the junction region of an STM leads to periodic heating and cooling of tip and sample. Since both are made of different materials, the result is a measurable potential difference. The first demonstration of this technique by the group of Wickramasinghe was applied to optical absorption measurements in the visible spectral range (Weaver et al., 1989). A similar Kelvin probe force microscope that is based on detection of photothermal signals by a modified AFM setup was applied also to this spectral range (Nonnenmacher et al., 1991; Nonnenmacher and Wickramasinghe, 1992). Apertureless photothermal near-field measurements in the IR range use a detection concept that is based on photoacoustic

Table 2. Setups for s-SNOM.

Illumination/ detection		Angle (beam-tip axis)	Feedback mechanism	Refs.
Side illumination/ reflection	Conventional microscope (NA 0.28 ... 0.55)	60°	AFM (tapping mode)	(Hillenbrand and Keilmann, 2000)
		60°	STM	(Barbara et al., 2005)
		80°	AFM (tapping mode)	(Akhremitchev et al., 2002)
	FTIR spectrometer and focusing mirror	(Not reported)	Shear-force feedback with optical detection	(Narita and Kimura, 2001)

spectroscopy. This technique takes advantage of the volume expansion caused by photothermal heating. The resulting acoustic or ultrasonic waves lead to surface deflections, which can be detected by AFM (Anderson, 2000a; Tomoda et al., 2003) or STM (Hida et al., 2001). These techniques not only add an additional optical contrast to SPM images, they also have great potential for the detection of buried subsurface structures as well as possible depth-resolved measurements.

2.4. *Tip fabrication*

The tip has direct effect on the degree of TERS enhancement, and consequently, the signal intensity and S/N ratio. Its fabrication procedure is thus of great importance for the performance of TERS as an analytical tool. The type of tip used is closely related to the method to control the tip–sample distance. Three types of SPM techniques are used in TERS, and they are shear-force feedback using a tuning fork, AFM with a cantilever, and the STM (Table 1). Almost all kinds of tips can be used in the tuning fork-based shear-force mode (Stöckle et al., 2000; Hartschuh et al., 2003b), but it can be tedious to attach the probe properly to the tuning fork to achieve a good feedback. The AFM is the most universal, versatile and robust method for surface analysis, with the advantage of being able to obtain complementary data such as friction force, electric force, magnetic force, or phase images simultaneously with the topography image.

Two types of methods have been used to metallize the AFM tips: the widely used physical deposition (sputter or vapor coating) (Hayazawa et al., 2000; Stöckle et al., 2000) and chemical deposition (Ag mirror reaction) (Wang et al., 2005). There are three disadvantages of the AFM-based TERS. First, while vapor-coated AFM tips are easy to prepare, the enhancement factors obtained in this fashion are known to vary a lot. This can be attributed to the fact that the physical deposition process is stochastic and the probability of obtaining a suitable "hot" particle on the tip apex is typically low (Anderson and Pike, 2002). Secondly, the metal coating can influence the performance of the cantilever. Heating of the cantilever during fast vapor deposition rates is known to lead to its distortion and makes it unsuitable to use for scanning (Hayazawa et al., 2002a). The adhesion of the metal coating to the silicon tip is also weak and peels off easily after several hours of scanning as shown in figs. 3(a) and 3(b) (Nieman et al., 2001). Finally, AFM tips are prohibitively expensive. The use of Ag-mirror coated tips has yet to be reported by other groups presumably due to its longer time and steps of preparation.

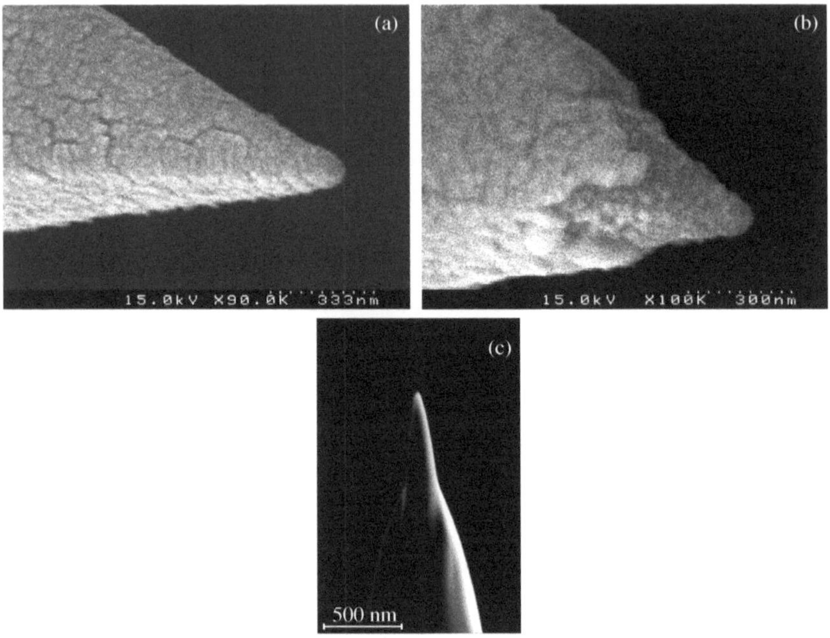

Fig. 3. SEM images of (a) an unused gold vapor-coated AFM tip and (b) the same tip after approximately 10 h of AFM imaging. It is evident that the gold wears away from the tip end with use, affecting the surface enhancement over time. (c) An electrochemically etched Ag tip sharpened by focused ion milling. Figures 3(a) and 3(b) reprinted from (Nieman et al., 2001), with permission from the American Institute of Physics. Figure 3(c) reprinted with permission from (Hartschuh et al., 2003b). Copyright (2003) by the American Physical Society.

Compared to the tuning fork scheme and AFM, the fabrication of the STM tip is easier and cheaper. The art of producing sharp metallic tips by electrochemical etching with optional focused ion beam milling (FIB) for tip sharpening have been developed in the past 20 years (fig. 3(c)) (Hartschuh et al., 2003b; Ren et al., 2004). The best Raman enhancement to date has been reported from STM-based TERS studies (Pettinger et al., 2004; Ren et al., 2005). Unfortunately, STM-based TERS requires conductive samples, which severely limits its application.

The material of the irradiated part of the tip is of great importance to whether it exhibits strong TERS enhancement with samples. This is related to the ability to create an enhanced electromagnetic field from the localized surface plasmon polariton at the tip apex during radiation by the excitation laser (Demming et al., 2005). Ag and Au are the two metals used with visible light excitation. The main advantage of Ag is that the imaginary part of its permittivity is much smaller than that of Au, which leads to a higher field enhancement. While Ag or Ag-coated tips have the advantage of being more compatible with a 488 nm laser, its enhancement is known to degrade rapidly due to oxidation (Watanabe et al., 2004). Au or Au-coated tips are used by some groups and are chemically durable for months (Mehtani et al., 2005b). It has been reported that the tips produced by the same etching procedure produce measurable enhancements that vary from negligible to several orders of magnitude (Hartschuh et al., 2003b; Pettinger et al., 2004). There has been contrasting information on the methods to make a Raman enhancement tip. For instance, both Au and Ag etched tips have been used with a He–Ne laser (633 nm) with similar Raman enhancement levels of 10^3–10^4 (Hartschuh et al., 2003b; Anderson et al., 2005). We believe that the ability to reproducibly fabricate a tip that exhibits consistently good TERS enhancement with a particular excitation wavelength is necessary for the widespread application of TERS for nano-scale chemical analysis. Theoretical investigations that provide guidelines for fabricating tips with strong Raman enhancement will be very useful. An alternative is to employ a tunable laser to tune into the correct surface plasmon resonance of the TERS-active metal particle.

§ 3. Enhancement factors and lateral resolution

The feasibility of a TERS application depends mainly on the extent of field enhancement that can be obtained. In several publications on TERS, enhancement factors for a certain combination of setup and sample are presented. Based on these values, comparisons between TERS and SERS measurements, experimental and theoretical data as well as comparisons between different TERS studies are possible. In the literature, several different ways of calculating TERS enhancement factors have been presented, which makes it difficult to directly compare the results. In this section we therefore discuss the various definitions of the TERS enhancement factor which are in use by different groups, and give an overview on published values and the basis for their calculation. This should help the readers to assess more critically these enhancement factors and to choose the appropriate way for estimation of enhancement factors in further TERS studies. Finally, a short comparison of the lateral resolution that has been achieved in apertureless near-field optical microscopy experiments is given.

3.1. TERS contrasts and enhancement factors

The calculation of TERS enhancement factors consists of two parts. First, the contrast between the near-field and far-field signals is derived from measured values. Second, based on this contrast, the enhancement factor is calculated, taking into account that the sources of far-field and near-field signal are of different size.

The contrast is defined as the ratio between near-field and far-field signal. It has been determined by measuring Raman signals with the tip in contact with the sample and with the tip retracted, respectively (Stöckle et al., 2000; Nieman et al., 2001; Bulgarevich and Futamata, 2004; Pettinger et al., 2004; Stokes et al., 2004).

$$\text{Definition (a)}: \qquad \text{Contrast} = \frac{I_{\text{with tip}}}{I_{\text{without tip}}}$$

A more precise definition takes into account that the far-field signal is present as a background also when the tip is in contact with the sample. Here, the near-field signal is defined as the difference between the signal intensities obtained with and without the tip (Hayazawa et al., 2000; Sun and Shen, 2001; Hartschuh et al., 2003a; Wang et al., 2003; Patanè et al., 2004; Mehtani et al., 2005b). This leads to a contrast value of

$$\text{Definition (b)}: \ \text{Contrast} = \frac{I_{\text{near field}}}{I_{\text{far field}}} = \frac{I_{\text{with tip}} - I_{\text{without tip}}}{I_{\text{without tip}}} = \frac{I_{\text{with tip}}}{I_{\text{without tip}}} - 1$$

In our opinion, definition (a) is of higher practical relevance because it is an indicator for the quality of TERS near-field images that can be achieved in a certain combination of setup and sample. On the other hand, definition (b) allows a more precise estimation of the field enhancement and is therefore more useful in comparisons between experimentally determined and theoretically calculated data. Since the difference between both definitions is 1, the method of calculation affects only small contrast values significantly. In the case of small near-field signals and/or high far-field background contributions, definition (b) can lead to contrasts that are <1. An increase of this value can be achieved by increasing the near-field enhancement by means of an appropriate tip, by minimizing the far-field background signal, or by the best possible compromise between both strategies.

As mentioned above, the enhancement factor also takes into account that the sources of the near-field and far-field signals are different. These signals have to be normalized to the corresponding illuminated areas (Definition (c)) or volumes (Definition (d)) (Hayazawa et al., 2000; Stöckle et al., 2000; Anderson, 2000b; Nieman et al., 2001; Sun and Shen, 2001; Wang et al., 2003; Hartschuh et al., 2003b; Bulgarevich and Futamata,

2004; Patanè et al., 2004; Pettinger et al., 2004; Stokes et al., 2004; Mehtani et al., 2005b).

$$\text{Definition (c)}: \quad \text{Enhancement factor} = \text{Contrast} \times \frac{A_{\text{focus}}}{A_{\text{tip}}} = \text{Contrast} \times \frac{d^2_{\text{focus}}}{d^2_{\text{tip}}}$$

$$\text{Definition (d)}: \quad \text{Enhancement factor} = \text{Contrast} \times \frac{V_{\text{focus}}}{V_{\text{near field}}} = \text{Contrast} \times \frac{d^2_{\text{focus}}}{d^2_{\text{tip}}} \times \frac{h_{\text{focus}}}{h_{\text{near field}}}$$

Here, d_{focus} and h_{focus} are diameter and height of the laser focus, d_{tip} is generally taken as two times the radius of curvature of the tip, and $h_{\text{near field}}$ is the estimated depth of the near field. In the case of the enhancement factor, the choice of the definition and the estimated values of the source dimensions sizes have a strong influence on the result. In the case of thin or opaque samples, calculations based on the focal height can lead to an overestimation of the far-field volume and thus to an overestimation of the enhancement factor. Therefore, the group of Sokolov proposed to multiply the focal area with the smallest of the following three quantities: the sample thickness, the light penetration depth, and the height of the focus (Mehtani et al., 2005b).

Estimation of the near-field source size is much more problematic, because this value cannot be measured directly. In most cases, the corresponding area is calculated using the relatively well-known curvature radius of the tip, even though theoretical studies and some experimental results suggest that the source of the near-field enhancement can be smaller than the area under the tip (Mills, 2002; Hartschuh et al., 2003a; Demming et al., 2005; Ren et al., 2005). If definition (d) is used, the height of the near field has to be estimated (Bulgarevich and Futamata, 2004; Sun and Shen, 2001). From evaluation of the TERS enhancement factor as a function of the thickness of CdS films, $h_{\text{near field}} = 20 \pm 5$ nm was obtained for an Ag-coated tip with radius 25 nm (Mehtani et al., 2005b). A plot of signal intensity versus distance between a tip (10–15 nm radius) and an SWNT sample resulted in a $h_{\text{near field}} = 11$ nm (Hartschuh et al., 2003b).

Since every combination of the contrast definitions (a) and (b) with the definitions of the enhancement factor (c) and (d) is possible, there are four ways to calculate the enhancement factor, which are in use by different groups. For convenience, we use the following abbreviations for these combinations: (a) + (c), (a) + (d), (b) + (c), and (b) + (d). An example should demonstrate how the definitions of the enhancement factor influence the calculation result. Sun and Shen (2001) published TERS measurements of silicon surfaces. The bulk sample was relatively thick compared to the thin films investigated in other studies. This results in an increased far-field background and thus in a lower contrast value. As mentioned above, small contrast values depend significantly on the definition that was used for their calculation. With the tip in contact, the intensity of the Raman signal was 1 a.u., whereas with the tip retracted the signal decreased to 0.65 a.u. For calculation of far-field and near-field volume, the authors estimated the following values: $d_{\text{focus}} = 3\,\mu\text{m}$, $h_{\text{focus}} = 500\,\text{nm}$, $d_{\text{tip}} = 100\,\text{nm}$, and $h_{\text{near field}} = 20\,\text{nm}$.

The height of the focus equals approximately to the light penetration depth of 515 nm radiation in silicon (Mehtani et al., 2005b), whereas the sample thickness was much higher. Thus, these values are in agreement with the method for estimation of the far-field volume mentioned above, which was proposed by the group of Sokolov (Mehtani et al., 2005b). When the same values are used, the four definitions of the enhancement factor lead to significantly different results:

$$\begin{aligned}
&\text{Definition (b) + (c)}: && \text{Contrast} = 0.54;\ \text{Enhancement factor} = 4.9 \times 10^2\\
&\text{Definition (a) + (c)}: && \text{Contrast} = 1.54;\ \text{Enhancement factor} = 1.4 \times 10^3\\
&\text{Definition (b) + (d)}: && \text{Contrast} = 0.54;\ \text{Enhancement factor} = 1.2 \times 10^4\\
& && \text{(reported by Sun and Shen)}\\
&\text{Definition (a) + (d)}: && \text{Contrast} = 1.54;\ \text{Enhancement factor} = 3.4 \times 10^4
\end{aligned}$$

In both the cases, contrast definition (a) leads to a value that is $\sim 3\times$ the result of definition (b). The enhancement factor is increased further, when definition (d) is used instead of (c) because of the additional factor $h_{\text{focus}}/h_{\text{near field}}$ that is $500/20\,\text{nm} = 25$ in this case. These two effects lead to a ratio of ~ 70 between the highest and lowest enhancement factor in this example, i.e. between (a) + (d) and (b) + (c). It should be pointed out that uncertainties of the estimated diameters and heights affect the result additionally. Since the enhancement factor depends on the square of the diameter, a slightly different value can change the result considerably. We conclude that estimated areas and volumes as well as the way of calculation could change the enhancement factor by approximately two orders of magnitude. Since most of the TERS enhancement factors published to date range over approximately two orders of magnitude from 10^2 to 10^4, a comparison of different studies is only possible if the method and estimated values that were used for the calculation are known. Therefore, comparison of contrasts and enhancement factors in Table 3 contains additional information on the calculation method and parameters such as d_{focus}, h_{focus}, d_{tip}, and $h_{\text{near field}}$.

Table 3. Comparison of TERS enhancement factors obtained with diverse tips, samples, and excitation wavelengths. (a)–(e) refer to different definitions of contrast and enhancement factor (see text).

Tip	Sample	Excitation wavelength (nm)	Estimated source size	Contrast	Enhancement factor	Refs.
AFM CM tip coated with Ag	Brilliant cresyl blue (BCB)	488	$d_{focus} = 300$ nm $d_{tip} = 50$ nm	$30^{(a)}$	$2 \times 10^{3(c)}$	(Stöckle et al., 2000)
Au tip (shear force)	Fullerenes	488	$d_{focus} = 300$ nm $d_{tip} = 20$ nm	$40^{(a)}$	$4 \times 10^{4(c)}$	(Stöckle et al., 2000)
AFM CM tip coated with Ag	Sulfur	785	(Not reported)	(Not reported)	$1 \times 10^{4(c)}$	(Anderson, 2000b)
AFM CM tip coated with Ag (TIR illum.)	Rhodamine-6G on Ag	488	$d_{focus} = 400$ nm $d_{tip} = 40$ nm	(Not reported)	$40^{(b)+(c)}$	(Hayazawa et al., 2000)
AFM CM tip coated with Au	Polydiacetylene *para*-toluene sulfonate (PDA-PTS)	488	$d_{focus} = 2,500$ nm $d_{tip} = 40$ nm	$3.4^{(a)}$	$2 \times 10^{4(c)}$	(Nieman et al., 2001)
AFM CM tungsten tip	Silicon	488	$d_{focus} = 3,000$ nm $h_{focus} = 500$ nm $d_{tip} = 100$ nm $h_{near\ field} = 20$ nm	$0.54^{(b)}$	$1.2 \times 10^{4(d)}$	(Sun and Shen, 2001)
AFM CM tip coated with Ag (TIR illum.)	Rhodamine-6G on Ag	532	$d_{focus} = 400$ nm $d_{tip} = 40$ nm	$1^{(b)}$	$1 \times 10^{2(c)}$	(Hayazawa et al., 2002a)
AFM CM tip coated with Ag (side illum.)	Rhodamine-6G on Ag	532	$d_{focus} = 1,800$ nm $d_{tip} = 40$ nm	$2^{(b)}$	$4 \times 10^{3(c)}$	(Hayazawa et al., 2002b)

Table 3 (*continued*)

Tip	Sample	Excitation wavelength (nm)	Estimated source size	Contrast	Enhancement factor	Refs.
AFM CM tip coated with Ag (TIR illum.)	Adenine	532	$d_{focus} = 400\,nm$ $d_{tip} = 30\,nm$	(Not reported)	$2.7 \times 10^{3(c)}$	(Hayazawa et al., 2003)
AFM CM tip coated with Au	Fullerenes	515	$d_{focus} = 1,000\,nm$ $d_{tip} = 100\,nm$	$2^{(b)}$	$2 \times 10^{2(c)}$	(Wang et al., 2003)
Au tip (shear force)	Single-walled carbon nanotubes (SWNT)	633	$d_{focus} = 330\,nm$ $w_{SWNT} = 20\,nm$ $h_{SWNT} = 10\,nm$ $d_{near\ field} = 26\,nm$ $w_{SWNT} = 20\,nm$ $h_{SWNT} = 10\,nm$	$1.3^{(b)}$	$16^{(d)}$	(Hartschuh et al., 2003a)
Ag tip (shear force)	SWNT	633	(Not reported)	(Not reported)	$1 \times 10^{3(d)}$	(Hartschuh et al., 2003b)
AFM TM tip coated with Au	Diamond microparticles	532	$d_{focus} = 240\,nm$ $h_{focus} = 520\,nm$ $d_{tip} = 100\,nm$ $h_{near\ field} = 20\,nm$	$5^{(a)}$	$7.5 \times 10^{2(d)}$	(Bulgarevich and Futamata, 2004)
Ag-coated silica fiber in micropos. Stage	BCB	633	$d_{focus} = 50\,\mu m \ldots 100\,\mu m$ $d_{tip} = 100\,nm$	$8^{(a)}$	$2 \times 10^{6} \ldots$ $8 \times 10^{6(c)}$	(Stokes et al., 2004)
Au STM tip	Chemisorbed CN^- on Au	633	$d_{focus} = 2,000\,nm$ $d_{tip} = 80\,nm$	$800^{(a)}$	$4 \times 10^{5(c)}$	(Pettinger et al., 2004)

Au STM tip	Malachite green isothiocyanate (MGITC) on Au	633	$d_{\text{focus}} = 2,000\,\text{nm}$ $d_{\text{near field}} = 70 \ldots 180\,\text{nm}$	4,000 … 14,000 (avg. 8,000)[a]	$1 \times 10^6 \ldots 7 \times 10^{6}$[e]	(Pettinger et al., 2004)
Au STM tip	MGITC on Au	633	$d_{\text{near field}} = 50\,\text{nm}$	(Not reported)	6.3×10^{6}[e]	(Pettinger et al., 2005)
AFM CM tip coated with Au	Silicon	515	$d_{\text{focus}} = 1,600\,\text{nm}$ $h_{\text{penetration depth}} = 500\,\text{nm}$ $d_{\text{tip}} = 40\,\text{nm}$ $h_{\text{near field}} = 20\,\text{nm}$	0.3[b]	1.2×10^{4}[d]	(Mehtani et al., 2005b)
AFM CM tip coated with Au	CdS film on aluminum	515	$d_{\text{focus}} = 1,600\,\text{nm}$ $h_{\text{sample thickness}} = 10\,\text{nm}$ $d_{\text{tip}} = 40\,\text{nm}$ $h_{\text{near field}} = 20\,\text{nm}$	6[b]	5×10^{3}[d]	(Mehtani et al., 2005b)
AFM CM tip coated with Au (optimized polarization geometry)	Silicon	515	$d_{\text{focus}} = 1,600\,\text{nm}$ $h_{\text{penetration depth}} = 500\,\text{nm}$ $d_{\text{tip}} = 40\,\text{nm}$ $h_{\text{near field}} = 20\,\text{nm}$	12[b]	5×10^{5}[d]	(Mehtani et al., 2005a)

CM, contact mode; TM, tapping mode; TIR, total internal reflection; *d*, diameter; *w*, width; *h*, height.

Pettinger et al. proposed an alternate way to calculate the TERS enhancement factor (Pettinger et al., 2004, 2005):

$$\text{Definition (e)}: \qquad \text{Enhancement factor} = g^{hrm4}$$

Here, the enhancement factor is derived from the electric field enhancement g, which is estimated by evaluating the temporal bleaching behavior of the sample. The main advantage of this method is its independence from estimated sizes of the involved signal sources. The resulting enhancement factor can also be used to calculate the near-field volume.

3.2. Comparison of TERS contrasts and enhancement factors

TERS enhancement factors reported in literature range from 16 to ~ 6×10^6 (Table 3). Whereas most of them lie between 10^2 and 10^4, recent studies have reported values in the 10^5–10^6 range. The highest enhancement factors have been achieved by the group of Pettinger using an Au STM tip with 60° side illumination in reflection mode ($\lambda = 633$ nm) (Pettinger et al., 2004, 2005). Here both definitions (a) + (c) and (e) resulted in enhancement factors ranging from 4×10^5 to 6×10^6. These enhancement factors are connected to contrasts that are more than two orders of magnitude higher than the corresponding values obtained in other studies. This huge effect can be attributed in part to highly intense gap mode plasmons as well as resonance Raman scattering, and carefully selected polarization and incidence angles of the exciting laser beam.

The group of Sokolov could raise the contrast that was achieved in TERS measurements of silicon surfaces from 0.3 up to 12 by optimizing the polarization of the excitation beam (Mehtani et al., 2005a, 2005b). In this case, polarization optics have been used for both, with improvement of the near-field enhancement and suppression of the far-field background signal. The authors selected a highly polarized Raman band of silicon for their experiments. The corresponding far-field signal can be suppressed when the measurements are performed in a depolarized geometry. On the other hand, a metallized TERS probe in contact with the sample enhances and depolarizes the Raman emission, which is therefore transmitted by the analyzer. The best compromise between strong Raman enhancement with polarization parallel to the tip axis and good suppression of the far-field background was found at polarization angles between 50 and 70° with respect to the tip axis. The optimum angle of the analyzer depended on sample material and orientation and was optimized additionally. In this way, the enhancement factor according to the definition (b) + (d) could be increased from 1.2×10^4 to 5×10^5.

It must be mentioned here that the sample quality will affect the Raman enhancement as well. For instance, for dye molecules spin-coated on glass slides, it is difficult to obtain a completely flat layer and there is a tendency for small crystals to form. If the tip is placed on an isolated crystal and TERS is measured, it is likely that the TERS enhancement be much higher. The same caution should be applied for systems such as SWNTs where it is impossible to form a uniform layer.

Surely, different studies with similar illumination schemes can be better compared based on TERS contrasts. The contrast is derived directly from measured signal intensities and is therefore independent from estimated parameters. Only the definition used, (a) or (b) has to be specified. Beside the better comparability, contrast values are of much higher practical relevance than enhancement factors (Mehtani et al., 2005b). The feasibility of TERS imaging in a given combination of setup and sample depends mostly on the contrast level that can be achieved. Similar to the signal-to-noise ratio, the ratio between near-field signal and far-field background has to be sufficiently high to obtain high-contrast and high-resolution TERS images. The importance of the enhancement factor lies more in its comparability to theoretical calculations of field and TERS enhancement. For example, Sun and Shen as well as Mehtani et al. published TERS investigations of bulk silicon samples and achieved a comparatively high enhancement factor of 1.2×10^4 (Mehtani et al., 2005b; Sun and Shen, 2001). On the contrary, the corresponding contrast values obtained, 0.54 and 0.3 using definition (b), reveal that TERS near-field imaging is hampered by a strong far-field background contribution.

When converted to definition (b), most published TERS contrast values so far are in the range from 0.3 to 2.4 (see Table 3). The groups of Futamata and Vo-Dinh obtained significantly higher values (Bulgarevich and Futamata, 2004; Stokes et al., 2004). The investigation of diamond microparticles and BCB resulted in contrasts of 4 and 7, respectively, when the published values are converted to definition (b). The huge difference in the corresponding enhancement factors of 7.5×10^2 in the first case and more than 10^6 in the latter can be explained by different far-field volumes that were illuminated. Since the group of Vo-Dinh used a single lens instead of a microscope objective for focusing of the laser beam, the far-field background was generated on a significantly larger area than in the experiments of Bulgarevich and Futamata.

Until now, the highest TERS contrast values have been achieved by the groups of Zenobi and Pettinger. In the first experimental demonstration of the TERS effect, Stöckle et al. obtained contrasts of 30 in the case of BCB and 40 in the case of C_{60} (Stöckle et al., 2000). For the measurements, Ag-coated AFM tips in contact mode and Au tips in shear-force feedback were used. The tips were illuminated in transmission mode by means of an inverted microscope ($\lambda = 488\,\text{nm}$). Pettinger et al. performed TERS measurements of chemisorbed CN^- on Au surfaces and obtained a contrast value of 800, which could be even improved by approximately one order of magnitude in the case of an MGITC film on Au leading to contrast values that ranged from 4,000 up to 14,000 (Pettinger et al., 2004). As already mentioned, this group applied an Au STM tip in side illumination geometry ($\lambda = 633\,\text{nm}$).

TERS measurements on silicon surfaces have to be discussed separately, because they are the only bulk samples in TERS experiments described up to now. In most other experiments, thin layers with a thickness of few tens of nanometers were analyzed. If a bulk sample is irradiated with light having a considerable penetration depth in this material, a far-field background signal is generated that is much higher than in the case of thin film samples and the contrast is decreased considerably. From this point of view, the contrast of 12 that could be reached by Mehtani et al. is remarkable (Mehtani et al., 2005a). As mentioned above, this value could be improved from 0.3 up to 12 by optimizing the polarization angles of excitation beam and analyzer. The setup was based on an Au-coated AFM tip in contact mode and 65° side illumination in reflection mode ($\lambda = 515\,\text{nm}$).

3.3. Lateral resolution in apertureless near-field microscopy

A good estimate for the lateral resolution that can be achieved in TERS imaging is the tip apex diameter which can be determined by SEM. The tip diameters listed in Table 3 range from 20 to 100 nm, allowing imaging with sub-wavelength resolution in every case. In several cases, etched metal tips that can be used in shear-force feedback or in STM setups enable higher lateral resolutions than metal-coated AFM probes because of the larger radii of curvature of the latter. Some experimental results suggest that the lateral resolution of a TERS image can be even better than the resolution of the corresponding topography image and thus have a smaller value than the tip diameter (Hartschuh et al., 2003a; Saito et al., 2005). Up to now, the highest spatial resolutions achieved in TERS experiments are in the range of 10–20 nm (Hartschuh et al., 2003b; Anderson et al., 2005; Saito et al., 2005). Some s-SNOM studies have shown a significantly increased lateral resolution. Whereas measurements in the

IR range ($\lambda = 10.6\,\mu m$) could be performed with a resolution of $\sim 17\,nm$ (i.e. $\lambda/600$) (Lahrech et al., 1996), similar measurements in the visible spectral range revealed the potential for a further improvement of this value. With excitation at $\lambda = 633\,nm$ s-SNOM imaging provided material-specific contrast with a lateral resolution of $\sim 10\,nm$ (Hillenbrand and Keilmann, 2002). Other studies claimed a lateral resolution as low as 1 nm (Zenhausern et al., 1995).

§ 4. Chemical applications

4.1. Dyes

In many cases, the starting point in tip-enhanced vibrational spectroscopy experiments is the study of thin films of dye molecules such as brilliant cresyl blue, rhodamine dyes, or MGITC (Hayazawa et al., 2000, 2002a, 2002b; Stöckle et al., 2000; Pettinger et al., 2004, 2005; Stokes et al., 2004). While there are no particularly interesting aspects of investigating spin-cast films of dyes with spatially resolving techniques such as TERS, dyes have two features that are advantageous from a practical point of view: (i) in many cases, the Raman enhancement can be estimated by comparing against the intensity of unenhanced Raman bands or with the fluorescence background; and (ii) often, one can profit from an additional resonance enhancement if the incident laser wavelength overlaps with the absorption band of the dye. In that sense, dyes are quite different from other organic compounds and may not be representative of the signal levels and enhancement factors one can expect for molecules lacking chromophores. A classic example is shown in fig. 4, the spectrum published by Stöckle et al. which is one of the first TERS experiments that has appeared in the literature (Stöckle et al., 2000).

4.2. Catalysis

Catalytic surfaces are attractive for investigation by TERS for a number of reasons. First, it is expected that chemical turnover takes place very locally, at "hot spots" of the catalyst surfaces. Second, catalytic reactions typically happen at elevated pressures and temperatures, i.e. conditions that are very far from those present in ultra high vacuum surface science. Much effort has been invested in overcoming this "pressure gap" for understanding catalysis at relevant conditions (Cremer et al., 1998; Somorjai and Rupprechter, 1999); TERS in principle has the capability to study catalysis at high

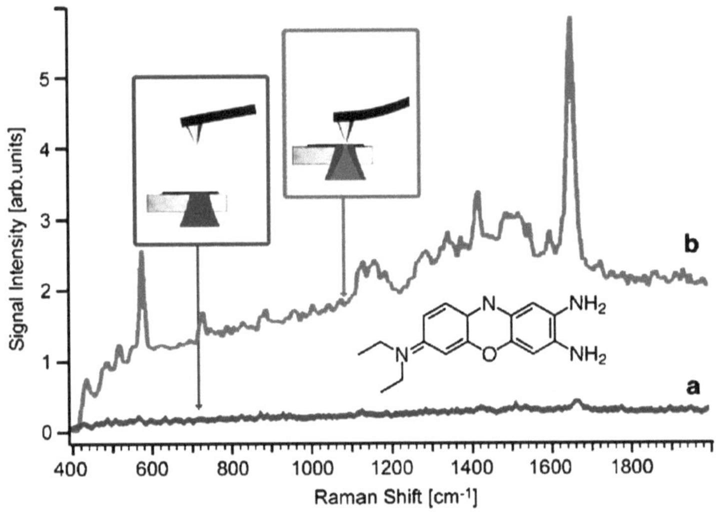

Fig. 4. Tip-enhanced Raman spectra of BCB dispersed on glass, measured with an AFM tip vapor-coated with silver. The two Raman spectra are measured with the tip retracted from the sample (a) and with the tip in contact with the sample (b). The acquisition time was 60 s per spectrum. The enhancement factor is estimated to be $\sim 2{,}000\times$, based on a 300 nm diameter illumination beam and a tip apex diameter of 50 nm. This particular example corresponds to the first entry (row) in Table 3. Adapted from Stöckle et al. (2000).

pressures and temperatures. Many industrially relevant catalysts are supported metal catalysts meaning that finely dispersed metal grains on an inert support such as silica, alumina, etc. are the active sites. These metal grains can lead to additional Raman enhancement, although this might complicate the data interpretation because transformations occurring on the support may not be observed with the same sensitivity.

Fokas and Deckert (2002) presented a study where aperture SNOM in combination with SERS was used to study Pd–Ag catalyzed hydrogenation of benzene to cyclohexane on a nanofabricated supported Pd–Ag/SiO_2 substrate (see fig. 5), with a resolution of $\sim$ 100 nm. The idea of studying model catalysts, albeit single crystal surfaces in this case, has also been mentioned in the context of TERS (Ren et al., 2005). The problem that was to be overcome in this case is that single-crystal metal surfaces show no or negligible SERS activity due to the absence of surface roughness, preventing SERS studies of adsorbates on single-crystal model catalyst surfaces. An external STM tip was therefore used to locally enhance the Raman scattered light from benzenethiol layers on Pt(1 1 0) and Au(1 1 0). A 20-fold difference in the Raman intensities on Au and Pt was found, which can be ascribed to the different optical properties of these

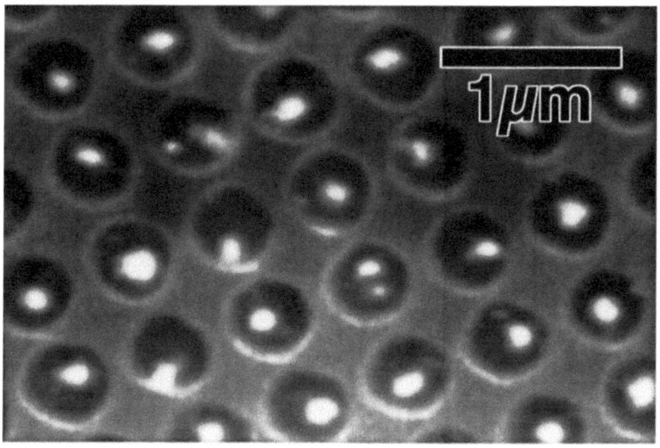

Fig. 5. SEM image of a nanofabricated model catalyst. The support consists of SiO_2 on Si which was etched by a photolithographic/interference technique. This step was followed by metallization with Pd, before removal of the remaining photoresist. By annealing, the metal that is left in the small indentations finally contracts to a small metal structure at the bottom of each indentation.

two metals. A control experiment was performed to verify the results and to exclude the presence of measurement artifacts.

We predict that the study of catalytically active surfaces – or at least model catalysts – with TERS will be a very fruitful area of application (Tian et al., 2002). If systems can be found whose corrugation is sufficiently small, and once the question how the heterogeneous nature of catalyst surfaces influences the local SERS enhancement has been properly addressed, the spatial resolution, chemical information content, and the possibility to operate at relevant pressures and temperatures of TERS has enormous potential to address important questions in catalysis.

4.3. Microfluidics and chromatography

Shear-driven chromatography (SDC) performed in miniaturized channels has been demonstrated to be a potential tool for rapid separation of small quantities of materials. Recently, mixtures of angiotensin peptides or coumarin dyes have been separated using such microfluidic devices at short distances of ≤ 2 mm (Clicq et al., 2004; Vankrunkelsven et al., 2006).

Detection of samples is a limitation of SDC. To overcome this problem, a TERS setup has been used by Anderson for the manipulation and analysis of low-volatility fluids flowing in open microchannels with groove widths and depths of 950 and 125 nm, respectively (Anderson, 2005). Contact AFM probes have been shown to function like shear plates to move liquids placed in these open grooves. Chemical identification of the liquids can be made by parking an Au-coated SERS-active AFM tip in the filled groove. In the experiment reported, ricinoleic acid placed in the channel migrated up onto the tip and cantilever by capillary action. The tip was then retracted and SERS spectrum measured at the cantilever shows the presence of the acid. If only very small amount of material can be removed, the SERS spectrum could be acquired at the tip end.

While this experiment would not be generally classified as TERS imaging, it does show the versatility of the instrument for different chemical applications. Issues on the practicality of this setup for chemical analysis do remain. For instance, in order to perform rapid identification of different types of liquids without cross-contamination, it is necessary to develop an in situ method to regenerate the surface of the SERS-active tip after each measurement. Otherwise, the tip has to be changed or cleaned by rinsing in solvents. Since the separation distance in SDC experiments would be at least an order of magnitude larger than the scan range of commercial AFMs, modification of the setups would be essential for real applications.

§ 5. Biological applications

5.1. Biopolymers

Reports in the field of apertureless near-field vibrational spectroscopy for imaging of biopolymers up to now have focused on DNA and the corresponding nucleobases. Near-field Raman spectra of nanocrystals of adenine have been determined by TERS measurements (Hayazawa et al., 2003; Watanabe et al., 2004). In this case, an inverted microscope was employed for epi-illumination ($\lambda = 532\,\mathrm{nm}$) of an Ag-coated AFM tip. Eight Raman bands which could be observed in the TERS spectra were assigned to certain vibrations of the adenine molecule. The most intense Raman signals are the in-plane ring breathing mode of the molecule (at $739\,\mathrm{cm}^{-1}$) and in-plane combination vibrations of C–H, N–H, C–N, and O–N bonds (at $1,328\,\mathrm{cm}^{-1}$), respectively. The TERS shifts differed from known Raman and SERS spectra. These differences could be explained by charge-transfer interactions between adenine and the metallized tip, which were in good agreement with corresponding density functional theory (DFT) calculations. A detailed discussion of such tip–sample interactions and their effect on TERS spectra can be found in Chapter 7 of this volume. The lateral resolution of the system was estimated to be 30 nm by conducting a one-dimensional scan over an adenine crystal. Taking into account the diameter of the illuminated area ($d_{\mathrm{focus}} = 400\,\mathrm{nm}$), calculation of the enhancement factor for the ring-breathing mode at $739\,cm^{-1}$ resulted in a value of 2.7×10^3 (see Table 3).

The feasibility of tip-enhanced CARS (TE-CARS) measurements of adenine was first demonstrated using immobilized nanoparticles for field-enhancement (Ichimura et al., 2003). Gold particles with a diameter of 60 nm were dispersed and fixed on a coverslip. Subsequently, this substrate was coated with adenine. CARS spectra were collected at isolated particles using a multi-focus CARS microscope (Kobayashi et al., 2002), where the frequency of the Stokes laser was tuned from $\omega_2 = 11,431.7\,cm^{-1}$ ($\lambda_2 = 874.8\,nm$) to $\omega_2 = 11,381.0\,cm^{-1}$ ($\lambda_2 = 878.7\,nm$) and the pump laser frequency was kept constant at $\omega_1 = 12,731.7\,cm^{-1}$ ($\lambda_1 = 785.4\,nm$). In this way, Raman shifts $\omega_{Raman} = \omega_1 - \omega_2$ in the range of $1{,}300–1,350\,cm^{-1}$ could be analyzed. This spectral range contains an intense Raman band at $1,328\,cm^{-1}$. The particle-enhanced CARS spectra were in good agreement with conventional Raman spectra and revealed enhancement factors of up to 2,000. Two attached particles showed a significantly higher enhancement of $\sim 6,000$.

In a further step, this principle was transferred to a scanning metallized tip (Hayazawa et al., 2004). Both, sample and tip were illuminated in transmission mode, using a 1.4 NA objective of an inverted microscope. This TE-CARS setup was applied to the imaging of DNA (Ichimura et al., 2004b). Nanoclusters consisting of poly(dA-dT) were cast and dried on a coverslip. The dimension of the nanoclusters was $\sim 100\,nm$ in diameter and $\sim 20\,nm$ in height. TERS spectra of such DNA samples revealed several peaks in the spectral range of $1{,}200–1,500\,cm^{-1}$. The most intense band at $1,337\,cm^{-1}$ could be assigned to the ring-breathing mode of the purin ring of adenine. TE-CARS imaging of the DNA sample was performed at this on-resonant frequency and at the off-resonant frequency $1,278\,cm^{-1}$. In the on-resonant TE-CARS image, the DNA nanoclusters could be clearly distinguished from the uncoated part of the coverslip. Here, the TE-CARS image was in good agreement with the simultaneously obtained AFM topography image. However, in the off-resonant scan the TE-CARS signal mostly vanished. Weak residual signals were due to off-resonant contributions from thymine and the DNA backbone. Additional possibilities for nonspecific background signals are related to the metallized probe tip (Ichimura et al., 2004a). Owing to local four-wave mixing, the tip can emit light with the same frequency as the CARS signal. In addition, after multi-photon excitation, noble metals show white light photoluminescence, which can contribute to the detected signal. Nevertheless, the capability of TE-CARS for spectrally resolved imaging of biopolymers has been demonstrated successfully.

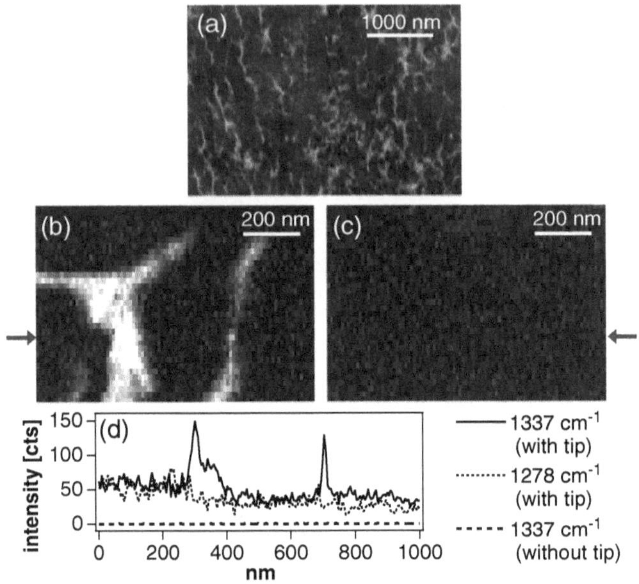

Fig. 6. TE-CARS imaging of the DNA network. (a) topographic AFM image, (b) TE-CARS image at on-resonant frequency of adenine ($1,337\,cm^{-1}$), (c) TE-CARS image at off-resonant frequency ($1278\,cm^{-1}$), and (d) cross sections along the line indicated by the arrows. Reprinted figure with permission from Ichimura et al. (2004a). Copyright (2004), by the American Physical Society.

In addition to nanoclusters, DNA was analyzed in form of network structures (Ichimura et al., 2004a). In this case, a solution of poly(dA-dT)-poly(dA-dT) was mixed with a Mg^{2+} solution and subsequently cast and dried on a coverslip. Here, the Mg^{2+} acts as a linker between DNA and negatively charged oxygen atoms of the glass surface.

The AFM topography image reveals a network consisting of single-DNA strands as well as bundles of up to 10 DNA molecules (see fig. 6(a)). The height of the structures is ~ 2.5 nm corresponding to the diameter of a single DNA double helix, whereas the width of the bundles varies between 2.5 nm and a few tens of nanometers. TE-CARS images obtained at the on-resonant and off-resonant frequency are shown in figs. 6(b) and 6(c), respectively. Again, the on-resonant image differs clearly from the corresponding off-resonant image, which can also be seen in the line profiles in fig. 6(d). Thus, the on-resonant image is based on a spectroscopic contrast, which allows the imaging of adenine-containing parts of the DNA sample. The active volume in these TE-CARS experiments was estimated to have a diameter of 20 nm and a height of 2.5 nm, which corresponds to the thickness of the sample. Taking into account this volume of approximately one zeptoliter, the estimated enhancement factor is ~ 100. Determination of the signal-to-noise ratio led to a smallest detectable DNA volume of ~ 1/4 zeptoliter. Thus, TE-CARS is a powerful tool for spectroscopically resolved imaging of biopolymer networks in their native state at the nanometer scale and provides vibrational spectral information from a sub-zeptoliter volume.

Similarly to TE-CARS, scattering SNOM (s-SNOM) in the IR range provides spectroscopically resolved imaging of DNA networks beyond the diffraction limit (Akhremitchev et al., 2002). Nanometer-scale imaging based on chemical contrast was demonstrated using patterned samples consisting of alternating areas coated with DNA and 1-hexadecanethiol, respectively. Therefore, monolayers of the two compounds were prepared in the form of 5-μm-wide stripes on Au-coated glass surfaces. Subsequently, 20 × 20 μm areas of the sample surface were scanned by a platinum-coated non-contact AFM tip. For s-SNOM measurements, the tip was side-irradiated by IR radiation from a tunable CO_2 laser with an ~ 80° angle of incidence. The radiation scattered at the tip–sample interface was collected in the same direction and detected by means of a mercury cadmium telluride (MCT) detector.

For s-SNOM imaging of DNA, the excitation wavenumber was tuned to the phosphate absorption band of DNA that is $\omega = 980\,\mathrm{cm}^{-1}$. In this way, DNA-coated areas of the sample could be clearly distinguished from hexadecanethiol stripes, whereas any significant contrast between these regions was visible in the corresponding topography image. The lateral resolution in the s-SNOM measurements was estimated to be around 200 nm by evaluating the signal behavior at the DNA-hexadecanethiol interface. With a sample thickness of ~ 2.5 nm, the sampled volume is ~ 100 zeptoliter, which is 100 times larger than in the TERS experiments mentioned above. Nevertheless, IR absorption signals from only ~ 5×10^5 or approximately one attomol phosphate groups were detected successfully and allowed to distinguish DNA from hexadecanethiol with a lateral resolution of ~ $\lambda/50$.

5.2. *Viruses and biological tissues*

First apertureless near-field optical spectroscopic measurements on biological material were performed using resonant optical scattering spectroscopy (REOSS) (Martin et al., 1996). Even though the measurements were not based on IR or Raman spectroscopy, this work should be mentioned here because it paved the way for vibrational spectroscopy and imaging of biological samples at the nanometer scale. In REOSS, a sharp AFM tip oscillates in non-contact mode close to the sample surface. Tip and sample are illuminated in transmission mode under TIR. Laser beams having wavelengths of 633 and 514 nm, respectively, were employed for excitation. The system provided a lateral resolution superior to 5 nm.

The tobacco mosaic virus (TMV) was stained using coomassie blue and adsorbed onto cleaved mica. The sample was imaged using non-contact AFM and REOSS. While scanning the sample, both resonant scattering of light and the variation of the tip height with respect to the laser focus affect the REOSS signal. The latter unwanted "liftoff" contribution could be corrected successfully using the independently determined AFM topography data. In this way, both techniques clearly revealed the cylindrical shape of the viruses. A comparison of REOSS measurements at two different wavelengths showed clear evidence for spectroscopic information achievable by this technique. The optical response at 633 nm was almost an order of magnitude stronger than at 514 nm, which is in good agreement with spectroscopic properties of coomassie blue. This result is very promising for further applications of apertureless near-field vibrational spectroscopic methods, i.e. IR and Raman spectroscopy, which should provide nanometer-scale imaging of biological samples without the need of staining.

The first demonstration of apertureless near-field Raman spectroscopy of an intact biological tissue was the application of TERS to the investigation of insect eye surfaces (Anderson and Gaimari, 2003). The compound eye of insects consists of hundreds of unit eyes called ommatidia. The ommatidial lens surfaces in several species of the order Diptera (flies) and the related order Mecoptera were imaged by AFM. The compound eye surfaces in most species under investigation showed a cerebral cortex-like pattern over the entire surface of each individual lens. Cross sections of the ommatidium of *Chrysopilus testaceipes* revealed periodic height variations of ~ 20 nm with a lateral distance of ~ 190 nm between ridges (see fig. 7).

AFM phase contrast imaging demonstrated variations in chemical composition between ridge and trough, which could be confirmed by TERS measurements. After acquisition of the AFM image, the metallized tip was parked on a ridge and a trough, respectively. Side illumination ($\lambda = 785$ nm) using a microscope objective with an angle of ~ 87° between laser beam and tip axis and detection through the same objective resulted in two Raman spectra having differences in spectral regions which are characteristic for aliphatic hydrocarbons (1,465 cm^{-1}), carboxylic acids (1,409 cm^{-1}; 1,425 cm^{-1}), esters (1,730 cm^{-1}), and proteins (1,658–1,672 cm^{-1}; 1,310 cm^{-1}). The ridges show more peptide bond

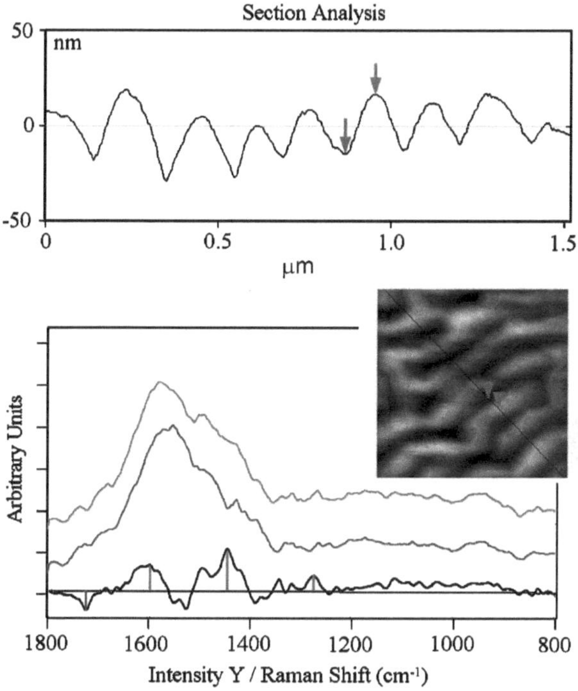

Fig. 7. TERS analysis of the ommatidal lens surface of a *Chrysopilus testaceipes* compound eye. Cross section of the ommatidium (upper graph), AFM phase image (insert), and Raman spectra (lower graph) taken from the top of a ridge (top curve, bottom panel) and the bottom of a trough (middle curve, bottom panel). The bottom curve in the lower graph is the subtraction of the trough spectrum from the ridge spectrum (2× enlarged). Reprinted from Anderson and Gaimari (2003), with permission from Elsevier.

functionality, whereas in troughs the content of fatty acid esters is increased. The control of surface wetting or anti-reflection coating are possible explanations for the morphological and chemical heterogeneities. This work showed the potential of TERS in the collection of phylogenetic data for a better understanding of evolutionary processes and in applications to biological tissues in general.

§ 6. Applications in materials science

6.1. Nanotubes

Single wall nanotubes were discovered by Iijima in 1991 (Iijima, 1991; Iijima and Ichihashi, 1993) and have generated intense interest due to their potential applications in nanotechnology. Numerous TERS investigations of SWNTs have been conducted since the near-field Raman imaging of an isolated semi-conducting SWNT by Novotny and co-workers, using an electrochemically etched Ag tip (Hartschuh et al., 2003b; Hayazawa et al., 2003; Anderson et al., 2005; Saito et al., 2005; Wang et al., 2005). Four Raman modes were observed in that significant work, the radial breathing mode (RBM), two graphitic bands (G, G′), and the disordered (D) band. Strong evidence that an isolated SWNT was detected instead of a bundle comes from the narrow full-width at half maximum (FWHM) of the Raman bands.

An interesting feature of the experiment is that the near-field resolution (23 nm) is superior to that obtained in the simultaneously collected topographical image (29 nm) (fig. 8). A similar experiment using a Si AFM tip functionalized with an 80 nm Ag particle has revealed that a high spatial resolution of 24 nm can be obtained (Wang et al., 2005). The near-field enhancement originating from an area much smaller than the size of the scattering particle is consistent with recent finite difference time domain (FDTD) simulations works by Richards and co-workers where at

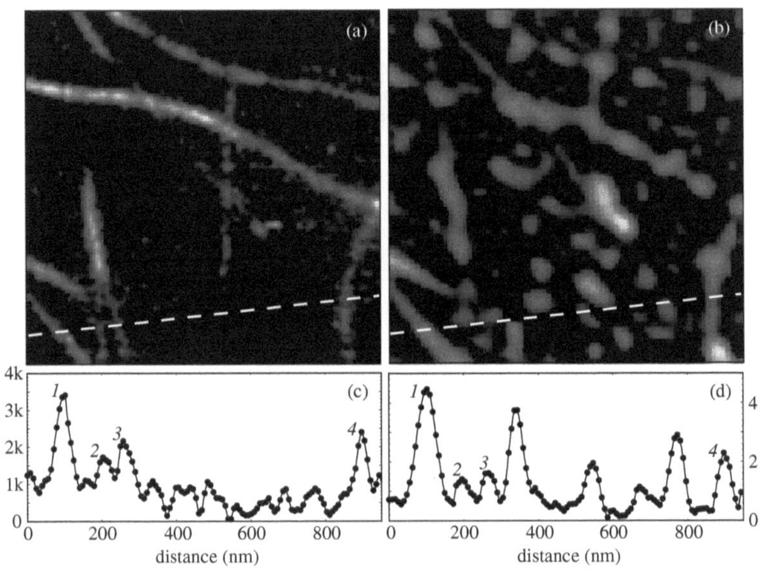

Fig. 8. (a) Near-field Raman image and (b) topographic image of SWNTs dispersed on glass. (c) and (d) cross-section taken along the dashed line in the Raman and topographic image. Reprinted with permission from Hartschuh et al. (2003b). Copyright (2003), by the American Physical Society.

resonance, an optical confinement far sharper than the tip radius has been predicted (Demming et al., 2005).

Defect-density imaging has also been demonstrated for an isolated SWNT produced by the arc discharge method (Anderson et al., 2005). Bumps of 5 nm in height have been observed on top of the SWNT in the topography image, and these have been assigned to Ni/Y catalyst particles. When localized TERS measurements were performed on spots of the SWNT close to or on the Ni/Y catalyst, the frequencies and intensities of the G and G′ bands changed. Differences in the local tube structure, which may be due to defects, junctions or interactions with the glass substrate may also be revealed from the TERS variation of the RBM frequency (diameter sensitive) and intensity.

The RBM and G-band of SWNTs have been found to exhibit opposite polarization dependences (Saito et al., 2005). The RBM and G bands exhibits greater TERS enhancement with p- and s-polarization, respectively. In addition, several symmetry components in the G-band show different enhancement efficiencies, which allow symmetry assignments to be made in the nanometer scale.

Recently, Hartschuh et al. presented the first simultaneous near-field photoluminescence and Raman imaging of isolated SWNTs with a spatial resolution better than 15 nm (Hartschuh et al., 2005). Highly localized and intense photoluminescence was observed from certain sections (20–30 nm) of an arc-discharge produced SWNT and has been assigned to the presence of localized excited states. These states may originate from localized chirality variations occurring from defects or from local environmental perturbations. However, for micelle-encapsulated SWNTs, the photoluminescence emission extends along the tube length (several hundreds of nm). By acquiring simultaneous near-field Raman and photoluminescence images, it is desired that a correlation between structural defects and the photoluminescence properties of individual SWNTs be obtained. More detailed information on carbon nanotube spectroscopy is available in Chapter 5 of this volume.

6.2. *Material-specific mapping*

Imaging with material-specific spectral contrast using local field enhancement by sharp metal tips can be done with excellent spatial resolution and relatively rapidly, as no spectral scanning or dispersion in a spectrometer is needed. This has been demonstrated for recognition of components in polymer blends, metal–dielectric composites, and on silicon surfaces.

Polymer blends are a very interesting and promising area of application for TERS and s-SNOM. These blends often show segregation of the polymer phases, forming domains with sizes in the 100 nm range and below. Fillers such as small silica or graphite particles and other additives add to the complexity of such samples, especially materials of industrial relevance. State-of-the-art Raman microspectrometry has a spatial resolution on the order of 1 μm, and is thus not capable to resolve polymer phase domains or filler particles (Zerda et al., 2003).

The group of Keilmann has applied IR s-SNOM at fixed wavelengths to image polymer mixtures, polymer films, and polymer beads with excellent spatial resolution. Several reviews about their work have recently appeared (Keilmann, 2002, 2004; Keilmann and Hillenbrand, 2004). The imaging is usually based on a theoretical calculation of the near-field extinction or scattering efficiency as a function of incident photon energy of the materials to be studied (Hillenbrand et al., 2002). A suitable wavelength is then chosen that gives maximum contrast between the compounds present on the sample surface, and s-SNOM images are recorded, often using double modulation methods (Taubner et al., 2003). High contrast can be achieved between metals and high-refractive index dielectric compounds (Knoll and Keilmann, 1999), and a spatial resolution of less than 10 nm has been found (Hillenbrand and Keilmann, 2002) even at IR wavelengths of $\lambda \approx 10\,\mu m$, corresponding to a resolution of $\lambda/1,000$. While the contrast is material dependent, the spatial resolution is not, but is determined by the properties of the tip alone. It is somewhat misleading, though, to talk about "spectroscopic imaging" and "identification" of materials in this context. While an image at fixed wavelength can be recorded in a few minutes (for the sake of argument, we assume 5 min), true spectroscopic imaging would take much longer. If a spectral range of $\Delta\omega = 2,000\,cm^{-1}$ should be covered with a resolution of $10\,cm^{-1}$, 200 scans would be necessary, requiring a total of 1,000 min, i.e. over 16 h. This is problematic due to long-term drifts of STM and AFM equipment. Suitable laser sources that are continuously tunable through the IR are also not readily available.

Figure 9 shows the imaging of a PMMA/PS polymer blend with a resolution of <70 nm carried out at different IR wavelengths in the 5.5–6 μm range (Taubner et al., 2004). This example demonstrates very nicely how spectroscopic contrast is achieved at an appropriate (fixed) wavelength, and how the contrast is completely reversed at another wavelength.

An interesting method was developed by these authors to prepare topographically flat polymer composite samples. Immiscible polymers, for example, polystyrene (PS) and polymethylmethacrylate (PMMA) were

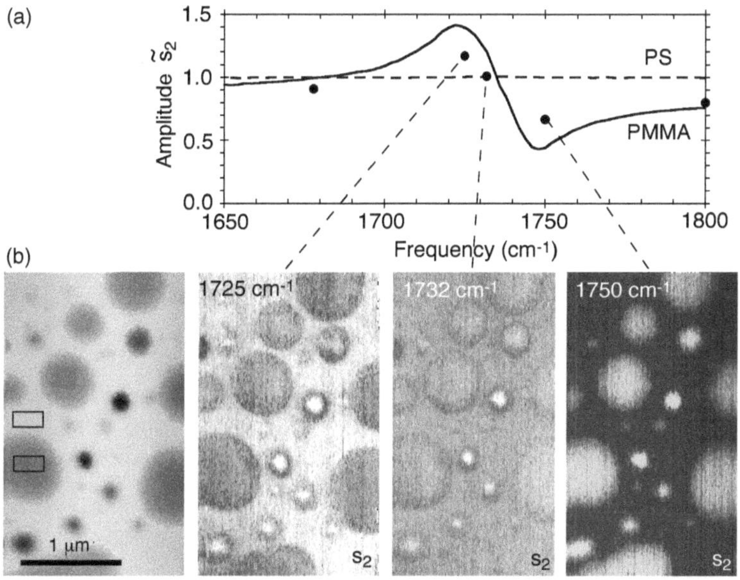

Fig. 9. s-SNOM analysis of PS in a PMMA matrix. (a) calculated s-SNOM amplitude spectra of PMMA (solid line) and of PS (dashed line). (b) topography (left) and IR amplitude (right) s-SNOM images of a 70 nm thick polymer blend film on Si at three different frequencies, as indicated in (a). The extracted IR contrast of PMMA relative to PS obtained by averaging in the regions indicated in the topography is shown as experimental data points in (a). Reprinted from Taubner et al. (2004), with permission from the American Institute of Physics.

sequentially dried on cleaved NaCl, followed by dissolving the NaCl in water. This yielded a nearly flat polymer surface with sharp material boundaries (Knoll and Keilmann, 1999). The same experimental approach can be used to study other materials. For example, Keilmann and co-workers have shown examples ranging from SiC partially covered with Au (Hillenbrand et al., 2002) to metal/Si/polymer three-component samples (Taubner et al., 2003).

Anderson demonstrated the feasibility of combining IR photoacoustic spectroscopy with AFM detection (Anderson, 2000a). Chopped radiation from an IR spectrometer was focused on the sample surface and surface deflections caused by photoacoustic pressure waves were detected by means of an AFM cantilever. The frequency of the generated acoustic waves equals the chopper frequency that was 3–4 Hz in this case. The ability to determine material-dependent signals from an area having a diameter of ~ 100 nm allowed the discrimination between fluorinated ethylene propylene (FEP) copolymer and polystyrene. The main advantage of photoacoustic detection compared to conventional s-SNOM lies in its capability of depth-resolved measurements and the detection of subsurface structures. Multi-layered samples consisting of Cr on SiO_2 or Cr and Al on SiO_2 were irradiated by chopped IR radiation (Tomoda et al., 2003). Since the chopper frequency was in the MHz range, ultrasonic waves were generated by the photoacoustic effect. For detection, ultrasound was converted into kHz waves by interference between photoacoustic waves and vibrations of the cantilever at a slightly shifted frequency. The authors called this "optical heterodyne force microscopy" (OHFM). A material-specific contrast with a lateral resolution of ~ 50 nm could be obtained, even in the case of buried structures, although data interpretation remains somewhat difficult.

6.3. *Semiconductors*

The possibility to conduct nanoscale stress mapping of silicon transistors has been gaining importance because of performance issues created by mechanical stress from miniaturization of these devices. The frequency, bandwidths, and band shapes of the phonon Raman bands are sensitive to the presence of stress, strains, and impurities.

Silicon surfaces have been studied by Sun and Shen (Sun and Shen, 2001, 2003a, 2003b; Sun et al., 2002). Near-field Raman mapping of a Si device consisting of 380 nm wide SiO_2 lines separated by 300 nm has been performed. 100 nm pitches in Si that was oxidized by thermal annealing have also been investigated with TERS using an Ag-coated quartz AFM tip (Poborchii et al., 2005). The Si–Si stretching frequency has been found to increase at the proximity of the pitch. This demonstrates that compression at the Si/SiO_2 boundary between the pitch and the rest of the substrate has occurred during the thermal oxidation process.

Lahrech et al. have demonstrated successfully that implanted boron lines in silicon can be detected by IR s-SNOM even in the absence of any topographical contrast (Lahrech et al., 1997). The setup was based on side irradiation of an oscillating tungsten tip by means of a CO_2 laser and provided lateral resolution of ~ 400 nm. Knoll and Keilmann have also performed near-field IR measurements on doped Si and have shown that subsurface mobile carriers can be probed by their response to an IR near-field with a yet-to-be fully optimized spatial resolution of 30 nm (Knoll and Keilmann, 2000).

Micro-SERS has been performed on $Si_{1-x}Ge_x$ for highly sensitive detection of the surface of strained Si using an epi-illuminescence setup (Hayazawa et al., 2005). A strained 30-nm-thick Si layer is grown on a Ge-doped Si substrate, and this side is placed facing the high NA objective and illuminated. A small shoulder at $513.8\,cm^{-1}$ is observed and attributed to the strained Si. When a similar substrate is covered with a 10 nm fractal-like Ag film, this signal is shifted to $515.3\,cm^{-1}$. This blue-shift is attributed to the presence of neighboring Ag atoms. Thus, the supposingly Raman-enhancing Ag layer affects the result of the analysis. This experiment has been suggested as a precursor for TERS using a sharpened metallized cantilever tip with side-on illumination. A better spatial resolution should be achievable. We suggest that for experiments of this nature, metallized silica fiber tips based on a tuning fork feedback or etched metal wires and STM feedback are better alternatives for two reasons. The first is to avoid perturbation from chemical bonding or pressure effects when the tip is in contact with the Si sample (Watanabe et al., 2004). The second is of a practical nature since the Si AFM tip would give very similar Raman signals that may interfere with the analysis of the strained Si bands.

Subsurface defects in semiconductors can also be detected by combinations of SPM techniques and photoacoustic spectroscopy (Hida et al., 2001). Chopped near-IR irradiation of GaAs semiconductors and detection of photoacoustic waves using an STM allowed the imaging of subsurface defect structures with nanometer resolution. The imaging contrast was in good agreement with the spectroscopic properties of defect and non-defect sites.

6.4. SERS substrates

Apertureless near-field optical microscopy is also a tool for investigating surface plasmon resonances, which provide the basis for both, SERS and TERS. While light intensity distributions on irradiated SERS substrates can be determined in aperture SNOM experiments (Klar et al., 1998; Krenn et al., 1999a, 1999b), apertureless s-SNOM techniques allow the measurement of local electric field enhancements with nanometer resolution (Adam et al., 2000; Hillenbrand and Keilmann, 2001; Hillenbrand et al., 2003). An s-SNOM setup based on an AFM operated in tapping mode and side illumination of the tip with visible laser radiation was applied to investigate surface plasmons on single isolated gold nanospheres (Adam et al., 2000). By tuning the excitation wavelength, plasmon resonance spectra of individual particles could be determined.

A screening of nanoparticles with regard to their SERS activity can be performed by scanning immobilized particles even in close aggregations at a fixed wavelength. Whereas the topography image reveals the whole aggregate, in the optical near-field amplitude image only the particles are visible which at least partially fulfill the resonance condition at the excitation wavelength under investigation with respect to size, shape, and orientation. This could be demonstrated successfully by the group of Keilmann by use of their s-SNOM setup with 633 nm side illumination and PtIr-coated Si tips (Hillenbrand and Keilmann, 2001). In addition to the signal amplitude mentioned above, evaluation of the phase signal enables the elucidation of the plasmon modes that are excited. Gap mode plasmons between densely packed gold nanoparticles could be found by scanning particle aggregates with sharper uncoated Si probes. Although gap mode plasmons were clearly visible in the s-SNOM images, the highest field enhancements are expected to be deeper inside the particle aggregates and are therefore not accessible to the tip. A possibility to solve this problem can be the use of carbon nanotubes attached to a conventional AFM probe. Such a supertip was applied to the visualization of surface plasmons on disk-shaped gold islands on a glass surface (Hillenbrand et al., 2003). A bundle of carbon nanotubes on an AFM probe resulted in a tip radius of ~ 25 nm, a value that could probably be improved down to a few nanometers by use of SWNT. Highly resolved s-SNOM images of the disks having a diameter of 91 nm resulted in a reproducible pattern consisting of two bright areas separated by a narrow dark zone showing evidence for a dipolar plasmon oscillation. Although the gold disk structures could be fabricated with good reproducibility, highly significant differences in the optical amplitude image between individual gold islands were observed. This finding was explained by shifts of the resonance frequency and underlined the strong dependence of surface plasmon resonances on small changes in size, shape, and environment of each particle.

The techniques described in this section can help to systematically investigate parameters that influence surface plasmon resonances and to better understand SERS and related effects. Thus, they can be useful tools in the improvement of SERS substrates as well as in the screening of nanoparticles that are potential probes for TERS experiments.

§ 7. Conclusions and outlook

We presented the use of tip-enhanced vibrational spectroscopy for the investigation of biological objects, nanostructured materials, and for the observation of highly localized chemical processes. The technology has not yet reached a state where routine nanoscale chemical analysis can be performed. This is evident from the scarcity of literature demonstrating applications to real-life problems. However, the technology is maturing, and we are confident that TERS and related methodologies have excellent potential for nanoscale molecular identification and spectroscopy.

Tens of different near-field vibrational spectroscopy and microscopy investigations have been published over the past 10 years. We like to make some general comments about these techniques to aid the reader in assessing their practical usefulness. From the point of view of nanoscale chemical analysis, the issues of spatial resolution, sensitivity, and selectivity are important. For practical reason, instrumental complexity and price are also considered. According to these criteria, TERS is probably the best choice. The setup is easy to build: it is only necessary to combine a Raman microscope and an SPM instrument, both of which have been commercially available for more than 10 years. The best spatial resolution of TERS is around 10–20 nm (Anderson et al., 2005) and the sensitivity is high enough to investigate the Raman spectrum from tens of molecules (Pettinger et al., 2004). In our opinion, TERS has the potential to detect Raman scattering from single molecules under optimum conditions. Near-field IR microscopy has also been very successful. For example, it allows to distinguish different polymer phases with a spatial resolution of 10 nm (Hillenbrand et al., 2002), which is slightly better than TERS. On the other hand, its drawbacks are obvious: real spectra are difficult to record due to the need to tune a laser through the IR spectral range. White light sources have also been combined with SPM and FTIR (Narita and Kimura, 2001), but both, signal-to-noise ratio and spatial resolution were poor because of the absence of a mechanism to enhance the near-field signal. Furthermore, from a practical view, there is no commercial instrument yet that can be used directly for near-field IR microscopy, perhaps because handling IR radiation is more difficult than visible light. Other s-SNOM methods also succeeded to a certain extent. The main problem of s-SNOM methods is that their chemical selectivity is poor compared to methods based on true vibrational spectroscopy, limiting their application to analytical and bioanalytical chemistry.

A key issue when employing tip-enhanced vibrational spectroscopy is that the near-field signal intensity that can be obtained is small, because it originates from a very small area and is easily overwhelmed by background from the total illuminated area. Ways to overcome this difficulty, and to develop TERS into a robust method have therefore received much attention in this chapter. A high enhancement factor not only implies greater signal, but also greater near-field to far-field contrast. A large contrast will eliminate the need to perform background subtraction and will ultimately permit a reasonable imaging speed.

Methods for estimating or calculating enhancement factors have to be carefully selected based on the type of sample (e.g. its thickness and uniformity). A comparison of different enhancement factors is only possible if the parameters of the calculation are well defined and precisely known. In our opinion, the near-field to far-field contrast is a more useful quantity for comparison, and is also of greater practical relevance than the enhancement factor. The contrast can be improved by optimizing the near-field enhancement (annular illumination, optimized polarization to increase the p-component, excitation by a tunable laser to hit the plasmon resonance), by suppressing the far-field background (high NA epi-illumination, annular illumination, far-field suppression by use of appropriate polarization geometry), or generally by the best possible compromise between both strategies.

The highest enhancement factors that have been reported to date are in the 10^6 range, the highest TERS contrast is $\sim 10,000$. The best lateral resolution achieved with these methods is on the order of 10–20 nm for TERS, 17 nm for IR s-SNOM, and 10 nm down to 1 nm for Vis s-SNOM. The main fields of application that are described in this chapter, are dyes, catalysis, biopolymers and biological samples, SWNTs, material-specific mapping, and semiconductors.

Finally, we would like to point out that vibrational spectroscopy is not the only method to gain detailed analytical chemical information on a lateral scale of nanometers. Near-field ultrasound measurements with high spatial resolution afforded by the presence of an AFM probe have started to appear in the literature (Moreau and Ketterson, 1992; Yamanaka et al., 1994; Blodgett and Spicer, 1999; Inagaki et al., 2000; Shekhawat and Dravid, 2005). In the future, related imaging methods with high spatial resolution and chemical contrast may become feasible. Several research groups have presented very interesting work on near-field or tip-enhanced laser surface melting and laser ablation (Jersch and Dickmann, 1996; Dickmann et al., 1997; Jersch et al., 1997; Boneberg et al., 1998; Huber et al., 1998; Lu et al., 1999; Huang et al., 2002; Chimmalgi et al., 2003, 2005a, 2005b). The interfacing of such methods with mass spectrometric detection of the ablation products is an equally fascinating prospect (Kossakovski et al., 1998; Stöckle et al., 2001).

Acknowledgments

We thank M.S. Anderson, Y. Inouye, F. Keilmann, R. Martinez, and L. Novotny for providing original figures for this chapter. Financial support for our work from the Deutsche Forschungsgemeinschaft (to TS), the ETH Zurich, and the Gebert Rüf Stiftung is greatly appreciated.

References

Adam, P.M., Benrezzak, S., Bijeon, J.L., Royer, P., 2000, J. Appl. Phys. **88**, 6919.

Akhremitchev, B.B., Sun, Y.J., Stebounova, L., Walker, G.C., 2002, Langmuir **18**, 5325.

Anderson, M.S., 2000a, Appl. Spectrosc. **54**, 349.

Anderson, M.S., 2000b, Appl. Phys. Lett. **76**, 3130.

Anderson, M.S., 2005, Anal. Chem. **77**, 2907.

Anderson, M.S., Gaimari, S.D., 2003, J. Struct. Biol. **142**, 364.

Anderson, N., Hartschuh, A., Cronin, S., Novotny, L., 2005, J. Am. Chem. Soc. **127**, 2533.

Anderson, M.S., Pike, W.T., 2002, Rev. Sci. Instrum. **73**, 1198.

Aravind, P.K., Metiu, H., 1983, Surf. Sci. **124**, 506.

Atesang, J., Geer, R., 2005, Proc. SPIE **5766**, 134.

Barbara, A., Lopez-Rios, T., Quemerais, P., 2005, Rev. Sci. Instrum. **76**, 023704.

Blodgett, D.W., Spicer, J.B., 1999, Appl. Phys. Lett. **74**, 3735.

Boneberg, J., Munzer, H.J., Tresp, M., Ochmann, M., Leiderer, P., 1998, Appl. Phys. A **67**, 381.

Bulgarevich, D.S., Futamata, M., 2004, Appl. Spectrosc. **58**, 757.

Chimmalgi, A., Choi, T.Y., Grigoropoulos, C.P., Komvopoulos, K., 2003, Appl. Phys. Lett. **82**, 1146.

Chimmalgi, A., Grigoropoulos, C.P., Komvopoulos, K., 2005a, J. Appl. Phys. **97**, 104319.

Chimmalgi, A., Hwang, D.J., Grigoropoulos, C.P., 2005b, Nano Lett. **5**, 1924.

Clicq, D., Vankrunkelsven, S., Ranson, W., De Tandt, C., Barn, G.V., Desmet, G., 2004, Anal. Chim. Act. **507**, 79.

Cremer, P.S., Su, X.C., Somorjai, G.A., Shen, Y.R., 1998, J. Mol. Catal. A **131**, 225.

Demming, A.L., Festy, F., Richards, D., 2005, J. Chem. Phys. **122**, 184716.

Dickmann, K., Jersch, J., Demming, F., 1997, Surf. Interface Anal. **25**, 500.

Fokas, C., Deckert, V., 2002, Appl. Spectrosc. **56**, 192.

Hartschuh, A., Anderson, N., Novotny, L., 2003a, J. Microsc.-Oxford **210**, 234.

Hartschuh, A., Qian, H., Meixner, A.J., Anderson, N., Novotny, L., 2005, Nano Lett. **5**, 2310.

Hartschuh, A., Sanchez, E.J., Xie, X.S., Novotny, L., 2003b, Phys. Rev. Lett. **90**, 095503.

Hayazawa, N., Ichimura, T., Hashimoto, M., Inouye, Y., Kawata, S., 2004, J. Appl. Phys. **95**, 2676.

Hayazawa, N., Inouye, Y., Sekkat, Z., Kawata, S., 2000, Opt. Commun. **183**, 333.

Hayazawa, N., Inouye, Y., Sekkat, Z., Kawata, S., 2001, Chem. Phys. Lett. **335**, 369.

Hayazawa, N., Inouye, Y., Sekkat, Z., Kawata, S., 2002a, J. Chem. Phys. **117**, 1296.

Hayazawa, N., Motohashi, M., Saito, Y., Kawata, S., 2005, Appl. Phys. Lett. **86**, 263114.

Hayazawa, N., Tarun, A., Inouye, Y., Kawata, S., 2002b, J. Appl. Phys. **92**, 6983.

Hayazawa, N., Yano, T., Watanabe, H., Inouye, Y., Kawata, S., 2003, Chem. Phys. Lett. **376**, 174.

Hida, A., Mera, Y., Maeda, K., 2001, Appl. Phys. Lett. **78**, 3190.

Hillenbrand, R., Keilmann, F., 2000, Phys. Rev. Lett. **85**, 3029.

Hillenbrand, R., Keilmann, F., 2001, Appl. Phys. B **73**, 239.

Hillenbrand, R., Keilmann, F., 2002, Appl. Phys. Lett. **80**, 25.

Hillenbrand, R., Keilmann, F., Hanarp, P., Sutherland, D.S., Aizpurua, J., 2003, Phys. Lett. **83**, 368.

Hillenbrand, R., Taubner, T., Keilmann, F., 2002, Nature **418**, 159.

Huang, S.M., Hong, M.H., Lu, Y.F., Lukyanchuk, B.S., Song, W.D., Chong, T.C., 2002, J. Appl. Phys. **91**, 3268.

Huber, R., Koch, M., Feldmann, J., 1998, Appl. Phys. Lett. **73**, 2521.
Ichimura, T., Hayazawa, N., Hashimoto, M., Inouye, Y., Kawata, S., 2003, J. Raman Spectrosc. **34**, 651.
Ichimura, T., Hayazawa, N., Hashimoto, M., Inouye, Y., Kawata, S., 2004a, Phys. Rev. Lett. **92**, 220801.
Ichimura, T., Hayazawa, N., Hashimoto, M., Inouye, Y., Kawata, S., 2004b, Appl. Phys. Lett. **84**, 1768.
Iijima, S., 1991, Nature **354**, 56.
Iijima, S., Ichihashi, T., 1993, Nature **363**, 603.
Inagaki, K., Kolosov, O.V., Briggs, G.A.D., Wright, O.B., 2000, Appl. Phys. Lett. **76**, 1836.
Jersch, J., Demming, F., Dickman, K., 1997, Appl. Phys. A **64**, 29.
Jersch, J., Dickmann, K., 1996, Appl. Phys. Lett. **68**, 868.
Kalkbrenner, T., Ramstein, M., Mlynek, J., Sandoghdar, V., 2001, J. Microsc.-Oxford **202**, 72.
Keilmann, F., 2002, Vib. Spectrosc. **29**, 109.
Keilmann, F., 2004, J. Electron Microsc. **53**, 187.
Keilmann, F., Hillenbrand, R., 2004, Philos. T. Roy. Soc. A **362**, 787.
Klar, T., Perner, M., Grosse, S., von Plessen, G., Spirkl, W., Feldmann, J., 1998, Phys. Rev. Lett. **80**, 4249.
Knoll, B., Keilmann, F., 1999, Nature **399**, 134.
Knoll, B., Keilmann, F., 2000, Appl. Phys. Lett. **77**, 3980.
Kobayashi, M., Fujita, K., Kaneko, T., Takamatsu, T., Nakamura, O., Kawata, S., 2002, Opt. Lett. **27**, 1324.
Kossakovski, D., O'Connor, S.D., Widmer, M., Baldeshwieler, J.D., Beauchamp, J.L., 1998, Ultramicroscopy **71**, 111.
Krenn, J.R., Dereux, A., Weeber, J.C., Bourillot, E., Lacroute, Y., Goudonnet, J.P., Schider, G., Gotschy, W., Leitner, A., Aussenegg, F.R., Girard, C., 1999a, Phys. Rev. Lett. **82**, 2590.
Krenn, J.R., Weeber, J.C., Dereux, A., Bourillot, E., Goudonnet, J.P., Schider, B., Leitner, A., Aussenegg, F.R., Girard, C., 1999b, Phys. Rev. B **60**, 5029.
Lahrech, A., Bachelot, R., Gleyzes, P., Boccara, A.C., 1996, Opt. Lett. **21**, 1315.
Lahrech, A., Bachelot, R., Gleyzes, P., Boccara, A.C., 1997, Appl. Phys. Lett. **71**, 575.
Lu, Y.F., Mai, Z.H., Qiu, G., Chim, W.K., 1999, Appl. Phys. Lett. **75**, 2359.
Martin, Y., Zenhausern, F., Wickramasinghe, H.K., 1996, Appl. Phys. Lett. **68**, 2475.
Mehtani, D., Lee, N., Hartschuh, R.D., Kisliuk, A., Foster, M.D., Sokolov, A.P., Maguire, J.F., 2005a, Proc. SPIE **5927**, 592707.
Mehtani, D., Lee, N., Hartschuh, R.D., Kisliuk, A., Foster, M.D., Sokolov, A.P., Maguire, J.F., 2005b, J. Raman Spectrosc. **36**, 1068.
Mills, D.L., 2002, Phys. Rev. B **65**, 125419.
Moreau, A., Ketterson, J.B., 1992, J. Appl. Phys. **72**, 861.
Narita, Y., Kimura, S., 2001, Anal. Sci. **17**, i685.
Nieman, L.T., Krampert, G.M., Martinez, R.E., 2001, Rev. Sci. Instrum. **72**, 1691.
Nonnenmacher, M., Oboyle, M.P., Wickramasinghe, H.K., 1991, Appl. Phys. Lett. **58**, 2921.
Nonnenmacher, M., Wickramasinghe, H.K., 1992, Ultramicroscopy **42**, 351.
Novotny, L., Bian, R.X., Xie, X.S., 1997, Phys. Rev. Lett. **79**, 645.
Novotny, L., Sanchez, E.J., Xie, X.S., 1998, Ultramicroscopy **71**, 21.
Patanè, S., Gucciardi, P.G., Labardi, M., Allegrini, M., 2004, Riv. Nuovo Cim. **27**, 1.
Pettinger, B., Picardi, G., Schuster, R., Ertl, G., 2003, J. Electroanal. Chem. **554**, 293.
Pettinger, B., Ren, B., Picardi, G., Schuster, R., Ertl, G., 2004, Phys. Rev. Lett. **92**, 096101.
Pettinger, B., Ren, B., Picardi, G., Schuster, R., Ertl, G., 2005, J. Raman Spectrosc. **36**, 541.
Poborchii, V., Tada, T., Kanayama, T., 2005, Jpn. J. Appl. Phys. **244**, L202.
Ren, B., Picardi, G., Pettinger, B., 2004, Rev. Sci. Instrum. **75**, 837.

Ren, B., Picardi, G., Pettinger, B., Schuster, R., Ertl, G., 2005, Angew. Chem. Int. Ed. **44**, 139.

Saito, Y., Hayazawa, N., Kataura, H., Murakami, T., Tsukagoshi, K., Inouye, Y., Kawata, S., 2005, Chem. Phys. Lett. **410**, 136.

Shekhawat, G.S., Dravid, V.P., 2005, Science **310**, 89.

Somorjai, G.A., Rupprechter, G., 1999, J. Phys. Chem. B **103**, 1623.

Stokes, D.L., Chi, Z.H., Vo-Dinh, T., 2004, Appl. Spectrosc. **58**, 292.

Stöckle, R., Setz, P., Deckert, V., Lippert, T., Wokaun, A., Zenobi, R., 2001, Anal. Chem. **73**, 1399.

Stöckle, R.M., Suh, Y.D., Deckert, V., Zenobi, R., 2000, Chem. Phys. Lett. **318**, 131.

Sun, W.X., Shen, Z.X., 2001, Mater. Phys. Mechan. **4**, 17.

Sun, W.X., Shen, Z.X., 2003a, J. Raman Spectrosc. **34**, 668.

Sun, W.X., Shen, Z.X., 2003b, Ultramicroscopy **94**, 237.

Sun, W.X., Shen, Z.X., Cheong, F.C., Yu, G.Y., Lim, K.Y., Lin, J.Y., 2002, Rev. Sci. Instrum. **73**, 2942.

Taubner, T., Hillenbrand, R., Keilmann, F., 2003, J. Microsc. **210**, 311.

Taubner, T., Hillenbrand, R., Keilmann, F., 2004, Appl. Phys. Lett. **85**, 5064.

Tian, Z.Q., Ren, B., Wu, D.Y., 2002, J. Phys. Chem. B **106**, 9463.

Tomoda, M., Shiraishi, N., Kolosov, O.V., Wright, O.B., 2003, Appl. Phys. Lett. **82**, 622.

Vankrunkelsven, S., Clicq, D., Cabooter, D., De Malsche, W., Gardeniers, J.G.E., Desmet, G., 2006, J Chromatogr A **1102**, 96.

Wang, J.J., Saito, Y., Batchelder, D.N., Kirkham, J., Robinson, C., Smith, D.A., 2005, Appl. Phys. Lett. **86**, 263111.

Wang, J.J., Smith, D.A., Batchelder, D.N., Saito, Y., Kirkham, J., Robinson, C., Baldwin, K., Li, G., Bennett, B., 2003, J. Microsc.-Oxford **210**, 330.

Watanabe, H., Ishida, Y., Hayazawa, N., Inouye, Y., Kawata, S., 2004, Phys. Rev. B **69**, 155418.

Weaver, J.M.R., Walpita, L.M., Wickramasinghe, H.K., 1989, Nature **342**, 783.

Wessel, J., 1985, J. Opt. Soc. Am. B **2**, 1538.

Wu, X., Wang, Y., Wang, J., Tian, Q., Yu, J., 2005, Proc. SPIE **5635**, 292.

Yamanaka, K., Ogiso, H., Kolosov, O., 1994, Appl. Phys. Lett. **64**, 178.

Zayats, A.V., 1999, Opt. Commun. **161**, 156.

Zenhausern, F., Martin, Y., Wickramasinghe, H.K., 1995, Science **269**, 1083.

Zerda, T.W., Song, G., Waddell, W.H., 2003, Proc. Spring ACS Rubber Div. Meeting **76**, 769.

Zumbusch, A., Holtom, G.R., Xie, X.S., 1999, Phys. Rev. Lett. **82**, 4142.

Chapter 5

Tip-enhanced optical spectroscopy of single-walled carbon nanotubes

by

A. Hartschuh, H. Qian, A.J. Meixner

Physikalische und Theoretische Chemie, Universität Tübingen, Tübingen, Germany

N. Anderson, L. Novotny

University of Rochester, The Institute of Optics, Rochester, New York 14627, USA

Tip Enhancement
Advances in Nano-Optics and Nano-Photonics
ISSN: 1871-0018

S. Kawata and V.M. Shalaev (Editors)
© *2007 Elsevier B.V. All rights reserved.*
DOI: 10.1016/S1871-0018(06)01005-3

Contents

§ 1. Introduction

Single-walled carbon nanotubes (SWCNTs) are quasi-one-dimensional quantum wires with great promise for application in photonics, opto- and nanoelectronics. Their widespread utilization in devices, however, requires a detailed understanding of the role of localized defects that are expected to alter the desired properties of idealized nanotubes. Alternatively, based on this understanding, intentional modifications such as doping could be envisaged as a strategy to improve device characteristics as in the case of conventional semiconductors. Tip-enhanced optical spectroscopy is ideally suited to address questions regarding local material properties providing high spatial resolution, essential signal enhancement and spectroscopic contrast (see e.g. Hartschuh et al., 2003b, 2004). The electromagnetic fields in the proximity of the nanoscale metal tip control both excitation and emission efficiencies giving rise to amplified optical responses depending on the local properties of the sample material. Visualization of localized phonon modes and luminescent states with nanoscale resolution is feasible.

Measurements on the single nanotube level using confocal microscopy already revealed variations of emission energies and excited state lifetimes from nanotube to nanotube indicating significant inhomogeneities on sub-diffraction length scales. Since both Raman scattering and photoluminescence can be detected from a single nanotube simultaneously (Hartschuh et al., 2003a, 2005b), SWCNTs are the ideal model system for the evaluation of tip-enhanced optical techniques.

In this chapter, we first describe our experimental setup for near-field optical imaging. The fundamental properties of SWCNTs are introduced in Section 3. In the following two sections, tip-enhanced spectroscopy is used to investigate Raman scattering and photoluminescence of SWCNTs with a spatial resolution better than 15 nm. Finally, the signal enhancement achieved for Raman scattering and photoluminescence on the same nanotube is compared and the different mechanisms are discussed.

§ 2. Experimental setup

The experimental setup is based on an inverted optical microscope with an x, y scan stage for raster scanning a transparent sample (fig. 1). Laser excitation light is provided either by short-pulse lasers to achieve high excitation densities or narrow-bandwidth cw-lasers for high spectral resolution. For the experimental results presented in this work, a cw-laser at 632.8 nm (HeNe-laser) is used. The laser beam is reflected by a non-polarizing beam splitter and focused by a high numerical aperture (NA) objective (1.4 NA) on the sample surface. A sharp gold tip is positioned in the focus of the beam and maintained above the sample surface at a distance of ≈ 2 nm by means of a sensitive shear-force feedback mechanism (fig. 1(c)) (Karrai and Grober, 1995).

Photoluminescence and Raman scattered light is collected with the same objective, transmitted by the beam splitter and filtered by a notch filter to remove the fundamental laser light. The optical signals are detected either by a combination of a spectrograph and a cooled charged coupled device (CCD) or by two single-photon counting avalanche photodiodes (APDs). In the beam path of the APDs, the light passes first through a dichroic

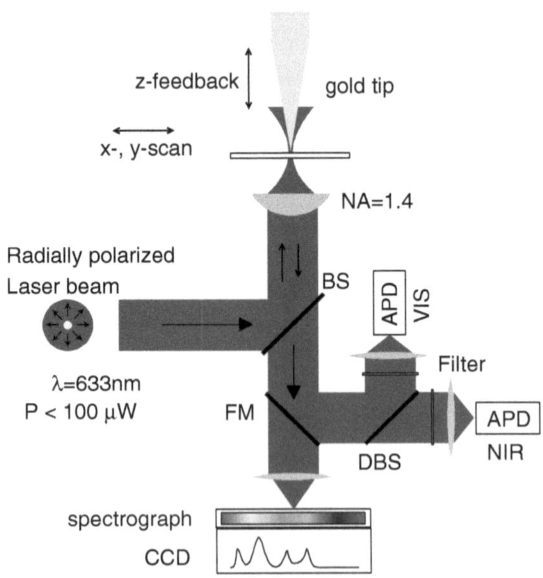

Fig. 1. Schematic of the experimental setup. A sharp metal tip is scanned through a tightly focused laser beam. The optical signal is detected either by APDs after spectral filtering or by a combination of spectrograph and a CCD. BS: Non-polarizing beam splitter. FM: Flip mirror. DBS to separate visible (VIS) and near-infrared light (NIR).

beam splitter (DBS) to separate visible light and infrared light before being detected by the two APDs independently. Narrow band-pass filters are used to select spectral windows to detect specific Raman or emission bands only. A near-field optical image is established by raster scanning the sample and simultaneously recording the optical signals. Sharp gold tips are produced by electrochemical etching in HCl.

In order to establish a strong-field enhancement at the tip, the electric field of the exciting laser beam needs to be polarized along the tip axis. To achieve this condition in our on-axis illumination scheme, we displace the tip from the center of the beam in the direction of polarization into one of the two longitudinal field lobes characteristic for strongly focused Gaussian beams or we use higher order laser modes such as the HG_{10} mode or the radially polarized mode with strong longitudinal fields in the center of the focus (Novotny et al., 1998; Quabis et al., 2000).

§ 3. Single-walled carbon nanotubes

SWCNTs are cylinders that consist of a single graphene sheet. The *multi-walled* form was first reported by Iijima (1991) 15 years ago. Two years later, SWCNTs were reported (Iijima and Ichihasi, 1993). They are typically between 0.5 and 2 nm in diameter, but up to millimeters in length. The structure of a nanotube is determined by the chiral vector $\mathbf{C}_h = n\mathbf{a}_1 + m\mathbf{a}_2$, that corresponds to the vector around the nanotube circumference in terms of the unit cell vectors of graphene $\mathbf{a}_1$ and $\mathbf{a}_2$. Thus, the tuple (n, m) uniquely defines the structure of a particular nanotube. In fig. 2(a), the structure of

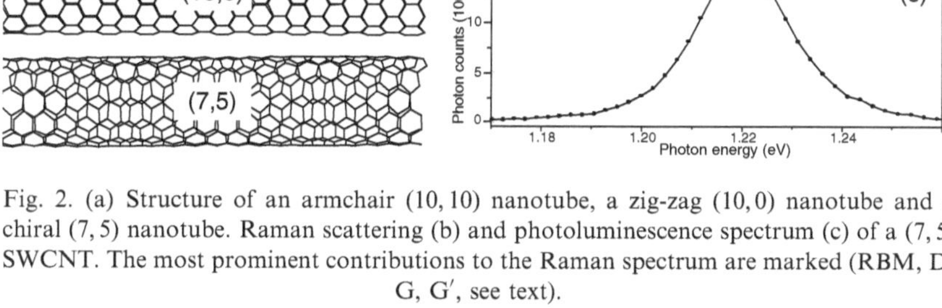

Fig. 2. (a) Structure of an armchair (10, 10) nanotube, a zig-zag (10, 0) nanotube and a chiral (7, 5) nanotube. Raman scattering (b) and photoluminescence spectrum (c) of a (7, 5) SWCNT. The most prominent contributions to the Raman spectrum are marked (RBM, D, G, G′, see text).

three different carbon nanotubes is shown. According to tight binding calculations (Saito et al., 1998) nanotubes with $(n-m) \bmod 3 = 0$ (n divisible by m) are metallic or semimetallic. All other nanotubes are semiconducting with a band gap in the near-infrared.

The growth of high-quality SWCNTs is still a very active field of research. So far, arc-discharge, laser ablation and chemical vapor deposition (CVD) are the three main methods for SWCNT production (see e.g. Dai, 2001; Liu et al., 2004). With the arc-discharge and laser methods, only powdered samples with broad diameter distributions can be produced. These powders contain large amounts of catalyst particles while most of the nanotubes are tangled into bundles requiring additional purification and separation procedures. CVD methods allow more control over the morphology and structure of the produced nanotubes and well-separated individual nanotubes either supported on flat substrates or suspended across trenches can be grown. The high-pressure catalytic decomposition of carbon monoxide (HiPCO) process, which is a floating catalyst CVD method, is used for the large-scale production of high-quality carbon nanotubes. In this chapter, we present experimental results obtained on nanotubes grown by arc-discharge, CVD and HIPCo methods.

Raman spectroscopy has been demonstrated to be of great value for the structural analysis of SWCNTs. The radial breathing mode (RBM) occurs in the frequency range $\nu = [100-500]\,\mathrm{cm}^{-1}$ and originates from vibrations transverse to the nanotube axis (see fig. 2(b)). Its frequency ν_{RBM} is directly related to the diameter d of the nanotube as the force needed for a radial deformation of the nanotube increases as the diameter decreases. Raman measurements using different laser excitation energies (Rao et al., 1997), showed a very strong, diameter-selective resonance effect. *Resonance Raman scattering* enhances the signal essentially and is the reason why a single nanotube can be observed in the first place. By combining the measured RBM frequency and considerations of resonant Raman scattering, the structural parameters (n, m) of an individual tube can be determined (Jorio et al., 2001).

The disorder-induced D-band appears at frequencies around $\nu = 1300\,\mathrm{cm}^{-1}$ and is also found in other carbons. The D-band results from phonons with large momentum for which momentum conservation is maintained by elastic scattering from defects. Its second order mode, the G′-band, occurs at frequencies around $\nu = 2600\,\mathrm{cm}^{-1}$ and is identified with two-phonon scattering processes arising from the creation of two phonons of equal and opposite momentum. In this case, no defects have to be present and the intensity ratio D-band/G′-band is discussed as a measure for the defect concentration (Reich et al., 2004).

Raman signals in the frequency range of $\nu = [1500-1600]\,\mathrm{cm}^{-1}$ constitute the G-band and originate from *C–C* stretching modes along the nanotube axis (G^+-component) and along the circumference (G^--component). While the assignment of these components is still under debate, it has been shown experimentally that there are characteristic differences between the G-bands of metallic and semiconducting nanotubes (Pimenta et al., 1998; Oron–Carl et al., 2005).

The **photoluminescence** of semiconducting SWCNTs occurs in the near-IR to IR spectral range and was only recently reported for the first time by (O'Connell et al., 2002) for an ensemble of SWCNTs encapsulated in micelles. The emission energy was found to be determined by the nanotube structure (n, m) (Bachilo et al., 2002) (see fig. 2(c)). At present, light emission from SWCNTs is expected to result from excitonic recombination (Wang et al., 2005; Perebeinos et al., 2004; Hagen et al., 2005; Ando, 1997).

For spectroscopists, SWCNTs are unique since both photoluminescence and Raman scattering can be observed simultaneously under the same experimental conditions (Hartschuh et al., 2003a). While consisting of carbon atoms only, electronic properties of carbon nanotubes are solely determined by their structural properties and symmetries. Any symmetry breaking, resulting from a localized topological defect for example, is thus expected to influence both, structural sensitive Raman spectra and electronically determined photoluminescence spectra. Measurements on the single nanotube level using conventional microscopy revealed that the spectroscopic properties can vary considerably from nanotube to nanotube. It was noticed that the emission energies are not completely uniform for a specific nanotube structure (n, m) but rather show a distribution centered at the energies observed in ensemble measurements (Hartschuh et al., 2003a; Htoon et al., 2004; Hartschuh et al., 2005b). Similarly, different excited state decay times have been observed for nanotubes of the same structure (Hagen et al., 2005). Because of these variations from nanotube to nanotube, inhomogeneous properties are also expected within individual nanotubes demanding a high-resolution near-field spectroscopic technique.

§ 4. Near-field Raman spectroscopy of SWCNTs

An essential improvement for the physical understanding of Raman scattering was achieved by going from bundles and nanotube ensembles to individual nanotubes thus avoiding *ensemble averaging*. Nanotubes of the same single (n, m) species were studied and chirality-dependent properties were identified. However, the spatial extent of the diffraction limited focus in conventional microscopy invariably leads to *spatial averaging* along individual nanotubes. Accordingly, structural defects, that are localized in nature, will perturb the observed spectra consisting of a superposition of different contributions. As an example, a single topological 5–7 defect can cause a sudden change of the chirality within a single nanotube (Yao et al., 1999). To avoid spatial averaging, high-resolution spectroscopy *within* individual nanotubes has to be performed. In the following, the superb spatial resolution achieved by tip-enhanced Raman spectroscopy is demonstrated. Examples outlining the capabilities of chemically specific near-field imaging are presented in which the technique is used to analyze the vibronic properties of SWCNTs.

In fig. 3(c) a near-field Raman image of an SWCNT on glass is shown together with the simultaneously acquired topography image of the same sample area in fig. 3(a). The Raman image was acquired by detecting the intensity of the G-band around 1600 cm^{-1} (Dresselhaus et al., 2005) after laser excitation at 632.8 nm using a radially polarized laser mode while raster scanning the sample. The optical resolution can be determined from the width of the signal full width at half maximum (FWHM) presented as line scan in fig. 3(d) to be about 13 nm, far below the diffraction limit of conventional microscopy. Raman spectra detected on top of this

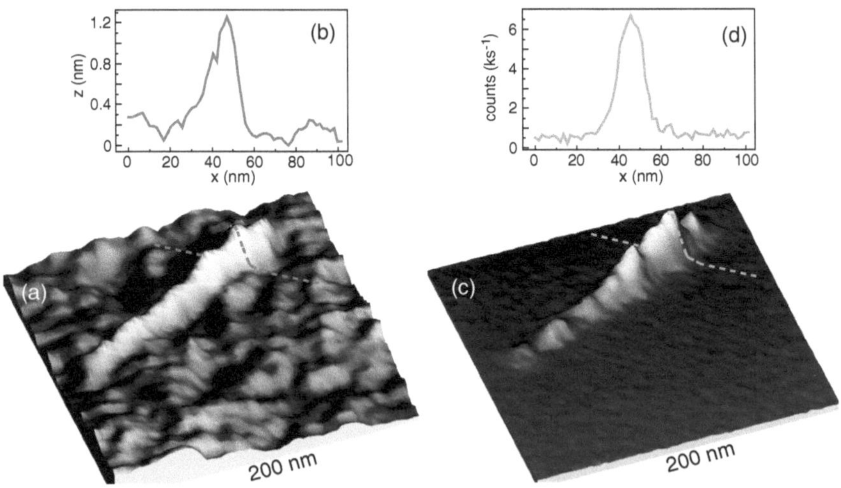

Fig. 3. (a) Topography image of an SWCNT grown by arc-discharge on glass. (c) Simultaneously acquired near-field Raman image established by detecting the G-band intensity around 700 nm after laser excitation at 632.8 nm. (b) and (d): Cross sections taken along the dashed line in the topographic (a) and optical image (c), respectively. The optical resolution is determined from the width of the cross section in (d) and is about 13 nm (FWHM). Besides the SWCNT, the topographic image reveals the roughness of the glass substrate.

nanotube (not shown) exhibit an RBM-frequency of $199\,\mathrm{cm}^{-1}$ corresponding to a nanotube diameter of $d \approx 1.2\,\mathrm{nm}$. The structure of the nanotube can be derived by considering the resonance Raman effect for the excitation wavelength of 632.8 nm and the determined diameter and is presumably $(n, m) = (14, 2)$.

In the next two examples, the vibrational properties along individual nanotubes are probed with high spatial resolution. Figure 4(a) shows the topography of SWCNTs grown by CVD on quartz. In fig. 4(b) a magnified optical view of the center area marked in (a) is presented. In this region, the nanotube did not grow straight and a bent part with a radius of curvature of about 300 nm can be seen. While the overall intensity of the G-band around $1585\,\mathrm{cm}^{-1}$ used to establish the image is found to be nearly uniform along the straight sections of the nanotube (positions 1, 2 and 4), the peak position of the G^+-component is shifted from $1575\,\mathrm{cm}^{-1}$ to $1582\,\mathrm{cm}^{-1}$ in the bent part (position 3). Cronin et al. (2004b) found that uniaxial stress in nanotubes strained by the tip of an AFM causes a considerable shift of the G-band position to lower frequencies by about $20\,\mathrm{cm}^{-1}$. The observed shift of the G^+-component to higher frequencies in the bent part in fig. 4(b) might thus result from non-symmetrical changes in the bond length.

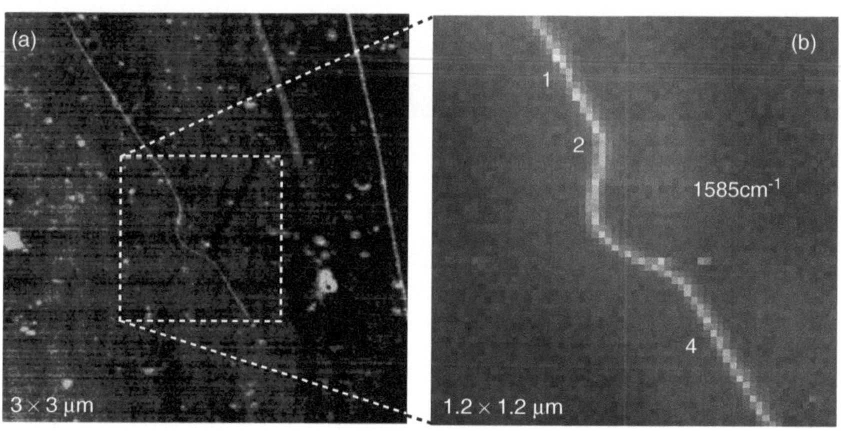

Fig. 4. (a) Topography image of SWCNTs grown by CVD on quartz showing mostly straight single nanotubes. Beside nanotubes, catalyst particles can be seen. (b) Near-field Raman image of the area marked in (a) established by detecting Raman spectra while raster-scanning the sample. The image is formed by integrating the intensity of the G-band around $1585\,\mathrm{cm}^{-1}$. Note that the catalyst particles render no optical signal demonstrating the chemical specificity of the near-field Raman technique. An analysis of the Raman spectra taken along the nanotube at positions 1–4 reveals uniform positions of the G^+-component for the straight ($1575\,\mathrm{cm}^{-1}$ at positions 1, 2 and 4) and a spectral shift of the G^+-component to $1582\,\mathrm{cm}^{-1}$ in the bent part around position 3.

Evidence for changes of the chirality (n, m) along individual nanotubes based on near-field Raman spectroscopy has been presented in Anderson et al. (2005). Here we discuss near-field images and spectra revealing a local transition from metallic to semiconducting characteristics. In the upper left section - positions 1 and 2 marked in fig. 5(a)–the nanotube features an RBM-frequency of $189\,\mathrm{cm}^{-1}$ as can be seen in the spectra in (d). While the intensity of the G-band remains constant along the nanotube (fig. 5(c)) the RBM-peak disappears at position 3 and reappears for position 4 indicating a local change in the structure of the nanotube. This is supported by closer inspection of the spectral contributions in the G-band region. Although the overall intensity remains nearly constant, the characteristics of the bands change considerably: spectra 1 and 2 feature a broad low-frequency G^--component around $1535\,\mathrm{cm}^{-1}$ characteristic for metallic nanotubes (Oron–Carl et al., 2005). The narrow peak at about $1590\,\mathrm{cm}^{-1}$ present in spectra 3 and 4, on the other hand, is

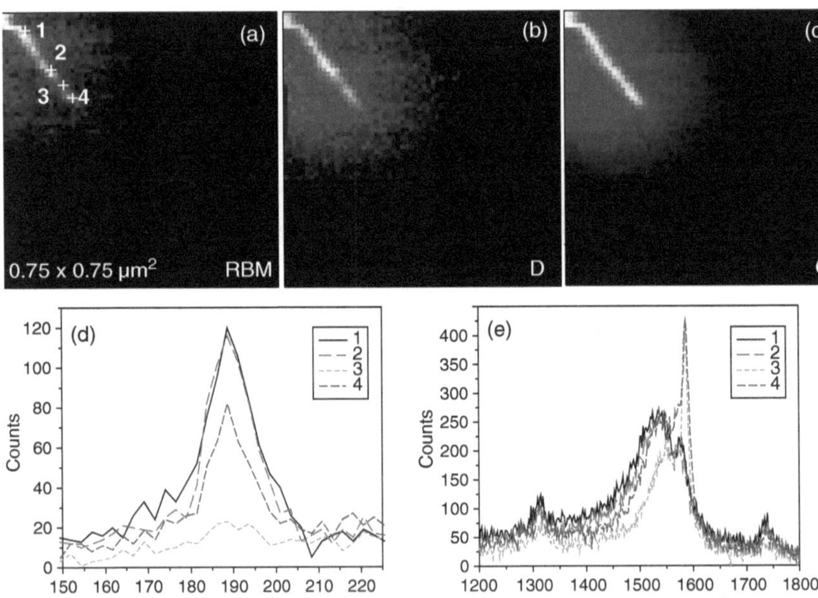

Fig. 5. Spectroscopic images of an SWCNT grown by the HiPCO method on glass. Scan range $0.75 \times 0.75\,\mu\mathrm{m}^2$. The images were acquired by detecting the intensity of the RBM around $189\,\mathrm{cm}^{-1}$ (a), the D-band around $1300\,\mathrm{cm}^{-1}$ (b) and the G-band around $1550\,\mathrm{cm}^{-1}$ (c), respectively. Raman spectra taken at positions 1–4 indicated in (a) are shown in (d), and (e). From the image in (a) and the spectra in (d), strong intensity variations of the RBM are evident indicating structural transitions along the nanotube. The spectral position and shape of the G-band in (e) clearly show that the nanotube is metallic at positions 1 and 2. Spectrum 3 and 4, on the other hand, reveal semiconducting G-band characteristics.

typical for semiconducting nanotubes. Remarkably, the disorder-induced D-band is pronounced close to the nanotube section where changes of the RBM-intensity occur (fig. 5(b)).

An alternative explanation for the localized transition from metallic to semiconducting G-band characteristics not involving changes in nanotube chirality is based on the presence of charged adsorbates. Electron transfer caused by charged species adsorbed to the nanotube sidewall is expected to induce changes in the occupation of electronic states thereby modifying both resonance Raman conditions and electron–phonon coupling (Strano et al., 2003). Spectro-electrochemical studies of SWCNTs films showed that the main vibrational features depend on the strength and sign of the applied external potential (Corio et al., 2003, 2004; Cronin et al., 2004a). It was suggested that a negative potential, corresponding to electron injection into the nanotube from the environment, moves the Fermi level of metallic nanotubes outside the conduction band leading to semiconducting G-band characteristics. Similarly, upon applying a positive potential, RBM and low-frequency G-band components disappeared while a sharper high-frequency G-band component increased.

The present example highlights the complexity of the information obtained in near-field Raman spectroscopy of SWCNTs. Local Raman spectra reflect both structural variations related to topological defects or changes in the nanotube chirality and electronic variations resulting from electron transfer involving charged defects.

§ 5. Near-field photoluminescence spectroscopy of SWCNTs

While Raman spectroscopy provides a wealth of structural information, information on electronic states is only obtained indirectly through considerations of the resonance Raman effect. Photoluminescence imaging, on the other hand, probes directly energies and spatial extent of luminescent states. Furthermore, photoluminescence intensities reflect the competition of radiative and nonradiative decay rates and give information on energy transfer processes.

Our investigations on SWCNTs using near-field Raman spectroscopy highlight the occurrence of structural inhomogeneities and structural defects. These localized perturbations can be expected to limit the spatial extent of delocalized excitonic states along a nanotube and thus reduce its photoluminescence quantum efficiency essentially.

Near-field photoluminescence and Raman images of micelle-encapsulated SWCNTs on mica are shown in fig. 6(a) and (b) together with the

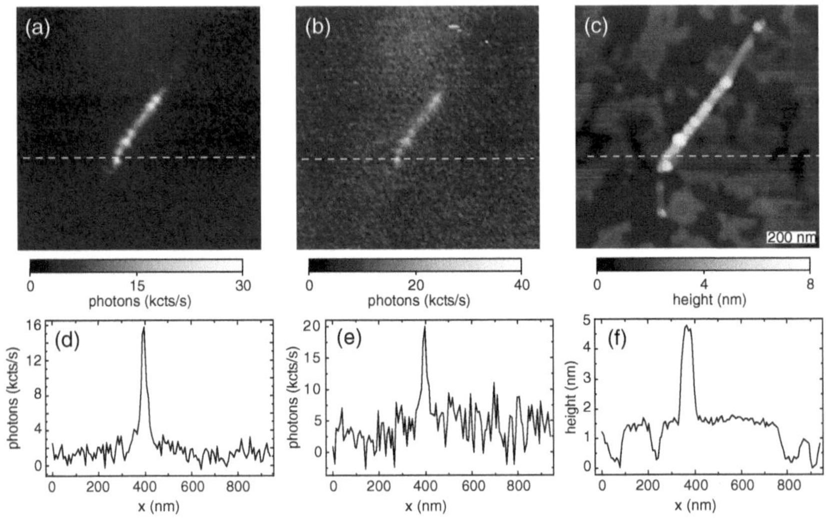

Fig. 6. Near-field photoluminescence (a), Raman (b) and topography image (c) detected simultaneously for micelle-encapsulated SWCNTs on mica. (d), (e) and (f) are optical and topographic cross sections taken along the dashed lines in (a), (b) and (c), respectively. The optical resolution achieved in the experiment is determined from the widths of the cross sections (FWHM) in (d) and (e). Similar optical resolution of about 15 nm is observed for photoluminescence and Raman scattering. (Figure reprinted with permission from ACS, A. Hartschuh et al., Nano Letters 5, 2310 (2005)).

simultaneously acquired topography image in (c) of the same sample area. The photoluminescence image was formed by detecting the light intensity around 950 nm upon laser excitation at 632.8 nm. Emission at 950 nm is expected to result from (8,3)-nanotubes (Bachilo et al., 2002). The Raman image represents the intensity of the G-band. In all three images, carbon nanotubes can be clearly observed extending from the upper right to the center. In the optical images (a) and (b) however, the upper section of the nanotubes is dark.

Possible explanations include the observation of a single nanotube that changes its structure and becomes non-luminescent or of a thin bundle in which the luminescent nanotube ends and only metallic nanotubes or nanotubes with emission energies outside the detection window remain in the upper section. In addition to nanotubes, the topography image also reveals two-dimensional features with a step height of about 1.4 nm as would be expected for a single layer of sodium dodecyl sulfate (SDS) surfactant that is used to form the micelles (O'Connell et al., 2002). The topographic height measured across the nanotubes is about 3.5 nm as can be seen from fig. 6(f). On the basis of the topography data, it is not

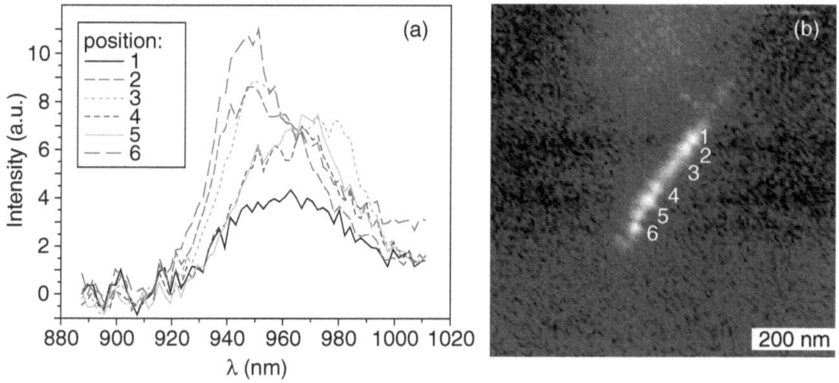

Fig. 7. (a) Near-field photoluminescence spectra taken along positions 1–6 in steps of 30 nm indicated in the near-field photoluminescenceimage of the SWCNT in (b). The spectral position and shape of the photoluminescence is found to vary significantly on the nanoscale. (Figure reprinted with permission from ACS, A. Hartschuh et al., Nano Letters 5, 2310 (2005)).

possible to distinguish between a thin bundle of nanotubes or an individual nanotube that is partially covered with SDS.

Photoluminescence is generally expected to be weak in bundles because of energy transfer from semiconducting nanotubes to non-luminescent metallic nanotubes that convert the energy non-radiatively. At present, the timescales for such processes is unknown. Our own experimental results however show that detectable emission occurs even for very thick bundles around 50 nm in diameter.

Remarkably, the photoluminescence in fig. 6 is extended on a length scale of about 500 nm, as would be expected for unperturbed and mobile excitonic states. For other nanotubes we observed a varying degree of localization of the emission signals (Hartschuh et al., 2005a).

The optical resolution achieved in the experiment is determined from the cross section shown in fig. 6(d) to be about 12 nm. In general, the optical resolution observed in near-field photoluminescence and Raman images is the same.

Near-field photoluminescence spectra taken along the nanotube in steps of 30 nm are presented in fig. 7. From the spectra it is clear that the observed photoluminescence emission is in fact composed of multiple emission bands with different emission energies and varying relative contributions. Owing to its limited spatial resolution, standard confocal spectroscopy would only show a single broadened luminescence peak resulting from a superposition of individual bands. The origin of these local energetic fluctuations are most likely changes of the dielectric function of the environment and the presence of charged adsorbates (Htoon et al., 2004; Perebeinos et al., 2004).

§ 6. Discussion of the signal enhancement and the image contrast

Tip-enhanced near-field optical spectroscopy makes use of the strongly-enhanced electric field components close to a sharp metal tip under laser illumination. The resulting enhancement for Raman scattering and photoluminescence is expected to be different. For Raman scattering, the electromagnetic enhancement of the signal M_{Raman} is caused by enhancement of both the incident field E_{incident} and the scattered field and can be approximated by $M_{\text{Raman}} \approx [E_{\text{local}}/E_{\text{incident}}]^4$ where E_{local} is the local electric field.

For photoluminescence, three different contributions can be distinguished.

- Enhanced excitation resulting from the locally enhanced electric field $M_{\text{PL}} = [E_{\text{local}}/E_{\text{incident}}]^2$.
- Modifications of the spontaneous emission rate due to variations in the local photonic mode density caused by the metal tip (Purcell effect).
- Additional non-radiative decay channels created by light absorption inside the metal tip.

Quantitative comparisons of the signal enhancements achieved simultaneously for Raman scattering and photoluminescencefor the same nanotube indicate that the tip-enhancement is in fact more efficient for photoluminescence in the case of SWCNTs (Hartschuh et al., 2005a). This observation can be explained by the extremely low photoluminescence quantum yield of SWCNTs ($\approx 10^{-3} - 10^{-4}$; O'Connell et al., 2002). Besides localizing the excitation field, the tip can also act as an antenna for the emitted radiation thus increasing the radiative-decay rate and the photoluminescence quantum yield (Thomas et al., 2004). This effect has already been used to study other molecular systems with low fluorescence quantum yield on rough metal films (Aslan et al., 2005). In the case of highly fluorescent dye molecules with quantum yields close to 1 typically studied in literature, a further increase of the radiative decay rate will play only a minor role such that quenching by the metal tip becomes more efficient (Frey et al., 2004; Farahani et al., 2005).

It is well known that the strongest field component is oriented along the tip axis giving rise to characteristic fluorescence patterns that can be utilized to retrieve the orientation of single fluorescent molecules (Frey

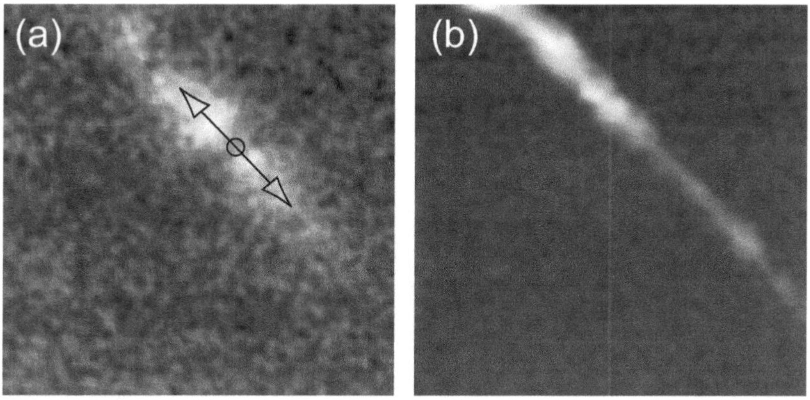

Fig. 8. Near-field photoluminescence (a) and topography image (b) of SWCNTs on mica. Scan range $200 \times 200\,\text{nm}^2$. The photoluminescence image exhibits a double lobe structure characteristic for a single transition dipole moment oriented within the sample plane as indicated by the arrow. The observed photoluminescence results presumably from a highly localized emissive state corresponding to a bound exciton.

et al., 2004). In the case of nanotubes however, the strongest optical transitions are polarized along the nanotube axis, i.e. perpendicular to the tip axis. Accordingly, only weak off-axis components of the enhanced fields contribute to the observed signal enhancement. In an idealized picture, the nanotube can be viewed as consisting of a string of transition dipole moments. The double lobe patterns, expected for single transition dipole moments (Frey et al., 2004), will add up to form a single line in optical scan images. For very short nanotubes, i.e. on the order of the optical resolution achieved in the near-field experiment, single double lobe patterns are expected and are in fact observed experimentally (see fig. 8).

The high optical resolution attained in tip-enhanced near-field optics indicates that the enhanced field is laterally confined to the size of the metal tip as expected from theoretical predictions (see e.g. Novotny et al., 1997). To demonstrate the confinement of the enhanced fields in the longitudinal direction, i.e. along the tip axes, the tip is positioned above an SWCNT and the Raman intensity is recorded as a function of tip–sample distance (fig. 9(a)). The enhanced signal is found to decay within $\approx 5\,\text{nm}$, corresponding to about half the width of the optical features observed for this particular tip in the optical scan images.

From our experimental results it is evident that the enhanced fields vary strongly within 5–10 nm. While this length scale exceeds typical molecular dimensions significantly, it is relevant for the electronic and vibronic states of SWCNTs (fig. 9(b)). At present, the spatial extent of localized phonon modes in nanotubes is unknown. Inelastic scanning-tunneling

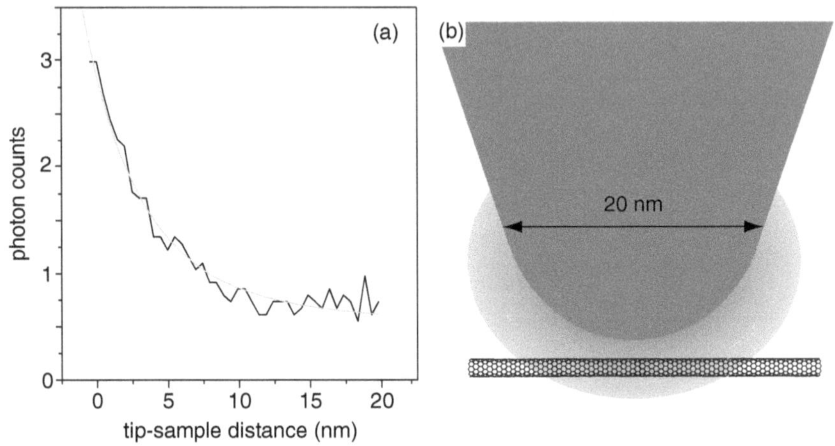

Fig. 9. (a) Dependence of the signal enhancement on the tip–sample distance. From the signal decay and the spatial resolution observed in the scan images, a three-dimensional field confinement on a length scale of 5–10 nm can be deduced. (b) Schematic of the experimental configuration illustrating the relative length scales of metal tip (apex diameter 20 nm), enhanced fields and carbon nanotubes.

spectroscopy indicates that a minimum length of the nanotube of about 3 nm is required to sustain an RBM (Vitali et al., 2004).

The size of excitons (Bohr radius) in semiconducting SWCNTs is discussed to be in the range of 1–5 nm depending on intrinsic properties such as nanotube diameter and chirality but also on extrinsic parameters such as the dielectric function of the environment (Perebeinos et al., 2004; Chang et al., 2004; Maultzsch et al., 2005). Accordingly, standard approximations leading to electric dipole transitions will be affected and optical selection rules could differ essentially involving additional multiplor transitions as predicted for quantum dots (Zurita and Novotny, 2002).

§ 7. Conclusions and outlook

Simultaneous near-field Raman and photoluminescence spectroscopy of SWCNTs with a spatial resolution better than 15 nm has been achieved providing a wealth of information on structural and electronic properties. Local spectroscopy reveals sample inhomogeneities that would remain hidden in conventional confocal microscopy measurements. Future experiments will allow us to establish a correlation between local structural and electronic properties derived from Raman and photoluminescence spectra, respectively. The quantitative study and comparison of the near-field signal enhancement for photoluminescence and Raman scattering, in particular their dependencies on the tip–sample distance, will help in refining our understanding of tip-enhanced optical techniques.

Acknowledgments

This work was funded by the DFG (grant ME 1600/5-1), the U.S. Department of Energy (grant DE-FG02-05ER46207) and the National Science Foundation (grant CHE-0454704).

References

Anderson, N., Hartschuh, A., Novotny, L., 2005, Nanoscale vibrational analysis of single-walled carbon nanotubes, J. Am. Chem. Soc. **127**, 2533–2537.

Ando, T., 1997, Excitons in carbon nanotubes, J. Phys. Soc. Jap. **66**, 1066–1073.

Aslan, K., Gryczynski, I., Malicka, J., Matveea, E., Lakowicz, J.R., Geddes, C.D., 2005, Metal-enhanced fluorescence: an emerging tool in biotechnology, Anal. Biotechnol. **16**, 55–62.

Bachilo, S.M., Strano, M.S., Kittrell, C., Hauge, R.H., Smalley, R., Weisman, R.B., 2002, Structure-assigned optical spectra of single-walled carbon nanotubes, Science **298**, 2361–2366.

Chang, E., Bussi, G., Ruini, A., Molinari, E., 2004, Excitons in carbon nanotubes: An ab initio symmetry-based approach, Phys. Rev. Lett. **92**, 196401.

Corio, P., Jorio, A., Demir, N., Dresselhaus, M.S., 2004, Spectro-electrochemical studies of single wall carbon nanotubes films, Chem. Phys. Lett. **392**, 396–402.

Corio, P., Santos, P.S., Brar, V.W., Samsonidze, G.G., Chou, S.G., Dresselhaus, M.S., 2003, Potential dependent surface Raman spectroscopy of single wall carbon nanotube films on platinum electrodes, Chem. Phys. Lett. **370**, 675–682.

Cronin, S.B., Barnett, R., Tinkham, M., Chou, S.G., Rabin, O., Dresselhaus, M.S., Swan, A.K., Ünlü, M.S., Goldberg, B.B., 2004a, Electrochemical gating of individual single-wall carbon nanotubes observed by electron transport measurements and resonant Raman spectroscopy, Appl. Phys. Lett. **84**, 2052–2054.

Cronin, S.B., Swan, A.K., Ünlü, M.S., Goldberg, B.B., Dresselhaus, M.S., Tinkham, M., 2004b, Measuring the uniaxial strain of individual single-wall carbon nanotubes: Resonance Raman spectra of atomic-force-microscope modified single-wall nanotubes, Phys. Phys. Lett. **93**, 2052–2054.

Dai, H., 2001, Nanotube growth and characterization, vol. 80 of Carbon Nanotubes, Topics in applied physics, Springer-Verlag, Berlin Heidelberg.

Dresselhaus, M.S., Dresselhaus, G., Saito, R., Jorio, A., 2005, Raman spectroscopy of carbon nanotubes, Phys. Reports **409**, 47–99.

Farahani, J.N., Pohl, D.W., Eisler, H.-J., Hecht, B., 2005, Single quantum dot coupled to a scanning optical antenna: A tunable superemitter, Phys. Rev. Lett. **95**, 017402–017405.

Frey, H.G., Witt, S., Felderer, K., Guckenberger, R., 2004, High-resolution imaging of single fluorescent molecules with the optical near-field of a metal tip, Phys. Rev. Lett. **93**, 200801–200804.

Hagen, A., Steiner, M., Raschke, M.B., Lienau, C., Hertel, T., Qian, H., Meixner, A.J., Hartschuh, A., 2005, Exponential decay lifetimes of excitons in individual single-walled carbon nanotubes, Phys. Rev. Lett. **95**, 197401–197404.

Hartschuh, A., Pedrosa, H.N., Novotny, L., Krauss, T.D., 2003a, Simultaneous fluorescence and Raman scattering from single carbon nanotubes, Science **301**, 1354–1356.

Hartschuh, A., Sánchez, E.J., Xie, X.S., Novotny, L., 2003b, High-resolution near-field raman microscopy of single-walled carbon nanotubes, Phys. Rev. Lett. **90**, 095503–095506.
Hartschuh, A., Beversluis, M.R., Bouhelier, A., Novotny, L., 2004, Tip-enhanced optical spectroscopy. Philosophical transactions: Mathematical, physical and engineering science, 807–819.
Hartschuh, A., Huihong, Q., Meixner, A.J., Anderson, N., Novotny, L., 2005a, Nanoscale optical imaging of excitons in single-walled carbon nanotubes, Nano Lett. **5**, 2310–2313.
Hartschuh, A., Pedrosa, H.N., Peterson, J., Huang, L., Anger, P., Qian, H., Steiner, M., Meixner, A.J., Novotny, L., Krauss, T.D., 2005b, Single carbon nanotube optical spectroscopy, Chem. Phys. Chem. **6**, 577–582.
Htoon, H., O'Connell, M.J., Cox, P.J., Doorn, S.K., Klimov, V.I., 2004, Low temperature emission spectra of individual single-walled carbon nanotubes: Multiplicity of subspecies within single-species nanotube-ensembles, Phys. Rev. Lett. **93**, 027401–027404.
Iijiima, S., 1991, Helical microtubules of graphitic carbon, Nature **354**, 56–58.
Iijima, S., Ichihasi, T., 1993, Single-shell carbon nanotubes of 1-nm diameter, Nature **363**, 603–605.
Jorio, A., Saito, R., Hafner, J.H., Lieber, C.M., Hunter, M., McClure, T., Dresselhaus, G., Dresselhaus, M., 2001, Structural (n,m) determination of isolated single-wall carbon nanotubes by resonant raman scattering, Phys. Rev. Lett. **86**, 1118–1121.
Karrai, K., Grober, R.D., 1995, Piezoelectric tip-sample distance control for near field optical microscopes, Appl. Phys. Lett. **66**, 1842–1844.
Liu, J., Fan, S., Dai, H., 2004, Recent advances in methods of forming carbon nanotubes, MRS Bulletin April, 244–250.
Maultzsch, J., Pomraenke, R., Reich, S., Chang, E., Prezzi, D., Ruini, A., Molinari, E., Strano, M.S., Thomsen, C., Lienau, C., 2005, Exciton binding energies in carbon nanotubes from two-photon photoluminescence, Phys. Rev. **B 72**, 241402(R)–241405(R).
Novotny, L., Bian, R.X., Xie, X.S., 1997, Theory of nanometric optical tweezers, Phys. Rev. Lett. **79**, 645–648.
Novotny, L., Sánchez, E.J., Xie, X.S., 1998, Near-field optical imaging using metal tips by higher-order hermite-gaussian beams, Ultramicroscopy **71**, 21–29.
O'Connell, M.J., Bachilo, S.M., Huffman, C.B., Moore, V.C., Strano, M.S., Haroz, E.H., Rialon, K.L., Boul, P.J., Noon, W.H., Kittrell, C., Ma, J., Hauge, R.H., Weisman, R.B., Smalley, R., 2002, Band gap fluorescence from individual single-walled carbon nanotubes, Science **297**, 593–596.
Oron–Carl, M., Hennrich, F., Kappes, M., v. Löhneysen, H., Krupke, R., 2005, On the electron-phonon coupling of individual single-walled carbon nanotubes, Nano Lett. **5**, 1761–1767.
Perebeinos, V., Tersoff, J., Avouris, P., 2004, Scaling of excitons in carbon nanotubes, Phys. Rev. Lett. **92**, 257402–257405.
Pimenta, M.A., Marucci, A., Empedocles, S.A., Bawendi, M.G., Hanlon, E.B., Rao, A.M., Eklund, P.C., Smalley, R.E., Dresselhaus, G., Dresselhaus, M., 1998, Raman modes of metallic carbon nanotubes, Phys. Rev. **B 58**, 16016(R)–16019(R).
Quabis, S., Dorn, R., Glöckl, O., Eberler, M., Leuchs, G., 2000, Focusing light to a tighter spot, Opt. Comm. **179**, 1–7.
Rao, A.M., Richter, E., Bandow, S., Chase, B., Eklund, P.C., Williams, K.A., Fang, S., Subbaswamy, K.R., Menon, M., Thess, A., Smalley, R.E., Dresselhaus, G., Dresselhaus, M., 1997, Diameter-selective Raman scattering from vibrational modes in carbon nanotubes, Science **275**, 187–191.
Reich, S., Thomsen, C., Maultzsch, J., 2004, Carbon Nanotubes: Basic Concepts and Physical Properties, Wiley-VCH, Berlin.
Saito, R., Dresselhaus, G., Dresselhaus, M.S., 1998, Physical Properties of Carbon Nanotubes, Imperial College Press, London.

Strano, M.S., Dyke, C.A., Usrey, M.L., Barone, P.W., Allen, M.J., Shan, H., Kittrell, C., Hauge, R.H., Tour, J.M., Smalley, R.E., 2003, Electronic structure control of single-walled carbon nanotube functionalization, Science **301**, 1519–1522.

Thomas, M., Carminati, R., Arias-Gonzalez, J.R., Greffet, J.-J., 2004, Single-molecule spontaneous emission close to absorbing nanostructures, Appl. Phys. Lett. **85**, 3863–3865.

Vitali, L., Burghard, M., Schneider, M.A., Liu, L., Jayanthi, C.S., Kern, K., 2004, Phonon spectromicroscopy of carbon nanostructures with atomic resolution, Phys. Rev. Lett. **93**, 136103–136106.

Wang, F., Dukovic, G., Brus, L.E., Heinz, T.F., 2005, The optical resonances in carbon nanotubes arise from excitons, Science **308**, 838–841.

Yao, Z., Postma, H.W.C., Balents, L., Dekker, C., 1999, Carbon nanotube intramolecular junctions, Nature **402**, 273–276.

Zurita, J.R., Novotny, L., 2002, Multipolar interband absorption in a semiconductor quantum dot. I. Electric quadrupole enhancement, J. Opt. Sco. Am. B **19**, 1355–1362.

Chapter 6

Scanning nano-Raman spectroscopy of silicon and other semiconducting materials

by

D. Mehtani, N. Lee, R.D. Hartschuh, A. Kisliuk, M.D. Foster, A.P. Sokolov

Department of Polymer Science, The University of Akron, Akron, OH 44325-3909, USA

J.F. Maguire

Materials and Manufacturing Directorate/MLBP, Wright-Patterson AFB, OH 45433-7750, USA

Tip Enhancement
Advances in Nano-Optics and Nano-Photonics
ISSN: 1871-0018

S. Kawata and V.M. Shalaev (Editors)
© 2007 Elsevier B.V. All rights reserved.
DOI: 10.1016/S1871-0018(06)01006-5

Contents

§ 1. Introduction

Tip-enhanced spectroscopies hold great promise for addressing the challenging task of characterizing the physical and chemical properties of nanoscale structures designed in today's and tomorrow's semiconductor industry. Current technology for silicon transistors can achieve structures 65 nm in size and the drive for ever higher density of microelectronic components will continue to motivate the development of still smaller structures. Unfortunately, the characterization of key properties of such structures remains very problematic and in many cases it is not possible with current spectroscopic techniques. As a single, prominent example, we point to the thermally induced local strains, found in some structures (Hopkins and Farrow, 1986), which play key roles in determining the performance of devices (Wolf, 1996) into which these structures are incorporated. Moreover, it appears that strained silicon (Si) works more effectively in fast chips and nanoscale characterization of strains in silicon structures is very important in current semiconductor science and technology.

Scanning probe microscopy (SPM) and electron microscopy can provide detailed information on the topography of nanoscale structures with very high (~ 1 nm) lateral resolution. However, these techniques do not provide information on chemical composition or conformational states and stresses in the materials. Spectroscopic techniques based on nuclear magnetic resonance (NMR), Raman scattering, IR-absorption/reflection and fluorescence are widely used for chemical characterization on larger length scales, and thus extensions of these techniques would be useful for nanoscale characterization. The last few years have seen efforts toward developing these techniques for nanoscale analysis.

Raman spectroscopy has been extensively used to map and understand crystal orientation (Hopkins and Farrow, 1986; Kolb et al., 1991; Wolf, 1996) and stresses (Kolb et al., 1991) in silicon structures with micron length scales. There is currently a strong demand for Raman spectroscopy techniques that can map these stresses with nanoscale spatial resolution. However, to use Raman spectroscopy for nanoscale analysis two challenges must be overcome. First, the desired spatial resolution necessitates overcoming the diffraction limitation of light. Secondly, the Raman signal, which is inherently weak even for macroscopic scattering volumes, will be dramatically reduced because of the drastic reduction in the characterization volume when going to the nanoscale. Thus, additional signal enhancement is required.

One route for possibly overcoming the first problem, the diffraction limit, that has already been extensively investigated, is the use of aperture-limited near-field optics (Dunn, 1999; Hecht et al., 2000). In this approach the spatial resolution is improved by illuminating the sample with a subwavelength aperture held at a distance $\sim$ 10–20 nm above the sample surface. Webster et al., (1998) used aperture-limited near-field Raman spectroscopy for mapping stresses in a damaged Si wafer. However, the low optical throughput of the apertured probes, usually $\sim 10^{-3}$–10^{-5} (Thio, et al., 2002), results in a prohibitively long imaging time of 9 h for an area of 4 μm^2 (60 s per point). Additionally, the spatial resolution is limited by the aperture size to $\sim$ 100–200 nm. Thus, although the aperture-limited near-field optics help to improve the resolution, they exacerbate the second problem – the weak signal coming from the small volume of interest.

An alternative approach that addresses both problems is apertureless near-field optics. This approach is based on the use of metal or metal-coated tips with a plasmon resonance at the apex that provides a strongly enhanced and localized signal (Courjon and Bainier, 1994; Patanè et al., 2004). The plasmon enhancement of optical signals is well known and has been used for a long time in surface-enhanced Raman spectroscopy (SERS) (Moskovits, 1985). The idea to apply this phenomenon to SPM and the use of metal or metal-coated tips for the local enhancement of optical signals was initially proposed by Wessel in 1985. However, this idea has received significant attention only during recent years. The technique of enhancing a local Raman signal using apertureless optics is most widely known by the name "tip-enhanced Raman spectroscopy" (TERS). However, when this capability is specifically combined with the means to scan the tip over a surface to collect a Raman image in one or two dimensions, the authors find the name "scanning nano-Raman spectroscopy" (SNRS) more descriptive, hence its inclusion in the title of this chapter. This terminology emphasizes the ability to construct an image, establishes the parallel to scanning micro-Raman spectroscopy and highlights the difference in length scale addressed by the two techniques. However, in keeping with the convention elsewhere in this volume, the term TERS is used predominantly in the remainder of this chapter.

One of the simplest approaches in TERS is to employ bottom-illumination optics (Stockle et al., 2000; Hayazawa et al., 2001; Hartschuh et al., 2003; Wang et al., 2003), in which an inverted microscope is used to illuminate the sample from below and the Raman signal is collected using the same optics in a backscattering geometry. A scanning probe microscope is placed on top of the inverted optical microscope and a modified tip with plasmon resonance at the apex strongly enhances the signal in the vicinity of the tip. Details of this approach are described in other chapters of this book. Using this technique Novotny and co-workers (Anderson et al., 2005) have demonstrated unprecedented lateral resolution of $\sim$ 14 nm for single-wall carbon nanotubes (SWNT) deposited on glass slides.

However, the bottom-illumination approach is limited to thin or transparent samples placed on a transparent substrate. One cannot use this approach for analysis of silicon structures (Hayazawa et al., 2005). Another disadvantage of this approach is the difficulty in aligning the incident beam polarization parallel to the tip axis (Novotny et al., 1998). It is known from simulations (Martin and Girard, 1997; Novotny et al., 1997; Sun, 2003a,b) that maximum enhancement can be achieved with the polarization of light parallel to the tip axis. In order to overcome these problems, another approach, side-illumination optics, has been proposed and used by a few groups (Anderson and Pike, 2002; Pettinger et al., 2004; Mehtani et al., 2005). In this geometry the objective of the optical system is placed above the sample surface at some angle with respect to the surface (fig. 1). The objective illuminates a μm-scale spot on the sample surface and collects the scattered light. A modified SPM tip brought into this spot provides a locally enhanced signal. This geometry enables the characterization of non-transparent samples and samples on non-transparent substrates. Anderson and Gaimari, (2003) have demonstrated a spatial resolution of $\sim$ 50 nm on intact biomolecular surfaces using this technique.

In this chapter, we overview developments in the side-illumination TERS technique and applications of TERS for the characterization of semiconducting materials and structures. We emphasize that even a tip

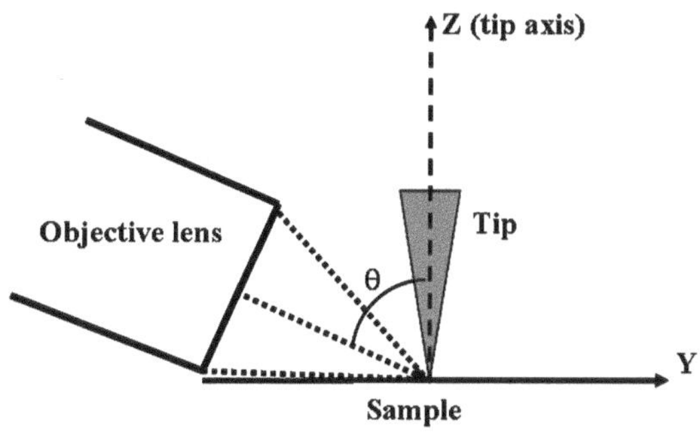

Fig. 1. Schematic of the objective placement for side-illumination optics.

enhancement factor of $\sim 10^4$ may not be sufficient for the characterization of some structures. We argue that the contrast between locally enhanced signal and the background far-field signal, and not the enhancement factor, is the most important parameter defining practical applications of TERS. Two different approaches of improving contrast, minimization of the far-field signal and maximization of the enhanced signal, are described and analyzed. Additionally, measurements of the optical properties of the apertureless tips are presented and discussed. Prospects and challenges for the side-illumination TERS technique are discussed in the summary.

§ 2. Side-illumination geometry and preparation of tips

The side-illumination optics have three major disadvantages in comparison to the bottom-illumination geometry: (i) no current commercial microscope can be used for this geometry; (ii) the image of the sample surface obtained through the objective is not clear because only a narrow portion of the image is in the focus of the oblique objective; and (iii) the SPM tip blocks part of the laser beam and shadows a portion of the sample, reducing the signal. Despite these problems, the side-illumination geometry seems to be the only choice for analysis of non-transparent samples and samples on non-transparent substrates (e.g. Si wafers). In addition, side-illumination geometry provides ease of aligning the incident beam polarization along the tip axis. This condition facilitates maximum enhancement (Martin and Girard, 1997) and is not easily achievable in the bottom-illumination geometry.

On the basis of this approach, a few groups have developed TERS with the objective placed at an angle θ varying between $\sim 60°$ (Pettinger et al., 2004; Mehtani et al., 2005) and $\sim 87°$ (Anderson and Pike, 2002) relative to the tip axis (fig. 1). Sun and Shen (2003a) used a modified top-illumination scheme with a mirror placed under the objective that provides side-illumination and collection of the signal. Simulations (Martin and Girard, 1997; Sun, 2003a,b) predict that for maximum enhancements the objective angle θ should be chosen in the range of 40–85° depending on the tip material and geometry. However, no detailed experimental studies of the enhancement factor as a function of θ have been performed yet. Pettinger's group (2004) used a scanning tunneling microscope (STM) in their setup. An advantage of STM is the possibility to hold a tip at a distance of $\sim$ 1–2 nm above the sample surface. STM requires, however, a highly conducting substrate. Other groups (Sun and Shen, 2001; Anderson and Pike, 2002; Mehtani et al., 2005) have used an atomic-force microscope (AFM) in contact mode.

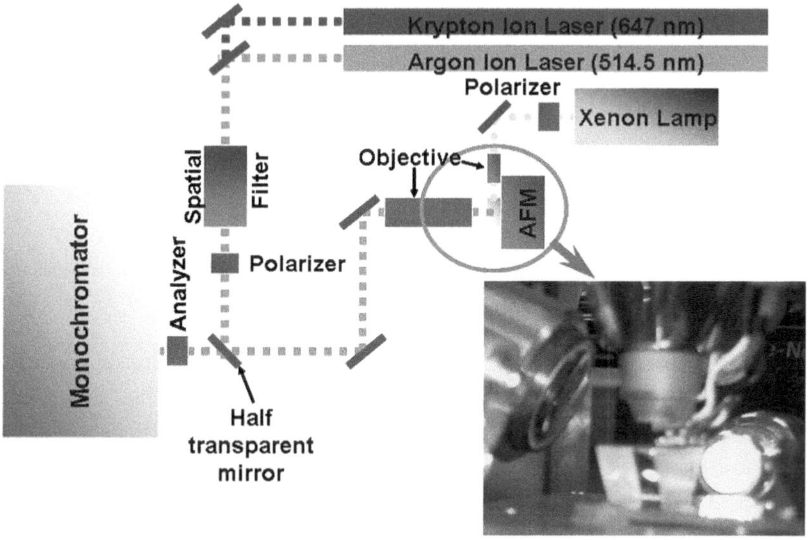

Fig. 2. Schematic of an SNRS system with side-illumination optics (Mehtani et al., 2005) with the inset photograph showing a specific configuration of the AFM head with the objective and a prism used for the measurement of optical properties of tips (reproduced from Mehtani et al., 2006, by permission).

Figure 2 shows a schematic of the nano-Raman spectrometer (described in detail in Mehtani et al., (2005)) developed in our group. It includes a Horiba Jobin Yvon Labram HR-800 Raman spectrometer optically coupled with a Quesant (QScope 250) AFM using a long-working-distance Mitutoyo (APO SL50) objective (50×, 0.42 NA). The sample is placed on a piezo-controlled XY-stage from nPoint. The objective is placed at an angle of $\sim 65°$ (relative to the vertical) and is fixed on an XY-stage controlled by step motors to position the objective with respect to the tip. The Raman signal is collected in the back-scattering geometry using the same objective. The 514.5 nm line of an argon ion laser and the 647 nm line of a krypton laser are used in our experiments, with an incident power of < 10 mW on the sample.

The nano-Raman system was modified to measure the optical properties of the tips using the principle of total internal reflection microscopy (details in Mehtani et al., 2006). A white light beam from a xenon arc lamp illuminates the side surface of a trapezoidal quartz prism (fig. 3) such that the beam is focused on the top surface under the condition of total internal reflection. This results in generation of an evanescent field on the top surface and a dark field (no propagating light) when seen by the objective of the collecting optics. When the apertureless tip is brought into the vicinity of the

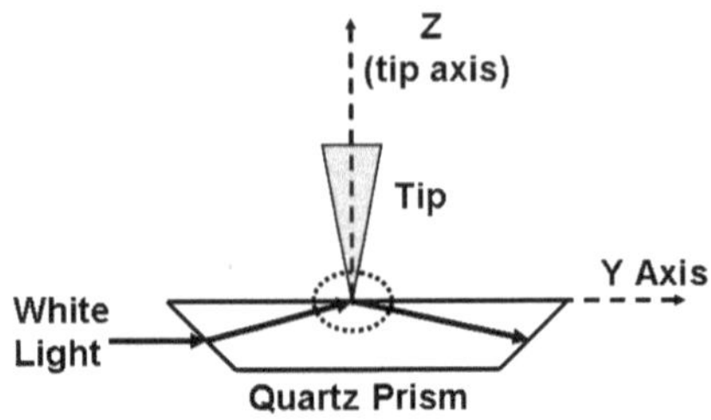

Fig. 3. Schematic of the total-internal reflection setup used to measure optical properties of the tips (reproduced from Mehtani et al., 2006, by permission).

surface, the evanescent field excites plasmons on the tip. The propagating light coming from these plasmons is measured by the objective of the side-illumination optics and analyzed to provide the optical spectrum of the tip.

Various tips have been used in the side-illumination geometry. Anderson and Pike (2002) used silver- and gold-coated conventional AFM tips prepared by plasma sputtering. Sun and Shen (2002) used self-made electrochemically etched tungsten tips coated with silver using RF sputtering. These tips were then bent and attached to a steel plate to enable their use in AFM tapping and contact modes. Pettinger and co-workers (Ren et al., 2004) used gold wire tips prepared by electrochemical etching.

In our studies we have tried commercially available Si and Si_3N_4 AFM tips and self-made tungsten tips. Tungsten tips were prepared using an etching solution of 2N NaOH and 2–3 V DC etching voltage (Ibe et al., 1990). All the tips were coated with either silver ($\sim$ 50 nm) or gold ($\sim$ 100 nm) using vapor deposition. In the case of gold-coated tips, an initial layer of chromium ($\sim$ 20 nm) was deposited on the tips for better adhesion. Metal was deposited at the rate of 0.3 Å/s in order to avoid distortion of the tips and the quantity of metal deposited was controlled using a Sycon (STM-100) monitor. The silver- and gold-coated Si_3N_4 tips were used in the contact mode of the Q250 AFM and the tungsten tips were attached to a tuning fork (Mehtani et al., 2006) to measure both the optical properties and the tip-enhanced Raman signal.

§ 3. Apparent enhancement and its localization

The signal in TERS consists of two components (fig. 4), the background signal coming from the entire laser-illuminated spot (we will call it the far-field signal) and the locally enhanced signal that comes from the nanoscale region in the vicinity of the tip (we will call it the near-field signal). The latter is the signal of interest and its spatial localization depends on the tip properties and geometry (Martin and Girard, 1997; Sun and Shen, 2003b).

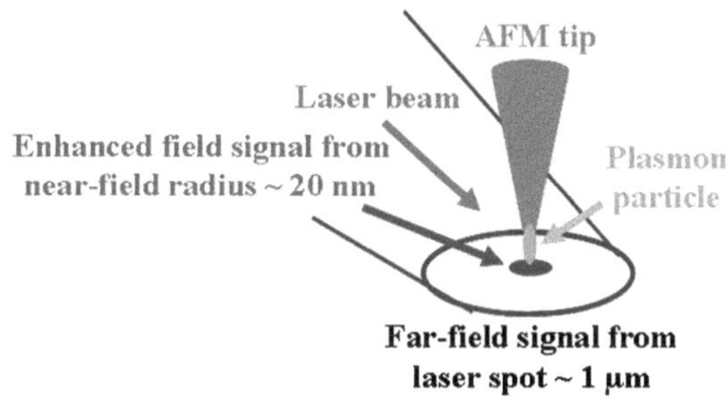

Fig. 4. Schematic of the laser-illuminated apertureless probe identifying portions of the illuminated area contributing to the far-field and near-field Raman signals in the case of side-illumination optics.

Tip enhancement of the Raman signal on different samples can be observed by comparing the Raman intensity without the tip (the far-field signal) to that when the tip is in contact with the sample (far-field plus near-field signals). Side-illumination TERS has been applied to Si-based structures (Sun and Shen, 2001, 2003a,b), dye and CN^- molecules (Pettinger et al., 2004), diamond particles (Anderson and Pike, 2002) and biological objects (Anderson and Gaimari, 2003), polymers, carbon nano-tube (CNT) and quantum dot material CdS (Mehtani et al., 2005). Apparent enhancement, defined as the ratio of the total signal, measured when the tip is in contact, to the far-field signal, measured when the tip is withdrawn, was significant for most of the samples studied. Figure 5 shows examples from the authors' work (Mehtani et al., 2005) of apparent enhancement for a conductive polymer blend of poly(3,4-ethylenedioxythiophene) and poly(styrenesulfonate) (PEDOT/PSS), SWNT, a 10 nm CdS film and a Si wafer. Among these examples, the smallest apparent enhancement is seen for Si. The enhancements obtained for Si and SWNT are close to the values reported by Sun and Shen, (2001) and Hartschuh et al. (2003). The results presented in fig. 5 demonstrate that the apparent enhancements achieved are sufficient for imaging thin films and layers. However, the enhancement is too weak in the case of Si because the background far-field signal remains significantly stronger than the tip-enhanced signal.

These measurements were carried out using different silver- and gold-coated tips to establish the reproducibility of the enhancement effect. The enhancement with silver-coated tips degraded after approximately 3 days while the enhancement provided by gold-coated tips remained stable over a couple of months (Mehtani et al., 2005).

A crucial question for the tip-enhanced signal is its localization. According to simulations, the localization depends on the tip radius and geometry (Martin and Girard, 1997; Sun and Shen, 2003b). The best lateral resolution

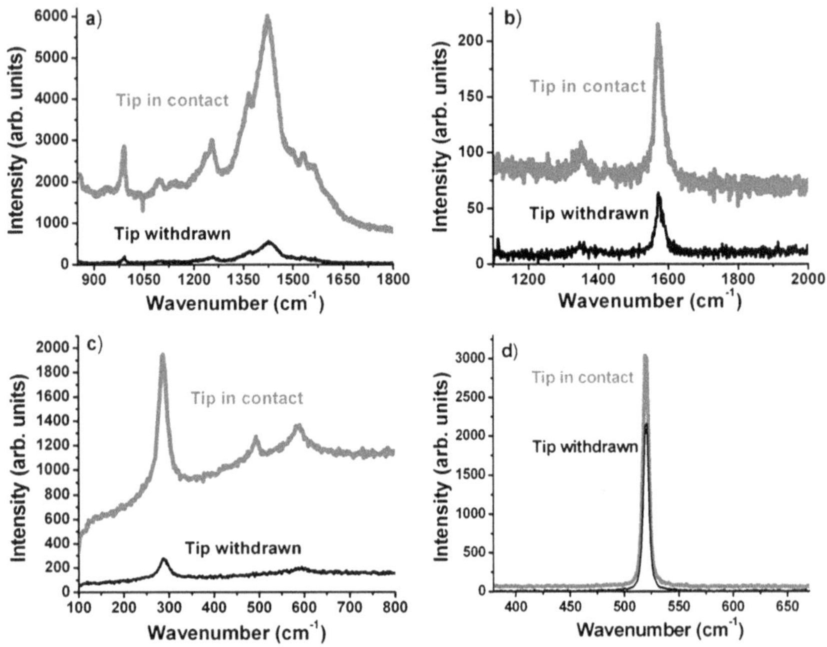

Fig. 5. Comparison of the spectra with tip withdrawn (far-field signal only) and tip in contact (far-field plus near-field) allows measurement of an apparent enhancement achieved with side-illumination optics and gold-coated tips: (a) Poly(3,4-ethylenedioxythiophene) poly(styrenesulfonate) (PEDOT/PSS) polymer film measured with $\lambda = 647$ nm, apparent enhancement ~ 9; (b) monolayer of SWNT measured with $\lambda = 514$ nm, apparent enhancement ~ 3; (c) 10 nm CdS film, apparent enhancement ~ 6, and (d) Si wafer measured with $\lambda = 514$ nm, apparent enhancement ~ 1.4 (data from Mehtani et al., 2005).

has been achieved in the bottom-illumination geometry using structures that are not very tall in order to minimize resolution degradation due to convolution of the true image with the tip shape. Approximately 14 nm resolution has been reported for SWNT (diameter ~ 1.5–2 nm) by the group of Novotny (Anderson et al., 2005) and ~ 15 nm resolution for DNA molecules using coherent anti-stokes Raman spectroscopy (CARS) has been reported by the group of Kawata (Ichimura et al., 2004). Anderson and Gaimari, (2003) reported ~ 50–100 nm lateral resolution on biomolecular samples using side-illumination geometry.

Localization of the enhancement effect in the depth direction was estimated using measurements of the apparent enhancement from CdS films with various thicknesses (fig. 6). Thicker samples gave lower apparent enhancement (fig. 6(b)). This lower enhancement is explained by the larger, thickness-dependent sample volume contributing to the far-field signal while the volume contributing to the near-field signal is limited and

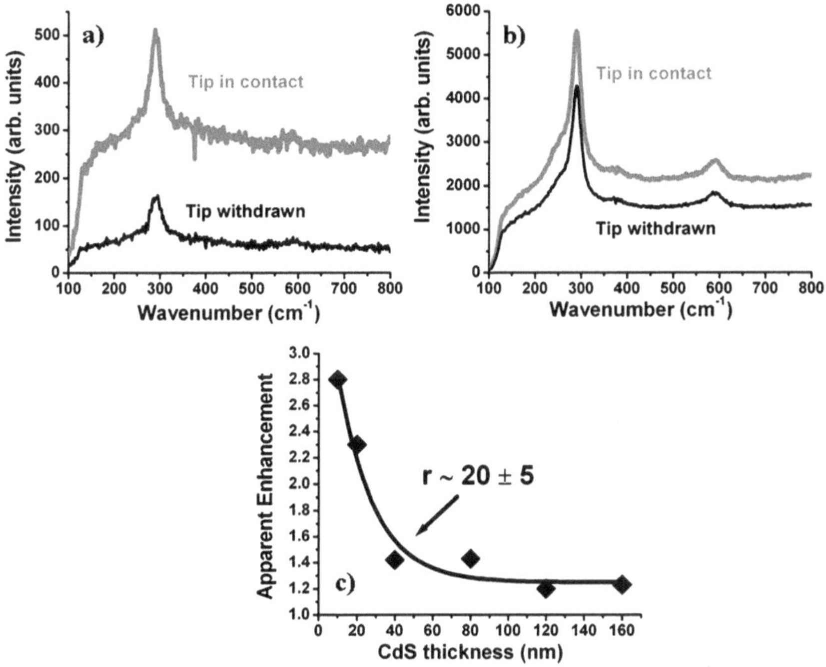

Fig. 6. (a) TERS spectra for 10 nm and (b) 160 nm CdS films. (c) Apparent enhancement as a function of film thickness for a series of films. The depth over which the enhancement is localized may be estimated at ~ 20 nm (data from Mehtani et al., 2005).

does not change with the film thickness beyond the localization depth. By fitting the data for apparent enhancement as a function of film thickness the localization depth was estimated to be ~ 20 ± 5 nm (fig. 6(c)) (Mehtani et al., 2005). This estimate agrees with predictions of simulations (Sun and Shen, 2003b) and with experimental estimates obtained in the bottom-illumination geometry using a curve of enhancement vs. tip-to-sample distance (Hartschuh et al., 2003), when the dependence of localization on tip radius is accounted for (Raschke and Lienau, 2003).

We also analyzed samples with ~ 200 and ~ 600 nm periodic structures of poly(methyl methacrylate) (PMMA) on top of a 20nm CdS film (figs. 7, 8). The PMMA lines with height ~ 30 nm provided a separation between the tip and the CdS, reducing the enhancement of the CdS signal. Gold-coated Si_3N_4 tips were used in AFM contact mode and the scanning was performed by moving the sample on a piezo-driven XY-stage. The full width at half maximum (FWHM) of the feature etched in the structure with 200 nm periodicity was 80 ± 10 nm. As can be seen from the 2-D image of topography in fig. 7(d), the 80 nm trenches were not well developed. Their analysis, however, provides the best test for the spatial

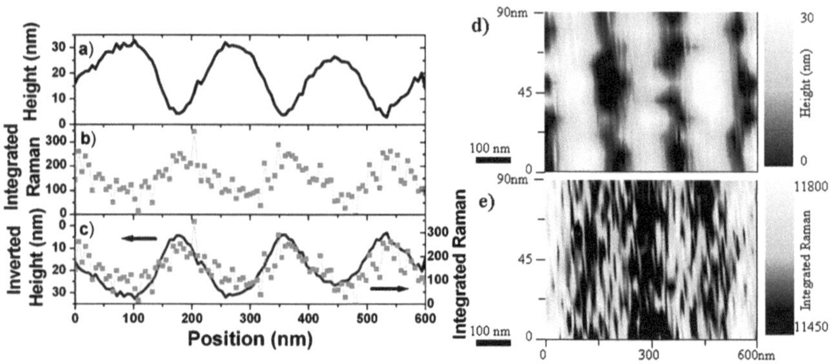

Fig. 7. (a) Line scan from the topographic image, (b) line scan created using the integrated Raman intensity, and (c) correlation between the inverted topography line scan (solid line) and integrated Raman intensity line scan (symbols) for a TERS image of PMMA on CdS structure with periodicity of 200 nm. (d) Two-dimensional AFM image and (e) TERS image of the same structures. Note that y scale is expanded for better visual comparison in (d) and (e).

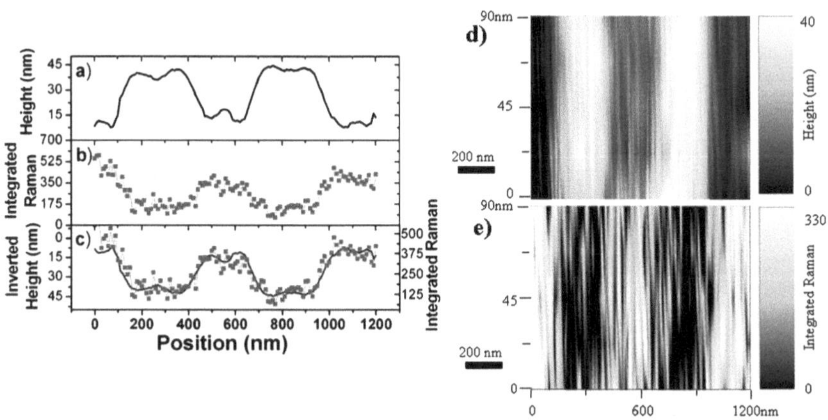

Fig. 8. The same as fig. 7, but for the PMMA on CdS structure with 600 nm periodicity: (a) topography and (b) integrated Raman intensity along the line scan, (c) correlation between the inverted topography (solid line) and integrated Raman intensity (symbols), (d) Two-dimensional AFM image, and (e) TERS image of the same structures. Note that y scale is expanded for better visual comparison in (d) and (e).

resolution. The topography (fig. 7(a)) and the Raman intensity (fig. 7(b)) were measured simultaneously for each of these structures. In this case, the CdS signal should correlate with the inverted topography (Fig. 7(c)), since a weaker signal is obtained when tip is on top of the PMMA lines. Reasonable correlation between the inverted topography and Raman intensity is observed for a 1-D scan in fig. 7(c) and for two-dimensional images shown in figs. 7(d), 7(e).

Similar analysis has been performed on structures with periodicity 600 nm (fig. 8). Excellent correlation between the inverted topography and the Raman intensity has been observed in this case. A comparison between the topography traces in figs. 7(d), 8(d) shows that the structures with 200 nm periodicity have more defects than the structure with 600 nm periodicity. For this reason, we used the structure with 600 nm periodicity to determine the lateral resolution of our system. Fits of the trench topographic and Raman profiles by Gaussian line shapes indicate an average FWHM of 250 ± 20 nm for both the topography signal and for the Raman signal. The TERS scan of PMMA lines on CdS indicate extremely high lateral resolution, probably comparable to the experimental uncertainty of 20 nm.

§ 4. Tip enhancement and contrast

It is obvious that an important parameter for apertureless near-field optics is the enhancement of the signal provided by the tip. Enhancement is an increase in the Raman signal intensity in the small (nanoscale) volume around the tip caused by the plasmon resonance of the tip. The enhancement factor (EF) is defined as the ratio of the near-field signal to the far-field signal per unit volume (Hayazawa et al., 2002; Hartschuh et al., 2003; Sun and Shen, 2003b; Bulgarevich and Futamata, 2004; Mehtani et al., 2005):

$$EF = \frac{I_{\text{near}}}{I_{\text{far}}} \frac{V_{\text{far}}}{V_{\text{near}}} = \left(\frac{I_{\text{total}}}{I_{\text{far}}} - 1\right) \frac{V_{\text{far}}}{V_{\text{near}}} \tag{4.1}$$

where, I_{near} and I_{far} are the near-field and far-field signal intensities, I_{total} is the total signal intensity when the tip is in contact, and V_{near} and V_{far} are the volumes contributing to the near-field and far-field signals, respectively.

The ratio of $V_{\text{far}}/V_{\text{near}}$ for thin films (thickness <20 nm) can be estimated as the ratio of the surface areas contributing to the far- and the near-field signals. Assuming that the laser spot on the sample surface has radius ~ 1 μm and that the localized radius of near-field enhancement, R, is ~ 20 nm, the ratio $V_{\text{far}}/V_{\text{near}}$ has a value of about 2,500. Thus, the apparent enhancement of 6 in the case of the 10 nm CdS film actually corresponds to an enhancement factor of $\sim 10^4$. On the other hand, the actual ratio of volumes must be used when calculating enhancement factors for thicker samples. With the same assumptions, V_{near} can be approximated as the hemispheric volume $2\pi R^3/3 \sim 2{*}10^{-5}$ μm^3. V_{far} for thick samples is defined by the surface area of the laser spot multiplied by the smallest of three parameter values: sample thickness, light penetration depth or the depth of focus of the confocal optics ($\sim 1\,\mu m$). The depth of penetration of 514.5 nm light in Si is $\sim 0.68\,\mu m$ (Feng et al., 1988). This gives an estimate of $V_{far} \sim 2\,\mu m^3$ for a Si wafer. With an apparent enhancement of $\sim$ 1.3–1.4 (fig. 5(d)), we estimate that the enhancement factor for Si is $\sim (3 - 4) \times 10^4$.

Enhancement factors in the range of 100–10^4 have been reported by several groups (Hayazawa et al., 2002; Hartschuh et al., 2003; Sun and Shen, 2003b; Bulgarevich and Futamata, 2004). Novotny and co-workers achieved enhancement factors between 100 and 1000 for TERS imaging of SWNT (Hartschuh et al., 2003; Anderson et al., 2005) using an etched gold-wire tip. This enhancement is sufficient for imaging when the number of nano-objects (SWNT in their case) placed in the laser illuminated spot is small. However, it will not be sufficient for the analysis of thin films, e.g. of CdS. Figure 5 shows that comparable enhancement factors ($\sim 10^4$) obtained for CdS and Si samples are sufficient for analysis of a 10 nm CdS film, but are not enough for good imaging of the Si surface. The near-field signal is weaker than the background far-field signal for Si (fig. 5(d)) and that background must be subtracted for good quality imaging.

These examples show that the tip-enhancement factor is actually not the most important parameter for practical applications of TERS. Mehtani et al. (2005) proposed to introduce the "contrast" between the near- and far-field signals as another important parameter. Contrast is defined as the ratio of the near-field signal to the far-field (background) signal:

$$C = \frac{I_{near}}{I_{far}} = \frac{I_{total}}{I_{far}} - 1 \tag{4.2}$$

The difference in the values of contrast observed for the thin CdS film (fig. 5(c)) and the Si wafer (fig. 5(d)) is attributed to the difference in V_{far} contributing to the background signal in these two cases. It is important to estimate the minimum enhancement factor required to obtain a reasonable contrast, say $C \sim$ 5–10, even for thick samples in which the light propagates to great depths. We have already estimated the near-field volume $V_{near} \sim 2 * 10^{-5}\,\mu m^3$, assuming the localized radius of enhancement $R \sim 20$ nm. V_{far} in a thick, transparent sample will be limited by the area of the laser spot on the sample defined by the quality of the confocal optics. Assuming the radius of the spot $\sim 1\,\mu m$, V_{far} can be estimated as $\sim 2\,\mu m^3$. Thus, in order to achieve reasonable contrast for any thick sample an enhancement factor of $\sim 10^6$ is required. This enhancement is two orders greater than that achieved so far in most groups developing TERS.

§ 5. Improving contrast for silicon

Sun and Shen (2001, 2003a,b) were the first to report tip-enhanced Raman spectra for silicon. They used modified top-illumination optics and observed an increase of approximately 50% in the Raman signal intensity at $\lambda = 488$ nm using silver-coated tungsten tips. They were able to image Si/SiO_x structures with a periodicity of about 700 nm. They demonstrated a strong difference between the far-field scan and a TERS scan and obtained reasonable correlation between the profiles of topography and TERS signals. The contrast (0.3–0.5) that was achieved is comparable to that seen in the data presented in fig. 5(d), but this contrast is not sufficient for effective TERS analysis of silicon structures, for example strain distributions. According to the definition in eq. (4.2), contrast can be maximized by maximizing the enhancement with a tip, minimizing the far-field signal or a combination of both. Poborchii et al. (2005) has recently proposed an interesting idea – to increase the contrast of the TERS signal from silicon by suppressing the background far-field signal using a depolarized scattering geometry. The Si Raman mode at 520 cm^{-1} is strongly polarized and the far-field signal can be suppressed using an analyzer set for a depolarized geometry. The authors used top-illumination optics and a quartz AFM tip with a silver particle at the apex. The tip was immersed in glycerol for refractive index matching. As a result, no shadowing by the tip was present and the silver particle at the apex was visible through the objective. The metal particle depolarizes the incident field in its vicinity, resulting in a partially depolarized enhanced near-field scattering signal. They achieved a contrast of ~ 2 times and a lateral resolution of ~ 100 nm, which was limited by the size of the silver particle.

We applied the depolarized scattering geometry to our side-illumination setup as well (Mehtani et al., 2005). This approach has a few advantages relative to the top-illumination scheme used by Poborchii et al. (2005): (i) the most enhanced signal under the tip (where the distance between tip and sample surface is smallest) can be collected in side-illumination optics, but is shadowed by the silver particle in the top-illumination scheme; (ii) the tip axis provides the preferential orientation for polarization in the side-illumination geometry, but not in the top-illumination scheme; and (iii) the side-illumination geometry should provide better lateral resolution due to the tip radius being smaller than the radius of the silver particles used by Poborchii et al. (2005).

It is known that the Raman scattering of Si depends on the polarization of the incident light relative to the crystallographic axes (Hopkins and Farrow, 1986; Nakashima and Mizoguchi, 1989). To achieve the highest contrast between the tip-enhanced and the far-field background signals, the polarization geometry should satisfy two criteria: (1) provide good tip enhancement and (2) provide an enhanced near-field signal with polarization that will not be suppressed by the analyzer (that is used to suppress the far-field signal). The maximum tip enhancement is expected for the polarization of the light parallel to the tip axis. Thus, the use of polarization perpendicular to the tip axis does not satisfy the first criterion of maximizing enhancement. Polarization parallel to the tip axis is also not optimal because the analyzer, set to transmit the depolarized light only, will suppress the enhanced signal. Hence, to maximize the contrast the polarization must be optimized to attain a balance between enhancement of the near-field signal and suppression of the far field signal.

We used a Si(100) wafer and started the optimization of the contrast by rotating the polarization of the incident beam in the absence of an analyzer for the scattered light. The optimum incident polarization appears at $\sim 70°$ relative to the tip axis (fig. 9). An analyzer was then mounted in the scattered beam and the analyzer rotated for a fixed angle of incident beam polarization. The optimum appears at a polarization of the incoming beam of $\sim 70°$ and an analyzer angle of $\sim 85°$–$90°$ (fig. 10(a)). A contrast of more than 10 has been achieved in this polarization configuration (figs. 10(a), 10(b)). The presented results show that the increase in contrast

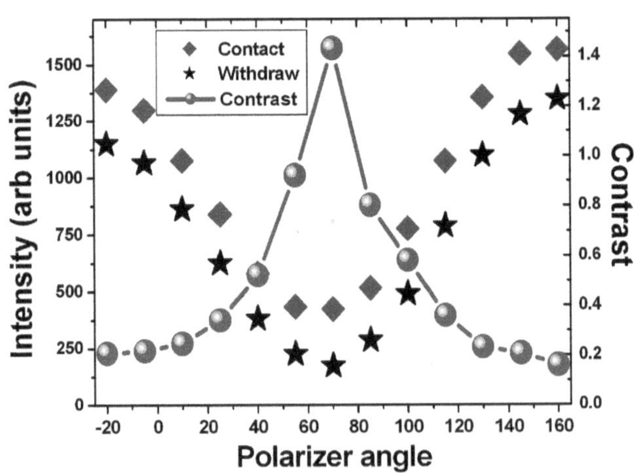

Fig. 9. Dependence of contrast on incident polarization for a Si wafer with a (100) face at $\lambda = 514$ nm with a gold-coated tip. Change in the signal when the tip is in contact and when the tip is withdrawn are also shown.

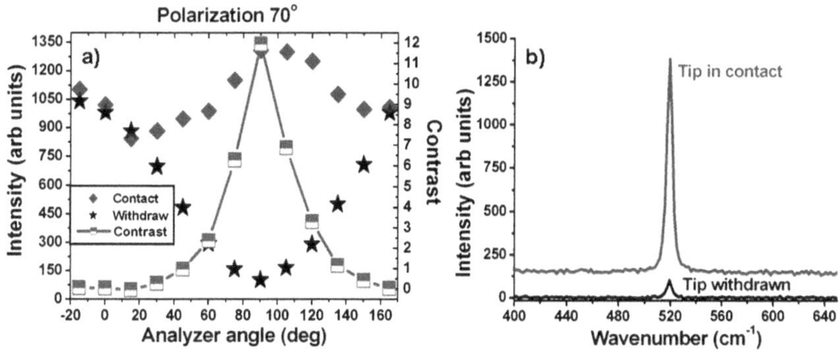

Fig. 10. (a) Dependence of intensity with the tip withdrawn (stars), intensity with the tip in contact (diamonds) and contrast (squares) on analyzer angle for an incident polarization of 70° (b) Comparison of the spectra measured for Si for an analyzer angle 90° demonstrate the contrast of ~ 12 than can be achieved with the depolarization geometry.

is achieved primarily by suppression of the far-field signal. A comparison between the contrast of 0.4 for silicon with conventional TERS measurements and the contrast of more than 10 using a depolarized geometry demonstrates convincingly the possibility of increasing TERS contrast by suppressing the far-field signal.

An important question that remains is how localized is the depolarized signal induced by the tip. To address this question, we measured TERS on Si wafers covered with SiO_x layers of various thicknesses. The SiO_x layer separates the tip from the Raman scatterer (Si), and thus reduces the enhancement of the signal. Theoretically, by varying the SiO_x thickness, d, from that of native oxide (~ 2 nm) to 350 nm we should obtain an estimate of the localization depth of the signal obtained in depolarized geometry (fig. 11). The contrast drops sharply for thicknesses of the SiO_x layer greater than 50 nm. However, its value is non-zero even for thicknesses greater than 150 nm, suggesting that part of the depolarized signal induced by the tip is not localized. Notice that the contrast increases for $d > 130$ nm and reaches a maximum at $d \sim 200$ nm. This is likely due to interference effects induced by the SiO_x layer. The non-localized portion of the depolarized TERS signal and the interference effects prevent detailed analysis of the near-field localization and necessitate alternative methods.

A more convincing demonstration of localization of the depolarized tip-enhanced signal has been obtained on a multilayer structure that consists of a 30 nm strained Si layer on top of a 300 nm SiO_x layer grown on a Si wafer (fig. 12 (inset)). Polarized measurements of the far-field signal using 514.5 nm laser light barely detect the contribution of the strained layer to the Raman signal because of the large (~ 700 nm) penetration depth of the

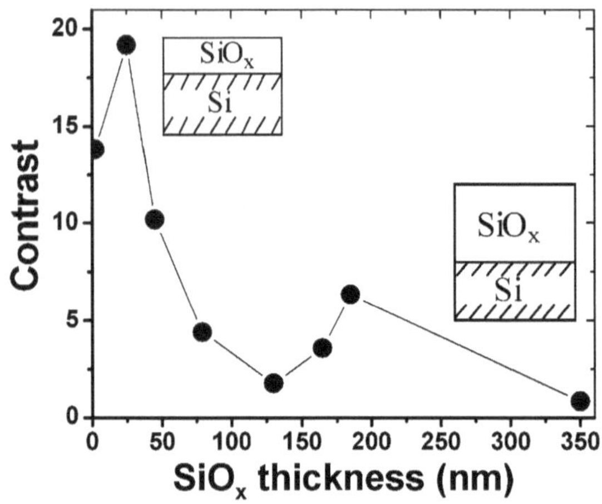

Fig. 11. Contrast between Raman signals with the tip in contact and tip withdrawn in the depolarized configuration for an SiO_x/Si structure as a function of SiO_x layer thickness.

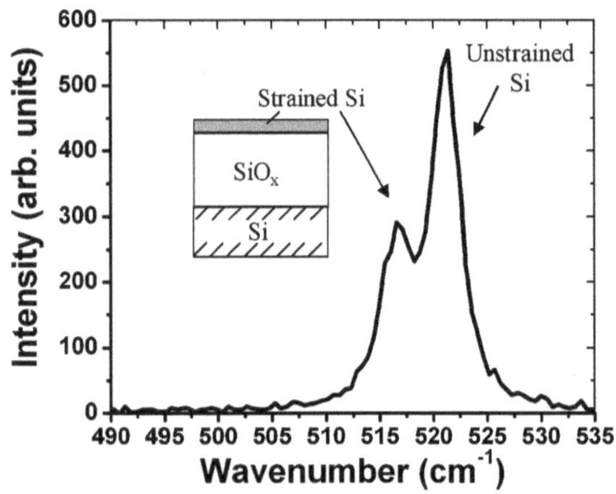

Fig. 12. Raman signal from a 30 nm layer of strained silicon atop a 300 nm thick SiO_x layer on a Si wafer. The inset shows the sample geometry.

laser light into Si (fig. 12). Selection of a particular polarization geometry allows one to make the intensities of the far-field Raman signals from the strained and unstrained Si comparable (fig. 13(a)). Bringing the apertureless tip in contact leads to enhancement of the Raman signal (fig. 13(b)) from the strained silicon by $\sim 350\%$, while an increase of only $\sim 90\%$ is observed for the signal from the unstrained silicon.

For the analysis of lateral resolution we used periodic SiO_x structures on a Si wafer (fig. 14(a)). Topographic and Raman signals were acquired

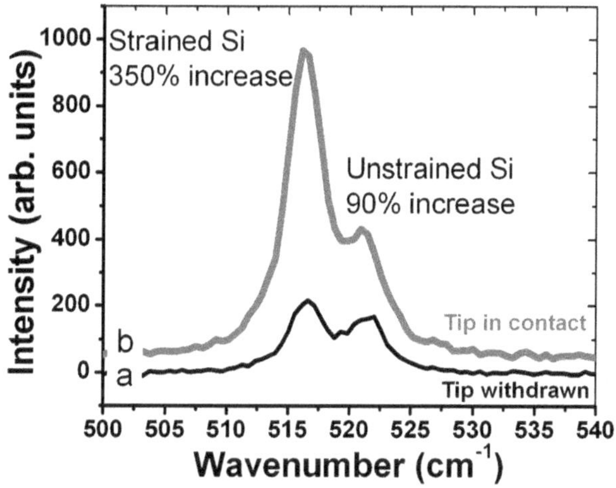

Fig. 13. Intensity plots demonstrating the relative contrasts for strained silicon at the surface and unstrained silicon ~ 300 nm below the surface. The polarization of the incident beam was chosen to be ~ 60° from the tip axis.

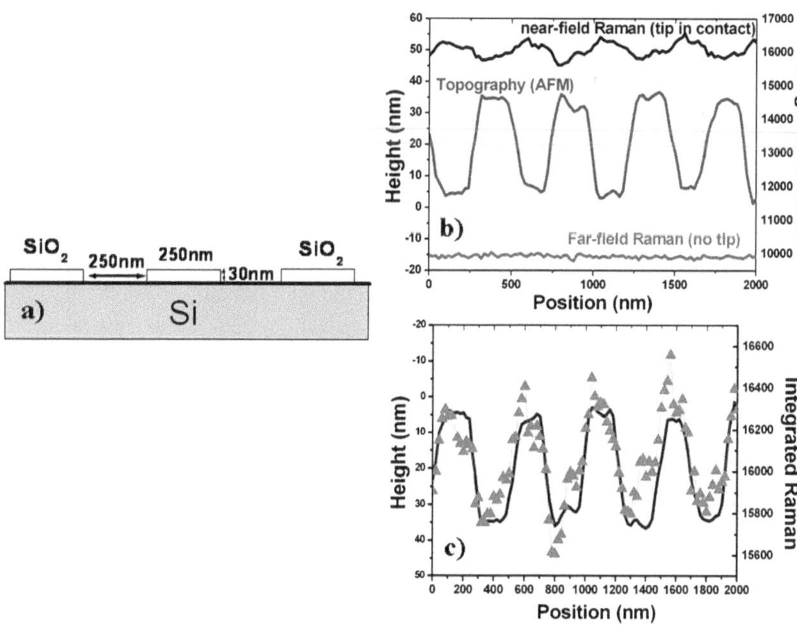

Fig. 14. (a) Schematic of the Si/SiO_x structures, (b) line scan comparing the integrated Raman intensity of the near-field signal and the far-field signal with the topography, and (c) correlation between the inverted topography line scan (solid line) and integrated Raman intensity line scan (▲).

simultaneously as the piezo-driven sample stage scanned in the XY plane with a step size of 20 nm and Raman acquisition 4 second per step. For this structure, as with the CdS/PMMA samples (fig. 7), the Raman signal from Si should be compared with the inverted topography. The near-field Raman signal follows well the inverted topography while no intensity modulation is visible in the far-field signal (figs. 14(b), 14(c)). Moreover, two-dimensional images also show excellent correlation between the topography of the sample (fig. 15(a)) and the variation in the integrated Raman intensity of the Si mode (fig. 15(b)). Fits of the profiles (fig. 14(c)) estimate an average FWHM of 216 ± 10 nm for the trench topography and 240 ± 20 nm for the Raman signal. These results once again indicate extremely high lateral resolution, comparable to the scanning step size 20 nm. The presented analysis demonstrates that the use of the depolarized geometry provides sufficient contrast and localization of the enhanced signal for TERS of Si structures. The contrast as achieved on Si

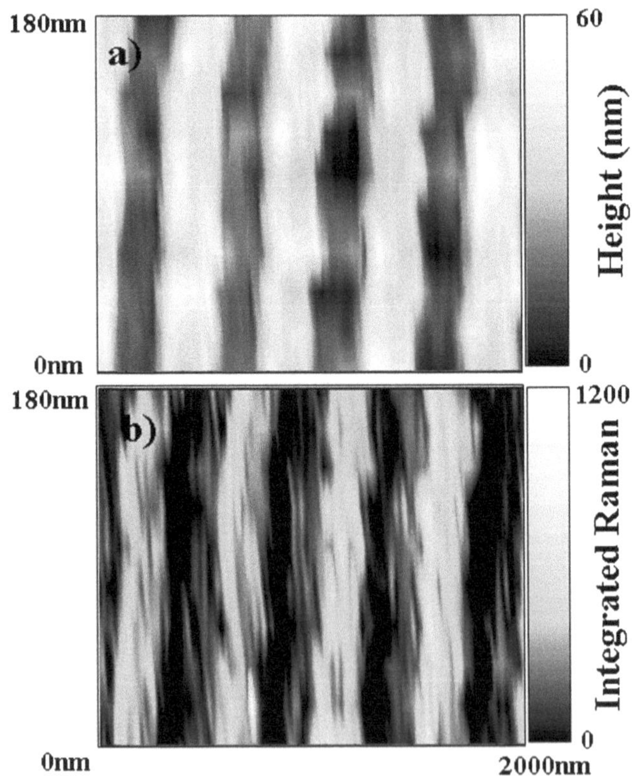

Fig. 15. (a) AFM and (b) TERS of Si two-dimensional images of the structures described in fig. 14. Note that y scale is expanded for better visual comparison.

should be sufficient for practical applications, e.g. for analysis of stress distributions in Si structures at the nanoscale.

§ 6. Optical properties of the apertureless tips

The best solution for improvement of contrast is an increase in the tip enhancement, which remains as the main challenge in apertureless near field optics. As has been mentioned above, most of the groups working with apertureless near-field optics have reported enhancement factors in the range of $\sim 10^2$–10^4. Estimates presented in Section 4 suggest that for a lateral resolution of ~ 20 nm, enhancement factors on the order of $\sim 10^6$ should be sufficient for the analysis of any sample. It has been shown theoretically (Demming et al., 1998; Sun and Shen, 2003b; Hartschuh et al., 2004) that achieving a local electric-field enhancement greater than 50 is feasible for TERS. This field enhancement corresponds to an enhancement factor $>10^6$ for the Raman signal. Moreover, SERS experimental studies have reported enhancements of up to $\sim 10^{12}$–10^{14}, in particular in the neighborhood of "hot spots" (Kneipp, 1997 et al., Nie and Emory, 1997). A possible theoretical explanation for this extreme enhancement was recently proposed (Li et al., 2005).

According to theoretical analyses (Denk and Paul, 1991; Novotny et al., 1997; Furukawa and Kawata, 1998; Zayats, 1999; Kottmann and Martin, 2001; Kottman et al., 2001; Porto et al., 2003), tip enhancement depends on its shape and geometry, the metal deposited, incident beam angle, wavelength and polarization. One of the most important adjustments to maximize the enhancement is to tune the incident laser wavelength to the resonance frequency of the metal plasmons. Richards and co-workers (Festy et al., 2004) have performed simulations of TERS using finite difference methods to show that there is an order of magnitude difference between the Raman enhancements for resonance and off-resonance conditions.

However, experimental realization of the optimum resonant enhancement conditions in TERS requires the measurement of the wavelength-dependent optical response of the tip apex. Since the tips have sub-wavelength dimensions, the optical measurement of the tip properties is not an easy task. Most research groups have relied solely on simulations of optical properties of the tips. A few groups (Aigouy et al., 1999, 2000; Bachelot et al., 2003; Kalkbrenner et al., 2004) have recently designed experimental techniques for the analysis of tip optical properties. Most of these techniques are indirect and complicated. One of the most straightforward approaches is to use the principle of total internal reflection microscopy, a technique used to measure the optical properties of metal nanoparticles (Sönnichsen et al., 2000). The application of this technique to measurements of apertureless tips is described in Section 2 above and presented schematically in fig. 3. Plasmons at the tip apex are excited by the evanescent field above a prism surface. The plasmons then radiate in the far-field and the radiation spectrum provides information on the wavelength of the plasmon resonance. Because the evanescent field decays very fast with distance from the surface, the plasmon properties at the very end of the tip are measured with this technique. Simulations of this experimental configuration (Mehtani et al., 2006) show that the wavelength of the maximum in the radiated spectrum measured in the far-field agrees with the wavelength for which the tip provides maximum enhancement of the electric field. Thus, this technique should provide reliable information on the tip enhancement properties. Raschke and co-workers (Neacsu et al., 2004) were the first to use this approach to measure optical properties of bulk gold and tungsten tips. Gold tips showed a wavelength-dependent response with a resonance maximum at ~ 600–700 nm, depending on the shape of the tip. On the other hand, tungsten tips did not show a strong wavelength-dependent response.

The authors used this technique to measure optical properties of gold- and silver-coated Si_3N_4 and tungsten tips (fig. 16) (Mehtani et al., 2006). Gold-coated tips exhibit a resonance maximum at 650–750 nm, silver-coated tips show a broad resonance in the range of 550–650 nm, while uncoated tips show a flat optical spectrum. The authors have compared the optical spectra of the tips to their TERS enhancements (Mehtani et al., 2006). Two different materials, namely PEDOT/PSS and Si, were used to

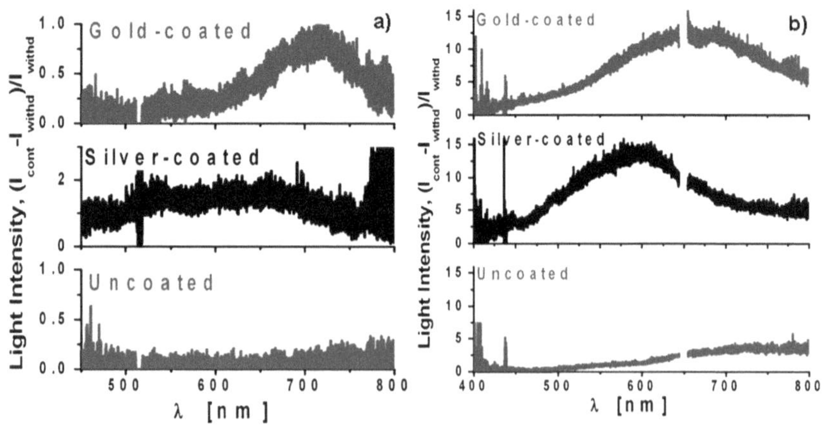

Fig. 16. Optical spectra of gold-coated, silver-coated and uncoated (a) silicon nitride and (b) tungsten tips (reproduced from Mehtani et al., 2006, by permission).

analyze correlations between the enhancement and optical properties of the tips. In the case of PEDOT/PSS (fig. 17(a)), silver- and gold-coated tips showed similar contrasts of ~ 1.5 for an incident wavelength of 514.5 nm.

When the gold-coated tip was excited with the 647 nm laser, the contrast for PEDOT/PSS increased to ~ 8. Comparing the contrast with the optical spectra in fig. 16(a), the five times higher contrast can be attributed to the fact that the resonant excitation of the gold-coated tips lies close to 647 nm. For silicon measurements (fig. 17(b)), the far-field signal was suppressed using an incident polarization of $\sim 70^{\circ}$ with respect to the tip. Under these conditions, the contrast was ~ 1.4 when silver-coated tips were excited with the 514.5 nm laser and ~ 0.6 when the tips were excited with the 647 nm laser. The contrast was ~ 1.75 for excitation of gold-coated tips with the 647 nm laser. A quantitative comparison of the resonance effect in this case requires taking into account the difference in the far-field scattering volumes. This variation in scattering volumes is primarily due to the changes in penetration depth, D, which can be estimated using $D = \lambda/4\pi k$, where k is the Si extinction coefficient for wavelength λ (Feng, 1988). Using the value of $k = 0.06$ ($\lambda = 514.5$ nm), the penetration depth is estimated to be 0.68 μm. For red light, $k = 0.017$ ($\lambda = 647$ nm) and $D = 3.0$ μm. So, the far field volume at 647 nm is ~ 4.5 times larger than that at 514.5 nm. Using this difference, the authors estimated that the enhancement factor with the silver-coated tip is ~ 2 times higher with the 647 nm laser than with the 514.5 nm laser. For the 647 nm laser, the gold-coated tip provides ~ 3 times stronger enhancement factor than does a silver-coated tip. Once again, correlation between enhancements factors and optical spectra of the tips is observed with gold-coated tips showing greater enhancement under resonant excitation at 647 nm. Silver-coated tips exhibit a broad resonance spectrum, with similar intensities at 514.5 nm and 647 nm, explaining the similar enhancements observed under these wavelengths for PEDOT/PSS and silicon. Thus, a good qualitative agreement between optical properties of the tips and their enhancement factors is observed. More detailed quantitative comparison will be required for a better understanding of the tip enhancements.

Analysis of the presented spectra suggests a possible explanation for the rather weak ($\sim 10^4$) enhancement provided by the tips. Optical spectra of the tips (fig. 16) show extremely broad spectral resonance. The reason for such a broad resonance may be the formation of many metal particles of different shapes and sizes during the vapor deposition. In any case, a broad spectral resonance cannot provide a very strong field enhancement. We expect that designing tips with sharper resonances might significantly increase the enhancements at the resonant wavelength.

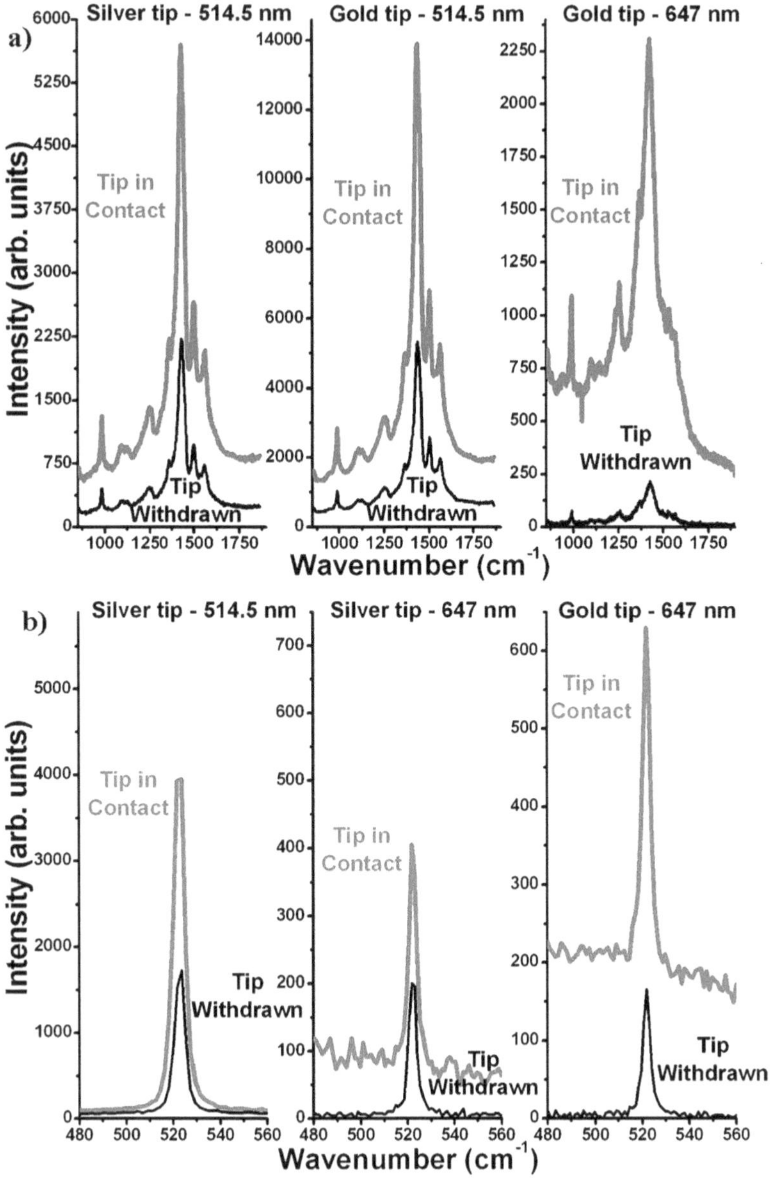

Fig. 17. Comparisons of Raman spectra measured with tip withdrawn and tip in contact showing the dependence of apparent enhancement on metal coating and laser wavelength for (a) PEDOT/PSS and (b) silicon (reproduced from Mehtani et al., 2006, by permission).

§ 7. Summary and outlook

Significant progress in the development of TERS techniques suitable to the study of silicon and other semiconducting materials, CNTs and biological molecules has been achieved in recent years and further advances are coming rapidly. The development of the technique is at an early, but very exciting, stage. A tip-enhanced signal has been obtained from many materials, including molecular and polymeric systems, carbon nanotubes, biological molecules and semiconducting materials. Localization of the tip-enhanced signal to length scales of $\sim 20\,\mathrm{nm}$ has been demonstrated. All of these results provide a "proof of principle", which offers excellent prospects for the TERS technique.

TERS in its current form can already be effectively applied to the analysis of thin films. However, the tip enhancements achieved to date do not generally provide sufficient contrast between the tip-enhanced signal and the far-field background signal for the analysis of thick samples. We emphasize that the contrast, not the tip enhancement, is the most important factor dictating the range of application of the TERS technique. In the specific case of silicon, it has been demonstrated that the increase in contrast that can be achieved by choosing an optimum polarization geometry is sufficient to make the analysis of Si nanostructures commercially viable.

Future work will focus necessarily on further increases in the tip enhancement factor. The enhancements of $\sim 10^2$–10^4 achieved so far will not generally be sufficient to provide effective contrast between the tip-enhanced local signal and far-field signal. We anticipate that future efforts will make critical use of the ability to measure the optical properties of the apertureless tips, in order to achieve sharper plasmon resonances tuned to the incident wavelength to maximize resonance effects. Achieving enhancements of order 10^6 will be necessary for the widespread application of TERS to samples of many types.

We emphasize that apertureless near-field optics based on tip enhancement can be applied not only to Raman measurements, but also to many other types of optical spectroscopy. Successful applications of the apertureless technique to CARS (Ichimura et al., 2004) and to fluorescence imaging (Sanchez and Novotny, 1999) have been demonstrated. Attempts to apply this technique to second-harmonic generation (SHG) (Zayats and Sandoghdar, 2001) have been reported. Apertureless optical microscopy for application to optical and infrared absorption imaging has been discussed in two publications (Zenhausern et al., 1994; Knoll and Kielmann, 1999). We expect that fabrication of apertureless tips with high enhancement factors will stimulate strong progress in all these directions.

The lateral resolution of $\sim 15\,\mathrm{nm}$, unprecedented for optical methods, which has already been achieved, puts apertureless techniques among the most promising future technologies for the chemical and physical analysis of materials at the nanoscale.

Acknowledgments

The authors acknowledge financial support from the Cooperative Center for Polymer Photonics, which is co-funded by the Air Force Research Laboratory, Air Force Office of Scientific Research and The University of Akron, the National Science Foundation (grant DMR-0215966) and the Ohio Board of Regents Action Fund. RDH thanks the NSF for Graduate Research Fellowship. We also thank Ivan Dolog, David Liptak and Laura Sowards for help with the preparation of CdS samples and Ed Evans for help with metal modifications of the tips.

References

Aigouy, L., Andréani, F.X., Boccara, A.C., Rivoal, J.C., Porto, J.A., Carminati, R., Greffet, J.J., 2000, Appl. Phys. Lett. **76**, 397.

Aigouy, L., Lahrech, A., Grésillon, S., Cory, H., Boccara, A.C., Rivoal, J.C., 1999, Opt. Lett. **24**, 187.

Anderson, M.S., Gaimari, S.D., 2003, J. Struct. Bio. **142**, 364.

Anderson, M.S., Pike, W.T., 2002, Rev. Sci. Instrum. **73**, 1198.

Anderson, N., Hartschuh, A., Cronin, S., Novotny, L., 2005, J. Am. Chem. Soc. **127**, 2553.

Bachelot, R., H'Dhili, F., Barchiesi, D., Lerondel, G., Fikri, R., Royer, P., 2003, J. Appl. Phys. **94**, 2060.

Bulgarevich, D.S., Futamata, M., 2004, Appl. Spectrosc. **58**, 757.

Courjon, D., Bainier, C., 1994, Rep. Prog. Phys. **57**, 989.

Demming, F., Jersch, J., Dickmann, K., Geshev, P.I., 1998, Appl. Phys. B: Lasers Opt. **66**, 593.

Denk, W., Pohl, D.W., 1991, J. Vac. Sci. Technol. B **9**, 510.

Dunn, R.C., 1999, Chem. Rev. **99**, 2891.

Feng, Z.C., Mascarenhas, A.J., Choyke, W.J., Powell, J.A., 1988, J. Appl. Phys. **64**, 6827.

Festy, F., Demming, A., Richards, D., 2004, Ultramicroscopy **100**, 437.

Furukawa, H., Kawata, S., 1998, Opt. Commun. **148**, 221.

Hartschuh, A., Beversluis, M.R., Bouhelier, A., Novotny, L., 2004, Phil. Trans. R. Soc. Lond. A **362**, 807.

Hartschuh, A., Sanchez, E.J., Xie, X.S., Novotny, L., 2003, Phys. Rev. Lett. **90**, 095503.

Hayazawa, N., Inouye, Y., Sekkat, Z., Kawata, S., 2001, Chem. Phys. Lett. **335**, 369.

Hayazawa, N., Inouye, Y., Sekkat, Z., Kawata, S., 2002, J. Chem. Phys. **117**, 1296.

Hayazawa, N., Motohashi, M., Saito, Y., Kawata, S., 2005, Appl. Phys. Lett. **86**, 263114.

Hecht, B., Sick, B., Wild, U.P., Deckert, V., Zenobi, R., Martin, O.J.F., Pohl, D.W., 2000, J. Chem. Phys. **112**, 7761.

Hopkins, J.B., Farrow, L.A., 1986, J. Appl. Phys. **59**, 1103.

Ibe, J.P., Bey, P.P., Brandow, S.L., Brizzolara, R.A., Burnham, N.A., Dilella, D.P., Lee, K.P., Marrian, C.R.K., Colton, R.J., 1990, J. Vac. Sci. Technol. A **8**, 3570.
Ichimura, T., Hayazawa, N., Hashimoto, M., Inouye, Y., Kawata, S., 2004, Phys. Rev. Lett. **92**, 220801.
Kalkbrenner, T., Hákanson, U., Sandoghdar, V., 2004, Nano Lett. **4**, 2309.
Kneipp, K., Wang, Y., Kneipp, H., Perelman, L.T., Itzkan, I., Dasari, R.R., Feld, M.S., 1997, Phys. Rev. Lett. **78**, 1667.
Knoll, B., Kielmann, F., 1999, Nature **399**, 134.
Kolb, G., Salbert, Th., Abstreiter, G.J., 1991, Appl. Phys. **69**, 3387.
Kottmann, J.P., Martin, O.J.F., 2001, Phys. Rev. B **64**, 235402.
Kottmann, J.P., Martin, O.J.F., Smith, D.R., Shultz, S., 2001, J. Microsc. **202**, 60.
Li, K., Li, X., Stockman, M.I., Bergman, D.J., 2005, Phys. Rev. B **71**, 115409.
Martin, O.J.F., Girard, C., 1997, Appl. Phys. Lett. **70**, 705.
Mehtani, D., Lee, N., Hartschuh, R.D., Kisliuk, A., Foster, M.D., Sokolov, A.P., Maguire, J.F., 2005, J. Raman Spectrosc. **36**, 1068.
Mehtani, D., Lee, N., Hartschuh, R.D., Kisliuk, A., Foster, M.D., Sokolov, A.P., Tsukerman, I., 2006, J Opt. A: App. Pure Opt., **5927**, S183–S190.
Moskovits, M., 1985, Rev. Mod. Phys. **57**, 783.
Nakashima, S., Mizoguchi, K., 1989, J. Appl. Phys. **65**, 2583.
Neacsu, C.C., Steudle, G.A., Raschke, M.B., 2004, Appl. Phys. B **80**, 295.
Nie, S., Emory, S.R., 1997, Science **275**, 1102.
Novotny, L., Bian, R.X., Xie, X.S., 1997, Phys. Rev. Lett. **79**, 645.
Novotny, L., Sánchez, E.J., Xie, X.S., 1998, Ultramicroscopy **71**, 21.
Patanè, S., Gucciardi, P.G., Labardi, M., Allegrini, M., 2004, Rivista Del Nuovo Cimento **27**, 1.
Pettinger, B., Ren, B., Picardi, G., Schuster, R., Ertl, G., 2004, Phys. Rev. Lett. **92**, 096101.
Poborchii, V., Tada, T., Kanayama, T., 2005, Japn. J. Appl. Phys. **44**, L202.
Porto, J.A., Johansson, P., Apell, S.P., López-Rios, T., 2003, Phys. Rev. B **67**, 085409.
Raschke, M.B., Lienau, C., 2003, App. Phys. Lett. **83**, 5089.
Ren, B., Picardi, G., Pettinger, B., 2004, Rev. Sci. Instrum. **75**, 837.
Sanchez, E.J., Novotny, L., Xie, X.S., 1999, Phys. Rev. Lett. **82**, 4014.
Sönnichsen, C., Geier, S., Hecker, N.E., Plessen, G., Feldmann, J., Ditlbacher, H., Lamprecht, B., Krenn, J.R., Aussenegg, F.R., Chan, V.Z-H., Spatz, J.P., Möller, M., 2000, Appl. Phys. Lett. **77**, 2949.
Stockle, R., Suh, Y.D., Deckert, V., Zenobi, R., 2000, Chem. Phys. Lett. **318**, 131.
Sun, W.X., Shen, Z.X., 2001, Mater. Phys. Mech. **4**, 17.
Sun, W.X., Shen, Z.X., 2002, Rev. Sci. Instrum. **73**, 2942.
Sun, W.X., Shen, Z.X., 2003a, J. Raman Spectrosc. **34**, 668.
Sun, W.X., Shen, Z.X., 2003b, Ultramicroscopy **94**, 237.
Thio, T., Lezec, H.J., Ebbesen, T.W., Pellerin, K.M., Lewen, G.D., Nahata, A., Linke, R.A., 2002, Nanotechnology, **13**, 429.
Wang, J.J., Smith, D.A., Batchelder, D.N., Saito, Y., Kirkham, J., Robinson, C., Baldwin, K., Li, G., Bennett, B., 2003, J. Microsc. **210**, 330.
Webster, S., Batchelder, D.N., Smith, D.A., 1998, Appl. Phys. Lett. **72**, 1478.
Wessel, J.J., 1985, Opt. Soc. Am. B **2**, 1538.
Wolf, IDe., 1996, Semicond. Sci. Technol. **11**, 139.
Zayats, A.V., 1999, Opt. Commun. **161**, 156.
Zayats, A.V., Sandoghdar, V., 2001, J. Microscopy, **202**, 94.
Zenhausern, F., O'Boyle, M.P., Wickramasinghe, H.K., 1994, Appl. Phys. Lett. **63**, 1623.

Chapter 7

Near-field optical structuring and manipulation based on local field enhancement in the vicinity of metal nano structures

by

R. Bachelot

Nanotechnology and Optical Instrumentation Laboratory, Delaunay Institute CNRS FRE 2848, University of Technology of Troyes (UTT), 12 rue Marie Curie, BP 2060 10010 Troyes cedex, France
e-mail: renaud.bachelot@utt.fr

Tip Enhancement | *S. Kawata and V.M. Shalaev (Editors)*
Advances in Nano-Optics and Nano-Photonics
ISSN: 1871-0018
DOI: 10.1016/S1871-0018(06)01007-7
© *2007 Elsevier B.V. All rights reserved.*

Contents

§ 1. Introduction: context and motivation

An important domain of the nanotechnology is the nanostructuration that involves numerous scientific and economic challenges (Gentili, et al., 2003; Bucknall, 2005). Especially, the innovative trend of modern technology lies in smaller, cheaper, faster and better performances. Industry must improve yield by increasing smaller instruments. For instance, cars, cameras and wireless telephone have combined many functions in a small box. Nanotechnology is an exact example of this trend representing complex technology in commodities. During the last few years, novel structures, phenomena and processes have been observed at the nanoscale from a fraction of nanometer to about 100 nm and new experimental, theoretical and simulation tools have been developed for investigating them. These advances provide fresh opportunities for scientific and technological developments in nanoparticles, nanostructured materials, nanodevices and systems. From the practical effect, the miniaturization of integration circuit and systems means the reduction of raw materials and energy waste. The smaller products are conducive to transportation and utilization, which proves to have more advantages of miniaturization over traditional products. In order to fit into the development of modern technology, the advances of nanotechnology are in urgent needs for its benefits. So the development of nanotechnology is driven both by science itself and market. This context explains why nanolithography and matter nanostructuration are important branches of nanotechnology envisioning various applications and research fields including ultrahigh density storage, nanoelectronics, nanomechanics and nanobiotechnology.

The current techniques of nanolithography are various and numerous. Exhaustive description and classification of these techniques is beyond the scope of this chapter. Most of them involve far-field illumination of the material to be modified. Figure 1 illustrates this class of lithographies. They are diffraction limited and the spatial resolution is theoretically not better than $\lambda/2n$ where λ is the wavelength of the source used and n is the refractive index of the medium. Among these techniques, let us cite X-ray and electron beam lithographies allowing for resolution in the 30–50 nm range.

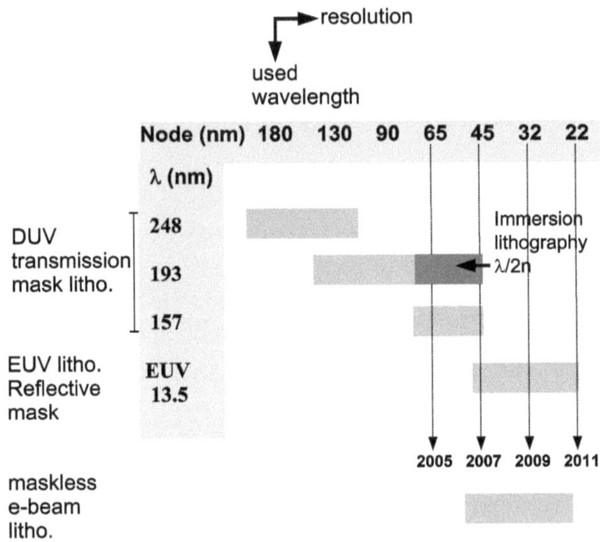

Fig. 1. Far-field lithographies.

Lithography using light (wavelengths included in the near-UV to near-IR domain) is particularly appreciated for several reasons of costs on simplicity. Presently, mask far-field optical lithography is the most widely used technique for pattern mass production in various fields such as microelectronics and microoptics (Sheats and Smith, 1998). This technology is optimized, and easy to implement compared with X-ray or ion-beam lithography. The photopolymers used are various, well known and have been optimized for several applications. However, the main limit of the optical lithography is its diffraction-limited spatial resolution. The typical resolution permitted by UV sources is currently about 150 nm. As shown in fig. 1, the current trend is to develop and use new small wavelength EUV sources, that would enable a resolution better than 50 nm in the near future. However, this approach requires expensive development of new technologies: new sources, new optics, and new photopolymers. As an example, the typical price for a 193 nm exposure tool is approximately 15 MDollars; the price of 157 nm exposure tools is expected to be as high as 25 MDollars and extreme ultra-violet (EUV) exposure tools may cost as much as 30 MDollars. Another approach consists in increasing the refraction index: immersion photolithography (using liquid) and solid immersion lens lithography (using a scanning microlens) allowed resolution to be improved by a factor 2, mainly limited by refractive index of the available materials (Sheats and Smith, 1998; Ghislain et al., 1999).

The near-field optical lithography (NFOL) is an alternative and elegant method of improving the resolution. NFOL relies on the use of spatially confined evanescent fields as optical sources. In the case of near-field illumination, the spatial confinement of the light–matter interaction is not limited by the light wavelength but rather by both source size and source-to-matter distance (Kawata et al., 2001; Courjon, 2003; Prasad, 2004). The advantage of NFOL compared with the other kinds of lithography is thus to be an optics-based technique without any $\lambda/2n$ resolution limit. NFOL actually relies on the use of lateral (parallel to material surface) wavevectors $k_{//}$ that are superior to k, the wavevector in medium n:

$$k^2 = k_{//}^2 + k_{\perp}^2, \tag{1.1}$$

where $k_{//} > k$.

Such high lateral wavevectors can be obtained by either total internal reflection (TIR) or diffraction by spatial frequencies $> 2n/\lambda$ (Goodman, 1996). Lateral wavevectors $> k$ involve evanescent waves ($k_{\perp}$, the component perpendicular to the surface is a pure complex in eq. (1.1), effective indexes $\gg 1$ (equivalent to a super immersion lithography !) and nanometer scale light confinement. Control, use and study of such electromagnetic waves constitute the near-field optics that has aroused large interest, an effort over the last two decades (Kawata et al., 2001; Courjon, 2003; Prasad, 2004). The advent of this science has opened a new field, and appreciation for, the control and manipulation of light at the nanoscale. Confined near-fields were used for optical microscopy and spectroscopy at the nanometer scale. Different chapters of the present book deal with these application.

Near-fields have been used as local optical sources for high resolution optical lithography too. Different types of near-field optical illumination have been used. They included solid immersion lenses and probes for scanning near-field optical microscopy. A general survey of the literature on NFOL can be found elsewhere (Royer et al., 2003). In the present chapter we would like to focus our attention onto the use of local optical fields of metal nano structures (MNS) for high resolution optical lithography. The principle of this approach is illustrated in fig. 2(a). It relies on the local optical interaction between MNS and photosensitive molecules. We will see as exhaustively as possible how scientists took advantage of the enhanced confined fields in the vicinity of MNS for modifying and manipulating matter with a subwavelength resolution. We shall show how rich and exciting this domain is, in particular how the

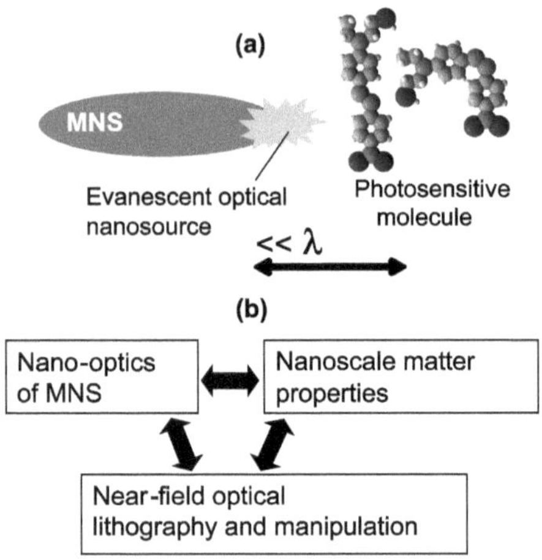

Fig. 2. Principle (a) and motivations (b) of using MNS or optical near-field lithography.

three following issues were addressed simultaneously (see fig. 2(b) as an illustration).

(*i*) *Knowledge of the nature of optical fields of MNS*. The interest here is twofold. First, this domain of research motivated several theoretical studies. Second, the photosensitive molecule can act as probes for the optical field, allowing for near-field photography of MNS field.

(*ii*) *Knowledge of the matter at the molecular scale*. The capability to illuminate the matter using a nanometer-sized optical source allows for better understanding of the photochemical properties of the molecules involved.

(*iii*) *Development of a novel powerful technique of optical lithography*.

As shown in fig. 2(b) all the above issues are linked to each other. For example, goal (*iii*) can not be achieved unless points (*i*) and (*ii*) are at a sufficiently advanced stage. As another example, many new information about optics of nanometals have been obtained (goal (*i*)) by attempting to achieve goal (*iii*).

The present chapter is divided into the following four sections. In Section 2, general consideration on optics of MNS are given. The physical effects, related to MNS, used and studied for NFOL are described. Obviously, a complete review of this area of science is beyond the scope of this chapter. The reader is referred to excellent recent reviews by Krenn et al. (2003); Zayats and Smolyaninov (2003); Pitarke et al. (2005). There are also very comprehensive books on optics of metals (Jackson, 1975; Bohren and Huffmann, 1982; Raether, 1988; Kreibig and Vollmer, 1996). Here we only remind and summarize the main involved phenomena. This reminder will be necessary for both commenting and appraising reported experiments. Section 3 is devoted to tip-enhanced optical lithography. In Section 4 we will describe the approaches based on localized surface plasmons. In Section 5 we will present NFOL optical mask-NFOL based on surface plasmons. Finally, we will be concluding and evoking some promising routes.

§ 2. General consideration on the optics of metal nanostructures

Three optical effects and phenomenons related to MNSs have been considered for NFOL and manipulation: local plasmon resonance, electromagnetic singularity and nanometer scale optical forces.

Figure 3 illustrates the first effect that is the local plasmon resonance. Plasmon excitations, i.e. collective oscillations of the conduction electrons, strongly influence the optical properties of metal nanostructures. Great interests and extensive works over the past 30 years were devoted to their studies (Jackson, 1975; Bohren and Huffmann, 1982; Raether, 1988; Kreibig and Vollmer, 1996). In particular, plasmon resonance found applications in several areas of science and technology including nanosensoring and integrated optics (Hutter and Fendler, 2004).

As most physical systems, the electron density present in metals structures can be modulated by resonances. Resonance parameters such as peak wavelength, and relaxation times depend mainly on charge density, electron effective mass and conductivity of the material. The Drude model is instrumental in predicting the dispersion of the permittivity $\varepsilon(\omega)$. For bulk metal, $\varepsilon(\omega)$ can be written as:

$$\varepsilon(\omega) = \varepsilon_\infty - \frac{\omega_p^2}{\omega(\omega + i\gamma_o)} \approx \varepsilon_\infty - \frac{\omega_p^2}{\omega^2} + i\frac{\gamma_o\omega^2}{\omega^3}, \tag{2.2}$$

where ω_p is the plasma frequency, γ_o the electron relaxation rate (related to damping due to collisions) and ε_∞ the dielectric constant at $\omega = \infty$ ($\varepsilon_\infty = 1$ if only free electron are considered). γ_o is the inverse of τ, the electron relaxation time (basically, the average time between two successive collisions) which depends on material properties:

$$\tau = \frac{\sigma m^*}{ne^2}, \tag{2.3}$$

where n, m^*, σ and e are the density, the effective mass of the conduction electrons, the material conductivity and the electron charge, respectively.

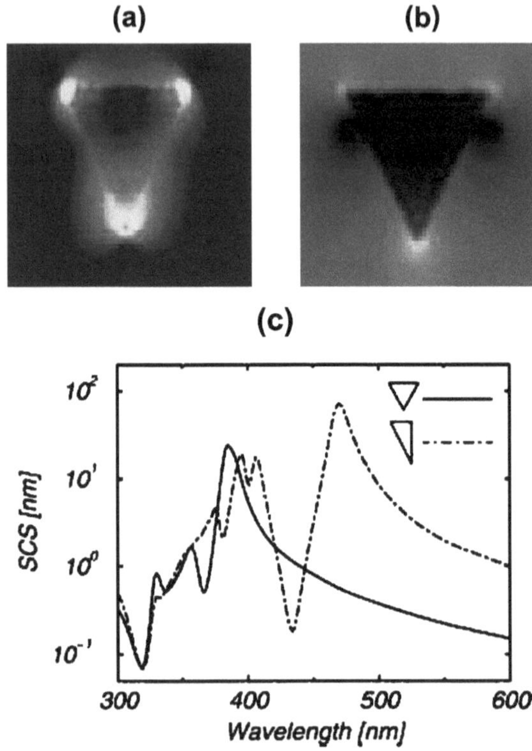

Fig. 3. Localized plasmon resonance of nanotriangles. From Kottmann et al. (2001). (a), (b) Calculated relative field intensity distribution for a 27 nm equilateral triangle. (a) $\lambda = 385$ nm (resonance). (b) $\lambda = 600$ nm (off resonance). (c) Calculated scattering cross section of nanotriangles as a function of λ.

ω_p depends on material properties too:

$$\omega_p = \sqrt{ne^2/\varepsilon_o m^*}, \tag{2.4}$$

where ε_o is the dielectric constant of the vacuum. Many properties of real metals have been shown to be well predicted by the above Drude model. It should be stressed that this model only takes into account free electrons within the conduction band and considers intraband electron transitions only. Correct description of nanometer-sized metal particles involves bounded electrons and interband transition from other bands to the conduction band. For bulk material, the electron resonance corresponds to a vanishing real part of the dielectric constant (see eq. (2.2)). Plasmon modes actually exist in a number of geometries and in various metals, with the strongest responses in noble metals such as gold and silver. In confined geometries such as metallic nanoparticles the plasmon resonance not only depends on materials parameters (σ, n, e and m^*) but also is affected by the shape of the object and its local environment as well as the condition of illumination (Hutter and Fendler, 2004). For example, the surface plasmon peak wavelength of 2-D arrays of gold disks is drastically affected by the size of the particle and by the near-field interaction between neighboring particles (Bouhelier et al., 2005a). For simple MNS geometries (spheres, oblate, prolate.), plasmon properties can be predicted by analytical theoretical models issued from the Mie Theory. Especially, the polarizability α of a spheroid takes the well-known following form (Bohren and Huffmann, 1982):

$$\alpha(\omega)_i = \varepsilon_0 V \frac{\varepsilon(\omega) - \varepsilon_m}{\varepsilon_m + L_i(\varepsilon(\omega) - \varepsilon_m)}, \tag{2.5}$$

where $\varepsilon(\omega)$, ε_0, ε_m are the complex dielectric functions of the MNS, the vacuum and the surrounding medium, respectively. V is the MNS volume and L_i describes the spheroid geometry along the axes i ($i = x$, y, or z). For example, for a sphere, $L_x = L_y = L_z = 1 \div 3$. Resonance of the MNS corresponds to a minimization of the denominator in eq. (2.5). This condition depends notably on the particle geometry, that is to say L_i, and on the dielectric environment, that is to say ε_m. In the case of nonregularly shaped MNS, numerical calculation are needed, as illustrated by fig. 3. Figure 3(c) shows the numerically calculated scattering cross section of two types of triangles as a function of the excitation wavelength. This figure reveals dipolar plasmons peaks. Figure 3(a) and fig. 3(b) shows the corresponding electric field intensity distribution in the case of resonance and off resonance respectively. For plasmon resonance, the local field intensity is much higher than for off resonance excitation. It is interesting to note that in both cases the field tends to be confined at the corners of the structure where radius of curvature of the object is small. This phenomenon is an off resonance optical effect similar to the well-known electrostatic lightning-rod effect. It constitutes the second near-field effect promising for NFOL.

The term lightning-rod effect refers to an electrostatic phenomenon where the electric charges on the surface of a conductive material are spatially confined by the shape of the structure. For a conductor with a nonspherical shape, surface charge density σ will vary from point to point along the shape. In the regions of high curvature, σ is locally increased, resulting in a large electric field just outside the material. If the object is at a high electric potential, the field can be large enough to trigger a dielectric breakdown of the surrounding medium. The confinement of the electric field around point of high curvature is routinely used to protect buildings from the destroying power of lightning. Sharply pointed poles ("lightning-rods") at an equipotential with the earth provide on one hand, a low-resistance electric path to a lightning discharge, and on the other hand, a mean to reduce the surrounding local electrostatic potential (Zeleni, 1932). Although this effect originates from macroscopic electrostatic considerations, a similar phenomenon can be observed in nanometer-sized metallic structures excited by an electromagnetic radiation. This point of view was discussed in detail by Van Bladel (1995).

The free electrons of the metal react to an electromagnetic excitation by inducing oscillating surface charges. When the surface presents a geometrical singularity such as a tip apex, the local surface charge density is drastically increased in this region. As a consequence, the electromagnetic field outside the tip is not only locally enhanced over the driving field but is also highly confined around the tip apex. This optical lightning-rod effect is represented in the calculations of fig. 4. Figure 4(a) shows field distribution at the end of cobalt-metallized silicon tip in air. Figure 4(b) shows field distribution at the end of gold tip in air. In both cases tips are illuminated by significant field component parallel to the tip axis. p-Polarization is actually suitable for exciting an electromagnetic singularity in the vicinity of metal tips, as shown in fig. 4(c) that represents field distribution at the extremity of a tungsten tip when illuminated by different polarization from p (incident field parallel to the incident plane) to s (incident field perpendicular to the incident plane). It should be stressed that this polarization effect is a consequence of the metal-dielectric boundary condition of the electric field as reminded in eqs. (2.6) and (2.7).

$$\mathbf{n}_{io} \times (\mathbf{E}_i - \mathbf{E}_o) = 0 \tag{2.6}$$

$$\mathbf{n}_{io} \cdot (\mathbf{E}_i - \mathbf{E}_o) = \frac{\sigma}{\varepsilon}, \tag{2.7}$$

where E_i and E_o are, respectively, the field within the metal and the dielectric medium (permittivity ε) at the metal surface and $\mathbf{n}_{io}$ is the unit vector perpendicular to the metal surface of charge density σ. Equation (2.6) says that the tangential field component is continuous at the metal surface: field vanishes (through the skin depth) both inside and outside the metal. Equation (2.7) says that the normal field component is discontinuous at the metal surface as a consequence of the Gauss Theorem: field vanishes inside the metal but can be high outside depending on σ. In the case of incident p-Polarization, the field is mainly parallel to the tip axis and normal to the metal just beneath the foremost tip's end, resulting in a

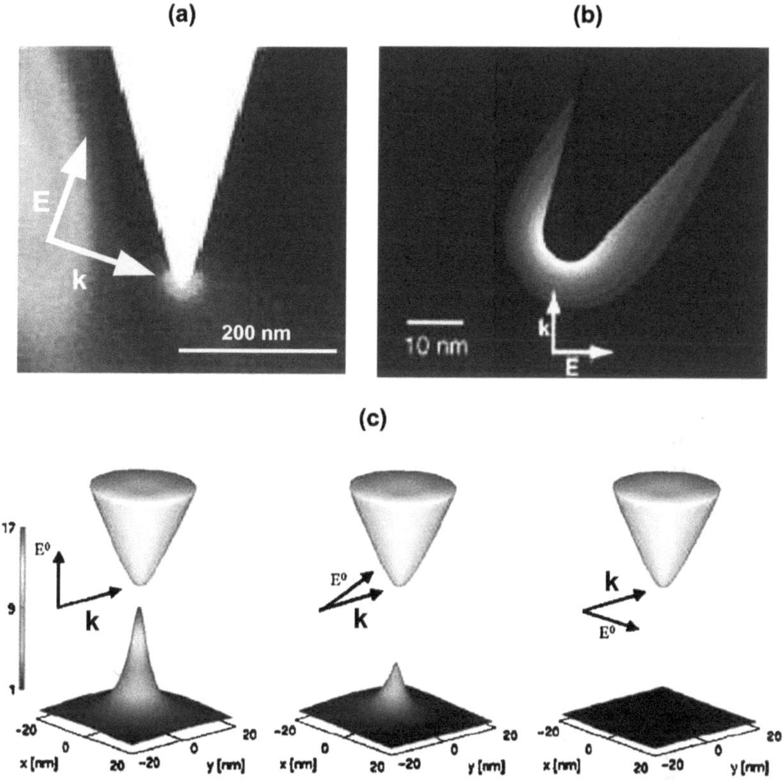

Fig. 4. Electromagnetic singularities excited at the extremity of metal tips (calculated intensity distribution). (a) Cobalt tip. From Bachelot et al. (2003). (b) Gold tip. From Sanchez et al. (1999). (c)Tungsten tip with different illumination condition. left: p-Polarization, right: s-Polarization, center: s–p Polarization. From Martin and Girard (1997).

local strong field. In the case of s-Polarization, field is tangential beneath the tip's end which remains uncharged. With incident s-Polarization only the tip's edges can present significant (nonlocalized) fields because the field is mainly normal at the edges. Nanosources such as these illustrated in fig. 4 have actually been excited by field components parallel to the tip axis. Such sources have been efficiently used for optical near-field microscopy and spectroscopy (Lahrech et al., 1997; Furukawa and Kawata, 1998; Sanchez et al., 1999; Hartschuh et al., 2003; Ichimura et al., 2004) and turns out to be powerful for NFOL, as it will be seen in Section 3.

It should be stressed that the above two effects are two aspects of the same physics: the physics of localized surface plasmon polaritons, i.e. localized surface plasmons excited by light (Zayats and Smolyaninov, 2003).

Figure 5 illustrates the third optical effect promising for nano matter manipulation and modification. Evanescent optical sources (field E) are

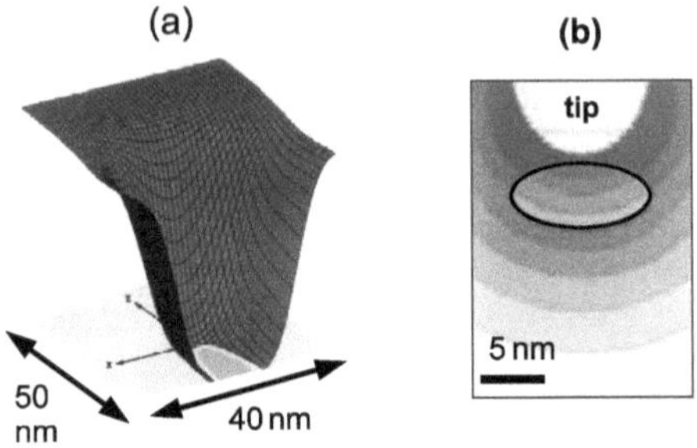

Fig. 5. Optical force induced by tip-enhanced field. (a) Calculated trapping potential of a 10 nm particle in the vicinity of a 5 nm radius gold tip. (b) Alignment of a prolate dielectric particle (black outline) beneath the tip (in white). Calculation from Novotny et al. (1997).

associated to forces that small particles of polarizability α are expected to experience (Novotny et al., 1997; Chaumet et al., 2002; Nieto-Vesperinas et al., 2004; Chaumet et al., 2005). Two different terms were found to determine such a force: the gradient force proportional to $Re[\alpha]\nabla|E|^2$, and the scattering force, proportional to $kIm[\alpha]|E|^2$, where k is the wavevector of the light involved (the scattering force corresponds to a radiation pressure). While the scattering term is weak for nanometer-sized particles, the gradient force term is expected to be strong due to the high field gradients characteristic of evanescent fields. For example, starting from Lorentz electrodynamic forces formalism applied to near-field optics, it was shown that a dipolar particle experiences, within an electromagnetic evanescent harmonic field, a time average longitudinal force that can be written as

$$\langle F_z \rangle \approx dRe[\alpha]\exp(-2zd), \tag{2.8}$$

where $Re(\alpha)$ denotes the real part of α, d is the penetration depth of the evanescent field along z, the longitudinal distance between the particle and the evanescent source (Nieto-Vesperinas et al., 2004). Figure 5(a) shows the calculated trapping potential of a dielectric 10 nm particle in the vicinity of a laser p-illuminated gold tip. Figure 5(b) illustrates the capability of the tip to trap a prolate particle. Such an effect is very promising for future molecule manipulation based on optical forces. This new approach could lead to a powerful tool for manipulation and engineering at the molecular scale. Compared to nanomanipulation based on STM (Dujardin et al., 1998), one could trap any kind of object including dielectric neutral and biological particles. Additionally, particle capture or release could be easily selected by tuning the incident light polarization, as discussed by Chaumet et al., 2005 (p for capture and s for release, see fig. 4(c)). So far no experimental demonstration was however carried out. That can be due to the fact that the involved expected forces are of few pN (Nieto-Vesperinas et al., 2004; Chaumet et al., 2005), which can be three orders of magnitude weaker than mechanical atomic force such as Van der Waals forces (Cao et al., 2002). Indeed, while micro-manipulation based on far-field laser trapping is a contactless method (Bayoudh et al., 2001), optical nanomanipulation using evanescent nano-source involves the presence of a material tip in interaction with the surface, leading to possible non-optical dominant forces. Furthermore, it was shown that the optical force can be either attractive or repulsive depending on particle material and used wavelength (as illustrated in eq. (2.8)), making presently somehow complicated the use of such forces. However, as it shall be shown in the following sections, molecular displacement based on local isomerization, was demonstrated with MNS optical near-fields.

§ 3. Tip-enhanced optical lithography (TEOL)

In this section we report and comment the use of metal tips for NFOL. In the following section this approach will be termed tip-enhanced optical lithography (TEOL) that belongs to the family of the scanning probe lithography techniques. An analysis of the literature on the subject reveals that mostly conical metal tips were used for TEOL. With this geometry, where the metal structure can be considered as semi-infinite, there is a subtle compromise between the establishment of a large surface plasmon resonance and a local electrostatic lightning-rod effect. Indeed, as discussed in Section 2, an electromagnetic singularity at a tips end can be efficiently excited for an incident polarization parallel to the tip's axis. However, for this polarization orientation, the main surface plasmon resonance of the tip is probably located in the infrared region of the spectrum. Furthermore, because of the semi-infinite geometry, the main resonance is likely to be broad and contributes marginally to the overall enhancement. This effect is illustrated by fig. 6 that shows dipolar plasmon resonance of an elongated gold particle. While the resonance of the small axis is in the visible region of the spectrum, the resonance of the major axis is clearly in the infrared. In the case of a semi-infinite tip, dipolar plasmon resonance should be in the far infrared where the photosensitive materials generally do not absorb electromagnetic radiation. Lateral tip resonance could be excited in the visible but in that case field is not confined at the tip end but, on the contrary, at its edges (see Section

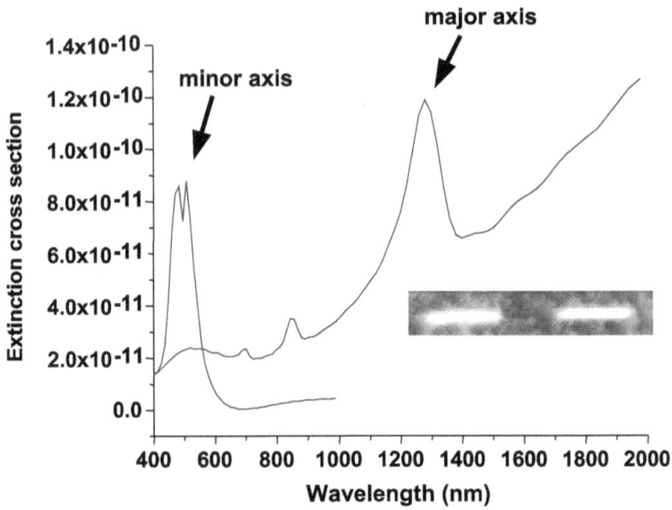

Fig. 6. Dipolar plasmon resonance of elongated MNS: Case of a gold ellipse. Extinction cross section calculated by the finite difference time domain method (A. Vial, LNIO-UTT). Inset: $3 \times 0.5\,\mu m^2$ AFM image of such ellipses.

2). As a consequence, optical sources within or close to the visible region of the spectrum were kept and tip plasmon resonance was not expected for TEOL. Actually, only off resonance tip electromagnetic singularities were used for tip-enhanced nanostructuring.

Yet, efforts have been dedicated to development of probes presenting single MNS at the very end of the probes (Kalkbrenner et al., 2000; Sqalli et al., 2002; Eah et al., 2005). In that case, tip's resonance is expected to be achieved in the visible. Moreover, finite-in-length metal probes were theoretically shown to present an intensity enhancement of up to 7,000 (Krug et al., 2002; Sun and Shen, 2003). Nevertheless, so far, no such tips were used as nanosources for TEOL.

TEOL was performed on inorganic surface and photopolymers.

3.1. TEOL on inorganic material

First experiments of TEOL were carried out on metal surfaces by the Dickmann's group in Germany (Jersh and Dickmann, 1996; Dickmann et al., 1997). Enhancement in the near-field of an STM tips was used to fabricate nanostructures on gold/palladium surfaces. A 10 nm resolution was achieved with tips illuminated by 5 ns pulses at $\lambda = 532$ nm issued from a Nd:YAG laser. Dickmann and co-workers named this approach "FOLANT" (focusing of laser radiation in the near-field of a tip). Figures

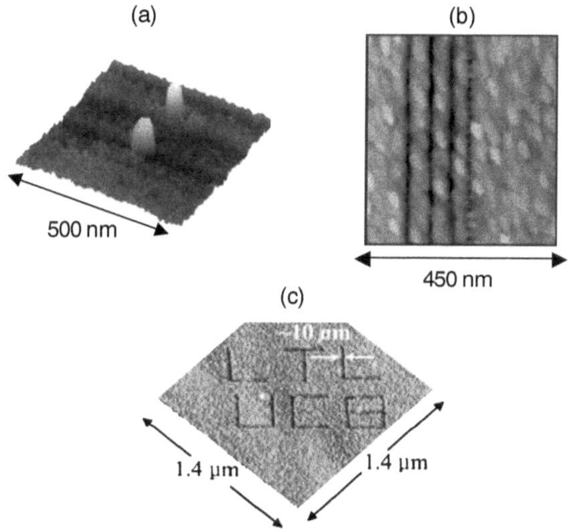

Fig. 7. TEOL on metal surfaces. AFM images of obtained structures (a) On gold/palladium substrate with a silver tip. (b) On gold/palladium substrate with a tungsten tip. (a), (b) are from Dickmann et al., 1997. (c) On gold film with a silicon tip (Chimmalgi et al., 2005).

7(a) and 7(b) show example of obtained structures. These figures illustrate two modes of TEOL found by Dickmann and co-workers. In case of use of silver-coated tungsten tips, hillocks were obtained (fig. 7(a)). Hillocks material was proved to be silver that has been transferred from the tip to the substrate. This suggested that transfer was triggered by excitation of the local plasmon resonance of silver aggregates at the tip's end. With plain tungsten tips, pits instead of hillocks were obtained (fig. 7(b)). The phenomenon was assumed to be caused by repulsing dipole–dipole interaction between surface gold grains, the dipole moment in the gold grains being induced by the tip-enhanced fields. Both tip and sample being metallic, it is also likely that cavity modes between tip and sample and associated field enhancement (Porto et al., 2003) played a significant role in dots formation. Although the above hypotheses were finally not proven, this first TEOL experiment lead to interesting discussion on nanophysics, opened the door to optical matter nanomodification and motivated several further researches involving multiphysics at the nanometer scale. In particular, it was proposed that indentation of the sample surface due to thermal expansion was the main reason for the formation of nanoscopic features (Huber et al., 1998). It was notably calculated that the effective temperature rise in the tip was in the 350–800 °C range yielding a tip expansion up to 5 nm (Huang et al., 2002). Photothermal laser ablation based on temperature increase, melding and evaporation, was also suggested (Geshev et al., 2000). More recently, it was proposed that surface gold ablation is due to a strong and fast temporal evolution of the temperature in the gold film locally illuminated by the tip's end (Chimmalgi et al., 2005). The same group confirmed results from Dickmann et al. (1997) by using AFM Si tips and femtosecond laser pulses, showing that an electromagnetic singularity can be also excited at the end of a semiconducting singularity. They obtained nanostructures with minimum dimensions of 10 nm (see fig. 7(c), as an example).

Semiconductor samples were also machined by TEOL. Lu and co-workers used this approach to locally induce nanooxidation on hydrogen-passivated Ge surface with an STM tip under laser illumination (Lu et al., 1999). The process of lithography was based on the local photo-breaking of the Ge–H bond by the nanosource excited at the tip's end (the sample was previously hydrogen passivated by HF etching). The involved sample zone oxidized instantaneously. The GeH/GeO_2 contrast was characterized by STM imaging. A 20 nm resolution was achieved with tungsten tips. Let us note that similar results were obtained by Herndon and co-workers, who used nano apertures in illumination mode to perform NFOL on H-passivated Si surfaces (Herdon et al., 1999). Apparently, these experiments were exceptional and authors did not go further into the experiments. However, it is likely that they will motivate future research because they remarkably illustrated the ability of TEOL to allow for surface chemistry at the nanometer scale.

3.2. TEOL on photopolymer

Following on from classical far-field optical lithography, TEOL was carried out on photopolymers by several groups. G. Wurtz and co-workers were the first to demonstrate that it is possible to induce locally one-photon absorption free-radical photopolymerization by tip-field enhancement (TFE) (Wurtz et al., 2000). This approach is illustrated in fig. 8(a). The liquid photopolymerizable formulation contains a sensitizer dye eosin absorbing green light. After absorption of actinic light by eosin the triplet state of the dye reacts with the amine Methyl Diethanolamine (MDEA) to form radicals. Radicals initiates the polymerization of the monomer Penta erythrilol triacrylate (PETIA). Owing to the monomer multifunctionality, the polymer quickly develops into a 3D network. Radical reactions are known to be quenched by oxygen, leading to a threshold energy below which no polymerization is possible (Espanet et al., 1999). TFE enabled the polymerization threshold dose to be locally exceeded at the tip's end while the threshold was not reached elsewhere.

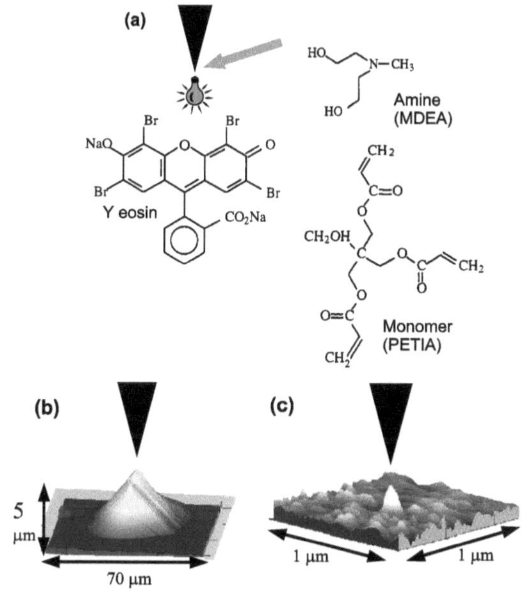

Fig. 8. TEOL on free radical photopolymerizable formulation (Wurtz et al., 2000). (a) Principle. (b) Polymerization with the far-field of a W tip. (c) Polymerization with the near-field of a W tip. (b), (c) are AFM images.

The authors reported a 70 nm resolution photopolymerization achieved with a tungsten tip illuminated by a p-polarized laser beam issued from a continuous He–Ne laser ($\lambda = 532$ nm, within the absorption band of eosin). Figure 8(c) shows an example of photopolymerized pitch. This experiment demonstrated for the first time the ability of a metal tip to induce a local photochemical process on polymers, confirming at the same time the existence of the TFE. Especially, polymerized parts were shown to be several orders of magnitude smaller than the total illuminated area. Additionally, while 1D nanometer confinement of polymerization was demonstrated in 1998 (Ecoffet et al., 1998), Wurtz et al. (2000) demonstrated that 3D encapsuling of the process (allowed by selective O_2 inhibition) was possible using 3D enhanced evanescent fields. Two important effects were observed. First, no local polymerization was possible with incident s-Polarization, as predicted theoretically (see Section 2). Second, the thermal origin of the polarization was ruled out. Indeed, as discussed previously, tip-enhanced field can be associated to high temperature increase (Huber et al., 1998; Chimmalgi et al., 2005). High temperature is expected to create radicals and trigger polymerization without the need of photo initiators. Red light ($\lambda = 632$ nm) was used as actinic source. At this wavelength eosin does not absorb light. On the other hand, at this wavelength, tungsten is more metallic and significant both TFE and temperature increase are expected. However, under these conditions, no polymerization was observed, confirming the pure photochemical origin of the obtained nanodots. This result was in agreement with previous prediction from Novotny et al. (1997) that tip's temperature increase is pretty modest in liquid. The tip's Fresnel far-field zone was also polymerized when the tip was put in a micronic volume of formulation. Figure 8(b) shows a micronic polymer object fabricated by TFE below a tungsten tip. The apparent cone should be viewed as a 3D molding of areas where the energy related to the tip's field was superior to the threshold energy ($\sim 10\,\mathrm{mJ\,cm^{-2}}$), illustrating the fact that tip's far-field is also enhanced by electromagnetic singularity (Aigouy et al., 1999).

Commercial photoresists were also structured by TEOL. In order to avoid both polymerization by tip's far-field (such as that shown in fig. 8(b)) and improve the resolution, near-field two-photon lithography on commercial negative tone (polymerizable) photoresists was performed by Yin et al. (2002). The authors used a metal-coated Si AFM tip illuminated by 120 *fs* laser pulses at 790 nm to induce locally two-photon polymerization. A 70 nm resolution was achieved. In that case, material was not liquid. It was a dry photoresist made of linear chains that developed more complex cross linking under optical illumination. Compared to experiments from Wurtz et al. (2000), who used a liquid formulation, the probe scanning was certainly more controllable and well-defined patterns were obtained as shown in fig. 9(b). The authors observed local two-photon

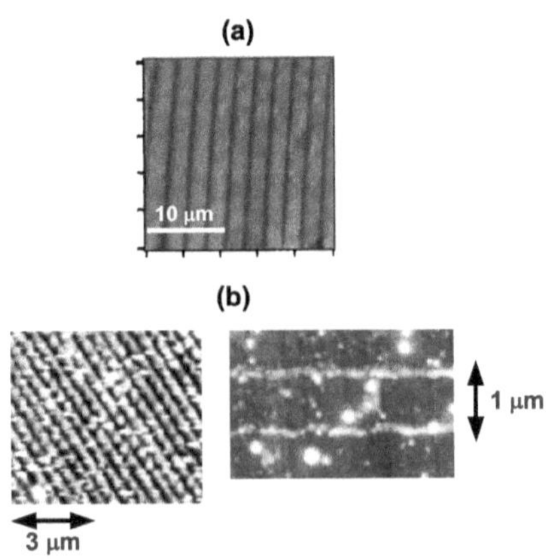

Fig. 9. TEOL on photoresist-type polymer (AFM images of produced structures). (a) One-photon TEOL on positive-tone photoresist with a Pt-coated silicon tip (from Tarun et al. (2002)). (b) Two-photon TEOL on negative-tone photoresist with a metal-coated silicon tip (from Yin et al. (2002)).

polymerization at intensities significantly lower than the threshold energy, revealing a field enhancement factor of about 7. This factor appears to be realistic and in relative agreement with recent calculation (Martin and Girard, 1997; Fikri et al., 2003; Micic et al., 2003). Additionally, experiments using a dielectric (silicon nitride) tip failed to produce observable structures at intensities below the threshold, confirming the role of surface charge density in TFE. This example remarkably shows how TEOL experiments allow study and quantification of tip-enhanced fields.

Positive-tone photoresists were also structured by TEOL. Tarun and co-workers successfully fabricated 100 nm wide lines using a metal-coated AFM tip illuminated by a laser beam ($\lambda = 402$ nm) through an immersion objective lens. Figure 9(a) shows an example of obtained subwavelength-sized pattern. The authors demonstrated that, in the case of low incident energy (below the threshold of photolysis), the tip's end can provide locally the amount of energy required to induce bond breaking in polymer, revealing again the existence of a local FE factor. A particularly interesting point here is that the authors successfully exited the tip's electromagnetic singularity by using the longitudinal electric field expected at the focus plane of a high numerical aperture (>1) objective lens, as predicted for example by Novotny et al. (1998).

As pointed out in Section 2, TEOL based on near-field optical forces was not experimentally demonstrated so far. However, matter migration induced by TFE has been recently reported. Metal tips were used to induce photoisomerization of azobenzene-dye molecules grafted to polymer matrix (H'Dhili et al., 2001, 2002; Bachelot et al., 2003; Royer et al., 2003). The photochemistry of this photopolymer is illustrated in fig. 10. Figure 10(a) shows azobenzene group grafted to a polymer matrix of polymethylmethacrylate (PMMA). The absorption band of the azo dye is centered in the visible (see absorption spectrum in inset). The absorption peak is typically situated around $\lambda = 500$ nm. This is why the dye is usually named DR1 (Dispersed Red One). Figure 10(b) shows schematically the process of isomerization. The stable state of the azobenzene molecule presented on the left side is the *trans*-isomeric configuration. The absorption in the visible range of a photon induces the transition to the *cis*-isomer. This state is metastable and the reverse transition to the *trans*-state takes place through either thermal activation or optical absorption. Therefore, a molecule absorbing a photon undergoes a complete *trans-cis-trans* isomerization cycle. Provided that there is a nonzero component of the light polarization vector along the intensity gradient, this transition induces a motion of the molecule and a related deformation of the matrix to which the molecule is grafted. The polarization selectivity lies in the fact

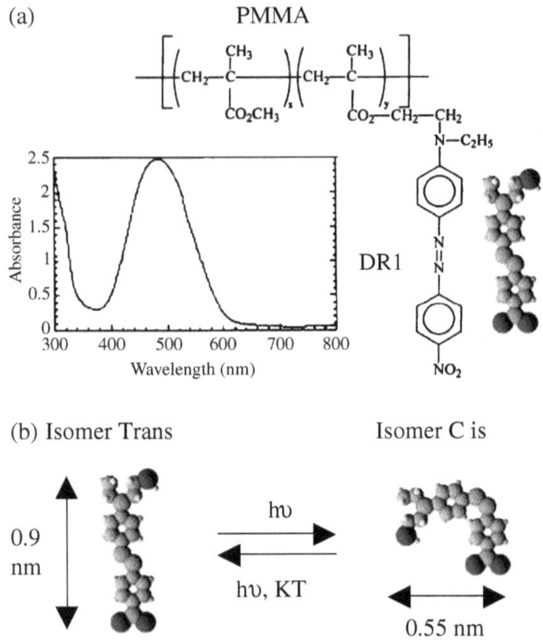

Fig. 10. Photochemistry of the PMMA-DR1. (a) Structure. (b) Photoisomerization. Inset: absorption spectrum of DR1.

that the azobenzene moieties in their *trans*-state are strongly anisotropic and preferentially absorb the light polarized along their main axis. The required illumination conditions to observe any matter migration are actually intensity gradient and a nonzero component of the polarization vector along the intensity gradient. As a consequence, the polymer is self-developing and after illumination its surface presents a topography related to the incident intensity distribution. Details on this material can be found in a recent comprehensive review paper by Natansohn and Rochon (2002).

The deformation properties of azo-containing polymer films under illumination were first used in the near-field to both map the electromagnetic field emitted by a metallized or uncoated fiber SNOM and perform NFOL (Davy and Spajer, 1996; Landraud et al., 2001). Below we describe TEOL on PMMA–DR1.

Azobenzene-containing films were illuminated with the presence of a metallized AFM tip at their surface. After illumination the optically induced topography was characterized in situ by AFM using the same tip. This approach enabled a parametric analysis leading to valuable

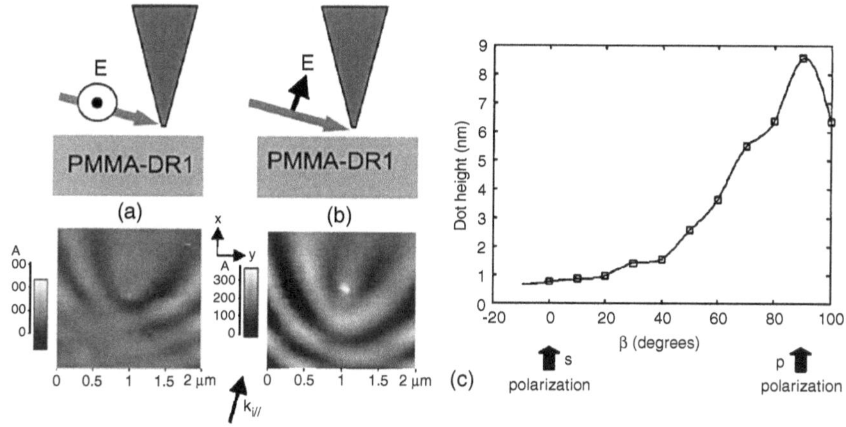

Fig. 11. TEOL on PMMA-DR1. Study of the influence of the incident polarization. The study was performed with a cobalt-metallized tip illuminated in reflection mode. (a) $2 \times 2\,\mu m^2$ tapping-mode AFM image obtained after s-Polarization exposure (see geometry above). (b) $2 \times 2\,\mu m^2$ tapping-mode AFM image obtained after p-Polarization exposure (see geometry above). $ki_{//}$ is the projection on the polymer surface of the incident wavevector. (c) Height of the obtained central singularity as a function of the polarization direction β. Data from Bachelot et al. (2003).

information on TFE (H'Dhili et al., 2001, 2002; Bachelot et al., 2003; Royer et al., 2003). In particular, the TFE was evaluated as a function of the illumination condition (state of polarization, angle of incidence,...) and tip's features (radius of curvature, material,...). As an example, fig. 11 shows the influence of the incident polarization on the TFE. Figure 11(b) shows the AFM image obtained after p-Polarization reflection mode illumination of a metal-coated Si tip in interaction with the PMMA–DR1 surface. The center of the image corresponds to the tip position during the exposure. The figure exhibits two different fabricated patterns: a far-field-type fringe system, which corresponds to diffraction by the tip cone, and a central nanometric dot due to the local enhancement of the electromagnetic field below the metallic tip. The far-field contrast vanished if the tip is illuminated by TIR (see fig. 12), providing valuable information about the suitable way of illuminating a tip for TEOL.

Figure 11(a) shows the obtained topography after s-Polarization illumination. The central dot appears to be low compared to fig. 11(b). The height of the central near-field spot was measured as a function of the incident angle of polarization (fig. 11(c)), confirming that the intensity of the local tip field is enhanced gradually from s-Polarization to p-Polarization (see fig. 4(c)). Figure 12 presents images showing the

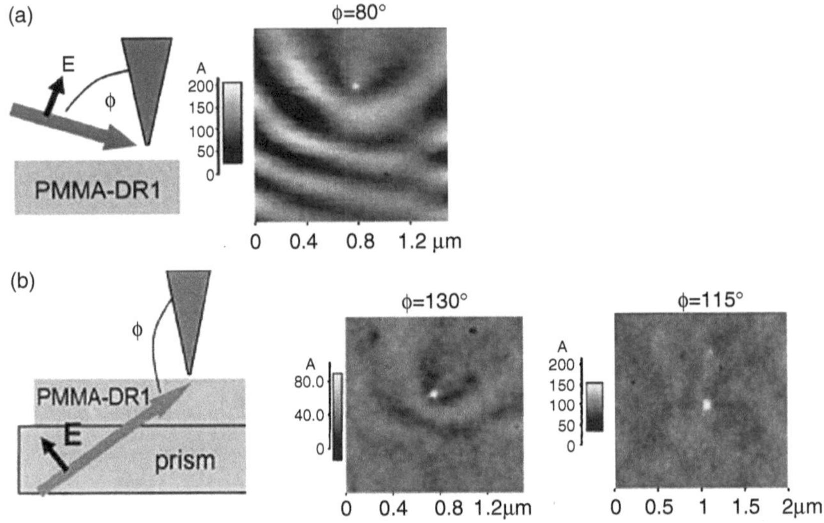

Fig. 12. TEOL on azobenzene-containing polymer films. Influence of the illumination condition. A Pt tip was illuminated by a p-polarized beam. (a) $1.4 \times 1.4\,\mu m^2$ Tapping mode AFM image obtained after a reflection mode illumination (see depicted geometry). (a) $1.4 \times 1.4\,\mu m^2$ and $2 \times 2\,\mu m^2$ Tapping mode AFM images obtained using the TIR illumination mode (see depicted geometry). Data from Bachelot et al. (2003).

influence of the illumination geometry on the field in the vicinity of the tip's extremity. This study was motivated by the presence of the fringes in the images of fig. 11. As pointed out before, these fringes are due to the diffraction caused by the tip cone. Hence, the fringe contrast was expected to decrease when the size of the illuminated tip decreases. Decreasing the illuminated tip zone was made possible by the use of a total internal reflection (TIR) configuration. This was confirmed experimentally as shown in fig. 12. Figures 12(a) and 12(b) present results of the experiment performed with an identical platinum-coated tip. Figure 12(a) was obtained in the same conditions as for fig. 11(b) while fig. 12(b) shows results of a TIR illumination carried out with a high-index prism. In fig. 12(b), we can note that the TIR configuration caused the fringe pattern to vanish while keeping the central nanometer sized dot. The residual fringes vanished when the penetration depth of the incident evanescent wave was decreased by increasing the angle of incidence. This experiment provided valuable information: considering the far-field diffraction caused by the tip, the TIR illumination mode is more suitable for tip-enhanced near-field microscopy and spectroscopy because background far-field effects are, in that way, minimized. It should be pointed out that this experiment

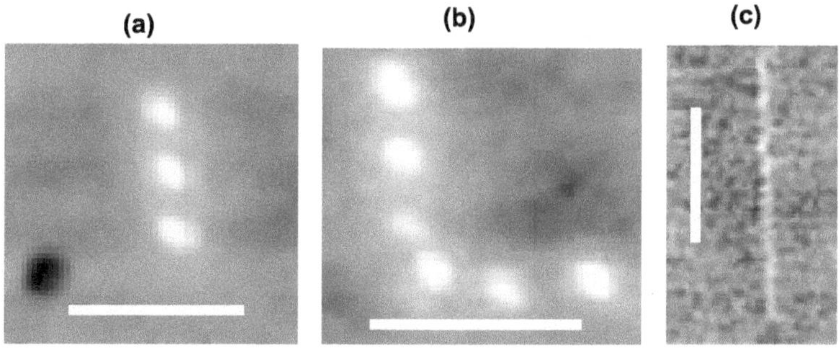

Fig. 13. TEOL on PMMA-DR1: fabrication of nanopatterns. (a), (b) successive exposures. (c) continuous exposure. White scale bars represent 500 nm. Data from Bachelot et al. (2003).

confirmed the efficiency of TIR illumination for FE, as theoretically predicted by Furukawa and Kawata (1998). These experiments were completed by studies of local migration of azo molecules within a diffraction spot locally perturbed by a metal tip (Gilbert et al., 2005). The above-cited near-field investigations lead to preliminary 30 nm resolution nanopatterning based on TFE (see fig. 13 as an illustration). All these experiments stimulated interesting discussions on the nature of the optical response of the molecule. Especially, it was shown that while lateral field components tend to make escape molecules from light along the polarization, longitudinal components lift the matter vertically, as a consequence of free space requirement from the molecule. This point of view was confirmed very recently by investigation of the effects of tightly focused higher-order laser beams on the photoinduced molecular migration and surface deformations in azo-polymer (Gilbert et al., 2006).

§ 4. NFOL based on localized 3-D surface plasmons

While plasmon resonant single particles were not used as sources for TEOL, localized surface plasmons of isolated particles were used for NFOL.

The migration of the azo-molecules (described in previous section) was shown to be possibly induced by field locally enhanced in the vicinity of noble metal nanostructures produced by electron lithography (Hubert et al., 2005).

A thin layer of PMMA-DR1 was deposited onto the metal structures. The sample was subsequently illuminated in normal incidence by a plane whose polarization state was controlled. Photoinduced mass transport created topographic modification at the polymer film surface that were

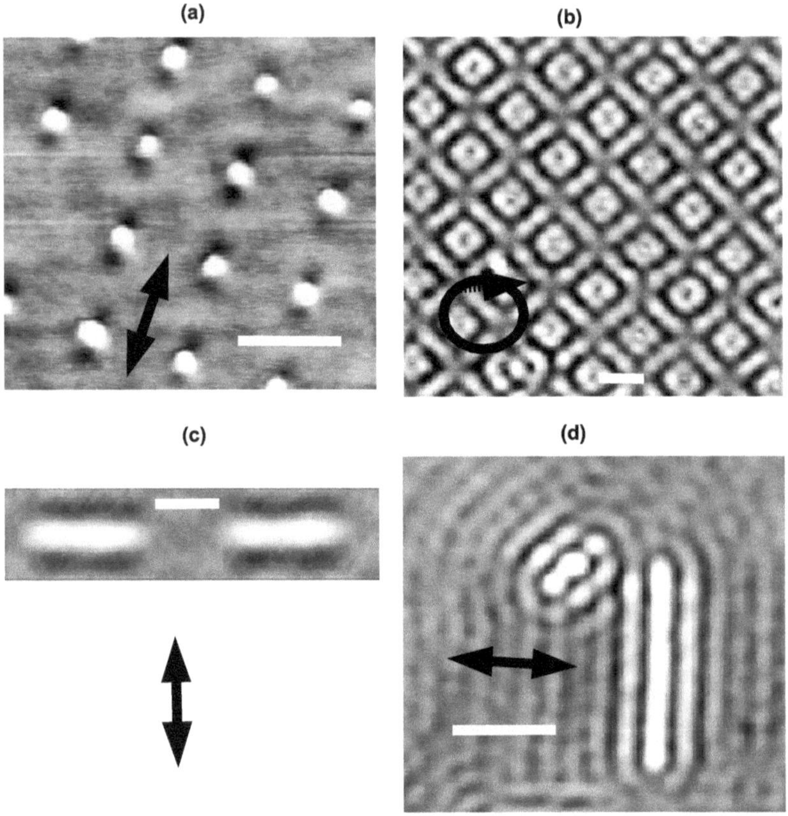

Fig. 14. NFOL with plasmonics structures. AFM images of the photomodified topography of an azo-polymer deposited on metal nanostructures. The incident polarization state is indicated by black arrows. White scale bars represent 500 nm. (a), (b) Silver dots. From Hubert et al. (2005). (c) Gold ellipses. From Hubert et al. (2005). (d) A gold dot in interaction with a gold nanorod (not published).

then measured by atomic force microscopy. Figure 14 shows a set of significant results. These are all AFM images taken after exposure (polarization shown in the images). Figure 14(a) shows the signature of dipolar near-field plasmon resonance of silver dots. The two holes parallel to the incident polarization were shown to be due to azobenzene molecules escaping from local enhanced field at the edges of the Ag particles under plasmon resonance. In fig. 14(b), an incident circular polarization was used and the result obtained shows that subwavelength scale complex polymer structuration is enabled by near-field optics. Figure 14(c) shows topography photoinduced by a gold ellipse excited along its minor axis. Resonance of this axis was excited at the wavelength used ($\lambda = 514$ nm, see fig. 6). The two observable dark stripes are the results of strong excitation of charge density at the ellipse's edges at resonance. Figure 14(d) is the result of the illumination of a gold dot in the vicinity of a gold rod. The interpretation of this image is underway, but it is likely to contain the signature of multipolar plasmon resonances and near-field coupling effects. AFM images were compared to numerical electromagnetic calculation. It was shown that the negative of calculated intensity basically corresponds to the obtained topography, revealing that molecules tended to escape light to minimize their internal energy. The interest of this study is clearly twofold: getting information on optics of nanometals and learning about properties of the molecules (see fig. 2). Since the azobenzene could be grafted to a large variety of matrix, this method may ultimately be applied to the nanooptical manipulation of photon-assisted molecular motors to produce nanoscale mass transport of a material. In that sense, data presented in fig. 14 can be considered as the first observation of matter manipulation induced by optical field enhancement in metal nanoparticles.

§ 5. Mask-based surface plasmon lithography

As seen in Sections 3 and 4, the use of metal tips or single particles allowed for nanometer scale matter modification. However, in an industrial point of view, these approaches are limited by both low speed and MNS's lifetime. This is why, with the purpose to surpass far-field UV mask lithography and to compete with X-ray lithography and nanoimprint lithography, mask-type approaches based on surface plasmons were introduced recently (Srituravanich et al., 2004a, 2004b; Luo and Ishihara, 2004). This approach is illustrated in fig. 15, it is generally termed "plasmonic nanolithography" or "surface plasmon nanolithography" (SPN).

SPN is a branch of optical contact lithography (OCL) that consist in illuminating the photosensitive material through a mask placed at a subwavelength distance from the material to be structured (see for instance Goodberlet and Kavak (2002)). In that way evanescent parts of the angular spectrum diffracted by the mask is used for lithography that is not diffraction-limited. The limit of the OCL is the extremely low transmission through the mask associated to poor field confinement, leading to reduced contrasts. SPN aims to enhance this transmission taking advantages of optical properties of metals. SPN is a very recent technique: first experiments were reported in 2004. They were notably motivated by discovery of extraordinary transmission through metal films perforated with a periodic array of subwavelength holes (Ebbesen et al., 1998). One of the main origins for the enhancement was found to be the surface plasmon resonance of the metal mask. SPN relies on the matching between

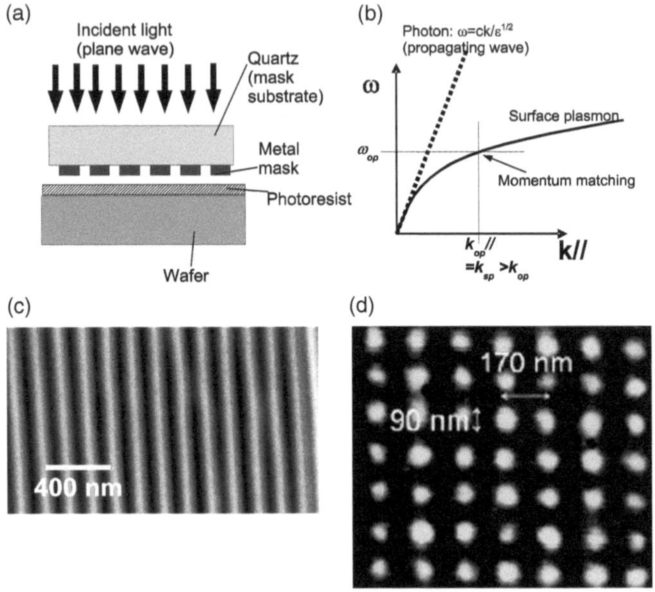

Fig. 15. NFOL based on surface plasmon. (a) Experimental arrangement. (b) Principle: surface plasmon and light dispersion. (c), (d) AFM images of subwavelength structures obtained on photoresists. (c) Array of lines. From Luo and Ishihara (2004a). (d) Array of dots. From Srituravanich et al. (2004b).

respective dispersion relation of surface plasmon and light. As shown in fig. 15(b), optical lateral wavevectors k_{op} cannot fit high SP wavevectors (k_{sp}) unless using evanescent waves (that are surface waves too) generated by diffraction by mask subwavelength structures (see eq. (1.1)). For a given optical pulsation ω_{op}, by selecting proper mask periodicity and proper dielectric constant one can get a main $k_{\mathrm{op}}// \sim k_{\mathrm{sp}}$ (momentum matching = conservation of wavevectors) leading to SP resonance and field enhancement (see fig. 15(b)). In other words, parts of the mask-diffracted angular spectrum are enhanced by coupling with the SP.

A resolution of about 100 nm was achieved on photoresists using near-UV light and gold masks made of periodic structures (see figs. 15(c),(d) as examples). So far experimental results were obtained with periodic structures exclusively. It is in accordance with the fact that, a priori, one single k_{op} (and thus one single period structure) is favored by enhancement (see fig. 15(b)). However, considering the relative flatness of the plasmon dispersion curve for high wavevectors, it is likely that a wide part of the mask-generated angular spectrum was plasmon-enhanced. Resolution is expected to be improved in the near-future by exploiting a branch of OCL, named "evanescent interferometric lithography". This approach relies on interference between evanescently decaying diffracted orders to create a frequency-doubled intensity pattern in the near field of the mask, leading to twice better resolution (Blaikie and McNab, 2001). This approach was recently proposed specifically for SPN and a resolution below 50 nm was predicted (Luo and Ishihara, 2004b; Liu et al., 2005). In the near future the main challenge of mask-SPL will be to produce specific structures that are of interest for industrial application, such as transistors. That will require production of complex specific mask geometries predicted by solving of near-field inverse problem (Fiddy, 1993; Pitman, 1992).

§ 6. Conclusion

This chapter presents the explored optical interaction between MNSs and photosensitive materials. Several examples have shown the ability of MNSs to induce physical and chemical processes at the molecular scale. These experiments not only allowed production of nanostructures using visible light but also opened the door to nano photochemistry based on MNSs. Moreover, they lead to progress in nano metals, in both experimental and theoretical points of view. Taking into account associated economical and scientific challenges the future of optical lithography based on MNS is likely to be successful. This technology certainly shall take advantage of the large variety of powerful physical effects in nano metals that are investigated since recently, such as second harmonic generation (Bouhelier et al., 2003), photoluminescence (Bouhelier et al., 2005b; Mühlschlegel et al., 2005), strong near-field coupling (Atay et al., 2004; Ghenuche et al., 2005), and multipole excitation (Krenn et al., 2000; Atay et al., 2004). All of these effects will permit higher resolution and better control of intensity and wavelength of the actinic light with regard to photosensitive materials.

Acknowledgment

The author would like to both cite and thank current and former co-workers for this subject of research (in alphabetical order): D. Barchiesi, A. Bouhelier, C. Carre, S. Chang, M. Derouard, C. Ecoffet, R. Fikri, Y. Gilbert, S.K. Gray, F. H'Dhili, C. Hubert, H. Ibn-El-Ahrach, M. Juan, S. Kostcheev, N. Landraud, G. Lerondel, D.J. Lougnot, J. Peretti, J. Plain, A. Rumyantseva, P. Royer, O. Soppera, C. Triger, A. Vial, G.P. Wiederrecht and G. Wurtz.

Interacting with them was (is) a great pleasure.

References

Aigouy, L., Lahrech, A., Gresillon, S., Cory, H., Boccara, A.C., Rivoal, J.C., 1999, Opt. Lett. **24**, 187.

Atay, T., Song, J.-H., Nurmikko, A.V., 2004, Nano Lett. **4**, 1627.

Bachelot, R., H'Dhili, F., Barchiesi, D., Lerondel, G., Fikri, R., Royer, P., Landraud, N., Peretti, J., Chaput, F., Lampel, G., Boilot, J.P., Lahlil, K., 2003, J. Appl. Phys. **94**, 2060.

Bayoudh, S., Mehta, M., Rubinsztein-Dunlop, H., Heckenberg, N.R., Critchley, C., 2001, J. Microsc. **203**, 214.

Blaikie, R.J., McNab, S.J., 2001, Appl. Opt. **40**, 1692.

Bohren, C., Huffmann, D., 1982, Absorption and Scattering of Light by Small Particles, Wiley, New York.

Bouhelier, A., Beversluis, M., Hartschuh, A., Novotny, L., 2003, Phys. Rev. Lett. **90**, 13903.

Bouhelier, A., Bachelot, R., Im, J.S., Wiederrecht, G.P., Lerondel, G., Kostcheev, S., Royer, P., 2005a, J. Phys. Chem. B **109**, 3195.

Bouhelier, A., Bachelot, R., Lerondel, G., Kostcheev, S., Royer, P., Wiederrecht, G.P., 2005b, Phys. Rev. Lett. **95**, 267405.

Bucknall, D.J. (Ed.), 2005, Nanolithography and patterning techniques in microelectronics, CRC.

Cao, J., Pham, D.K., Tonge, L., Nicolau, D.V., 2002, Smart Mater. Struct. **11**, 767.

Chaumet, P.C., Rahmani, A., Nieto-Vesperinas, M., 2002, Phys. Rev. Lett. **88**, 123601.

Chaumet, P.C., Rahmani, A., Nieto-Vesperinas, M., 2005, Phys. Rev. B **71**, 045425.

Chimmalgi, A., Grigoropoulos, C.P., Komvopoulos, K., 2005, J. Appl. Phys. **97**, 104319.

Courjon, D., 2003, Near-field Microscopy and Near-field Optics, Imperial College Press, London.

Davy, S., Spajer, M., 1996, Appl. Phys. Lett. **69**, 3306.

Dickmann, K., Jersh, J., Demming, F., 1997, Surface and interface analysis **25**, 500.

Dujardin, G., Mayne, A., Robert, O., Rose, F., Joachim, C., Tang, H., 1998, Phys. Rev. Lett. **80**, 3085.

Eah, S.-K., Jaeger, H.M., Scherer, F., Wiederrecht, G.P., Lin, X.-M., 2005, Appl. Phys. Lett. **86**, 31902.

Ebbesen, T.W., Lezec, H.J., Ghaemi, H.F., Thio, T., Wolff, P.A., 1998, Nature **391**, 667.

Ecoffet, C., Espanet, A., Lougnot, D.J., 1998, Adv. Mater. **10**, 411.

Espanet, A., Ecoffet, C., Lougnot, D.J., 1999, J. Polym. Sci., Part A: Polym. Chem. **3**, 2075.

Fiddy, M.A., (Ed.), 1993, Inverse Problems in Scattering and Imaging, Proceedings of SPIE volume, 1767, SPIE Press.

Fikri, R., Barchiesi, D., H'dhili, F., Bachelot, R., Vial, A., Royer, P., 2003, Opt. Comm. **221**, 13.

Furukawa, H., Kawata, S., 1998, Opt. Comm. **148**, 221.

Gentili, M., Giovannella, C., Selci, S., (Eds.), 1994, Nanolithography: A borderland between STM, EB, IB, and X-ray lithographies, NATO Science Series E, Springer.

Geshev, P.I., Demming, F., Jersh, J., Dickmann, K., 2000, Thin Solid Films **368**, 156.

Ghislain, L.P., Elings, V.B., Crozier, K.B., Manalis, S.R., Minne, S.C., Wilder, K., Kino, G.S., Quate, C.F., 1999, Appl. Phys. Lett. **74**, 551.

Ghenuche, P., Quidant, R., Badenes, G., 2005, Opt. Lett. **30**, 1882.

Gilbert, Y., Bachelot, R., Royer, P., Bouhelier, A., Wiederrecht, G.P., Novotny, L., 2006, Opt. Lett. **31**, 613.

Gilbert, Y., Bachelot, R., Vial, A., Lerondel, G., Royer, P., Bouhelier, A., Wiederrecht, G.P., 2005, Opt. Expr. **13**, 3619.

Goodman, J.W., 1996, Introduction to Fourier optics, McGrawHill, New York.

Goodberlet, J.G., Kavak, H., 2002, Appl. Phys. Lett. **81**, 1315.

Hartschuh, A., Sanchez, E.J., Xie, X.S., Novotny, L., 2003, Phys. Rev. Lett. **90**, 95503.

H'Dhili, F., Bachelot, R., Lerondel, G., Barchiesi, D., Royer, P., 2001, Appl. Phys. Lett. **79**, 4019.

H'Dhili, F., Bachelot, R., Rumyantseva, A., Lerondel, G., Royer, P., 2002, J. Micros. **209**, 214.

Herdon, M.K., Collins, R.T., Hollingswouth, R.E., Larson, P.R., Johnson, M.B., 1999, Appl. Phys. Lett. **74**, 41.

Huang, S.M., Hong, M.H., Lu, Y.F., Lukyanchuk, B.S., Song, W.D., Chong, T.C., 2002, J. Appl. Phys. Lett. **91**, 3268.

Huber, R., Koch, M., Feldmann, J., 1998, Appl. Phys. Lett. **73**, 2521.

Hubert, C., Rumyantseva, A., Lerondel, G., Grand, J., Kostcheev, S., Billot, L., Vial, A., Bachelot, R., Royer, P., Chang, S., Gray, S.K., Wiederrecht, G.P., Schatz, 2005, Nano Lett. **5**, 615.

Hutter, E., Fendler, J.H., 2004, Adv. Mater. **16**, 1685.

Ichimura, T., Hayasawa, H., Hashimoto, M., Inouye, Y., Kawata, S., 2004, Phys. Rev. Lett. **92**, 220801.

Jackson, J., 1975, Classical Electrodynamics, 2nd ed., Wiley, New York.

Jersh, J., Dickmann, K., 1996, Appl. Phys. Lett. **68**, 868.

Kalkbrenner, Th., Ramstein, M., Mlynek, J., Sandoghdar, V., 2000, J. Microsc. **202**, 72.

Kawata, S. (Ed.), et al., 2001, Near-field Optics and Surface Plasmon Polaritons. Springer-Verlag Berlin and Heidelberg GmbH & Co. K.

Kottmann, J.P., Martin, O.J.F., Smith, D.R., Shultz, S., 2001, J. Micros. **202**, 60.

Krenn, J.R., Leitner, A., Aussenegg, F.R., 2003, Metal Nano-Optics, in Encyclopedia of Nanoscience and Nanotechnology by H.S. Nalwa (Ed.), American Scientific Publishers.

Kreibig, U., Vollmer, M., 1996, Optical Properties of Metal Structures, vol. 25 of Springer Series in Materials Science, Springer, Berlin.

Krenn, J.R., Schider, G., Rechberger, W., Lamprecht, B., Leitner, A., Aussenegg, F.R., Weeber, J.C., 2000, Appl. Phys. Lett. **77**, 3379.

Krug, J.T., Sanchez, E.J., Xie, X.S., 2002, J. Chem. Phys. **116**, 10895.

Landraud, N., Peretti, J., Chaput, F., Lampel, G., Biolot, J.P., Lahlil, K., Safarov, V.I., 2001, Appl. Phys. Lett. **79**, 4562.

Lahrech, A., Bachelot, R., Gleyzes, P., Boccara, A.C., 1997, Appl. Phys. Lett. **71**, 575.

Liu, Z.-W., Wei, Q.-H., Zhang, X., 2005, Nano Lett. **5**, 957.

Lu, Y.F., Mai, Z.H., Qiu, G., Chim, W.K., 1999, Appl. Phys. **75**, 2359.

Luo, X., Ishihara, T., 2004a, Jap. J. Appl. Phys. **43**, 4017.

Luo, X., Ishihara, T., 2004b, Opt. Expr. **12**, 3055.

Martin, O.J.F., Girard, C., 1997, Appl. Phys. Lett. **70**, 705.

Micic, M., Klymyshyn, N., Suh, Y.D., Lu, H.P., 2003, J. Phys. Chem. B **107**, 1574.

Mühlschlegel, P., Eisler, H.-J., Martin, O.J.F., Hecht, B., Pohl, D.W., 2005, Science **308**, 1607.

Natansohn, A., Rochon, P., 2002, Chem. Rev. **102**, 4139.

Nieto-Vesperinas, M., Chaumet, P.C., Rahmani, A., 2004, Phil. Trans. R. Soc. Lond. A **362**, 719.

Novotny, L., Bian, R.X., Xie, X.S., 1997, Phys. Rev. Lett. **79**, 645.

Novotny, L., Sanchez, E.J., Xie, X.S., 1998, Ultramicroscopy **71**, 21.

Pitarke, J.M., Silkin, V.M., Chulkov, E.V., Echenique, P.M., 2005, J. Opt. A: Pure Appl. Opt. **7**, S73.

Pitman, A.G.R., 1992, Multidimensional inverse scattering problems, Monographs and Surveys in Pure and Applied Mathematics, vol. 51, Longman Scientific Technical, Harlow.

Porto, J.A., Johansson, P., Apell, S.P., Lopez-Rios, T., 2003, Phys. Rev. B **67**, 85409.

Prasad, P.N., 2004, Nanophotonics, Wiley, Hoboken, New Jersey.

Raether, H., 1988, Surface Plasmons on Smooth and Rough Surfaces and Gratings, Springer, Berlin.
Royer, P., Barchiesi, D., Lerondel, G., Bachelot, R., 2003, Phil. Trans. R. Soc. Lond. A **362**, 821.
Sanchez, E.J., Novotny, L., Xie, X.S., 1999, Phys. Rev. Lett. **82**, 4014.
Sheats, J.R., Smith, B.W. (Eds.), 1998, Microlithography Science and Technology, Marcel Dekker Inc., New York.
Sqalli, O., Utke, I., Hoffmann, P., Marquis-Weible, F., 2002, J. Appl. Phys. Lett. **92**, 1078.
Srituravanich, W., Fang, N., Durant, S., Ambati, M., Sun, C., Zhang, X., 2004a, J. Vac. Sci. Technol. B **22**, 3475.
Srituravanich, W., Fang, N., Sun, C., Luo, Q., Zhang, X., 2004b, Nano Lett. **4**, 1085.
Sun, W.X., Shen, Z.X., 2003, J. Opt. Soc. Am. A **20**, 2254.
Tarun, A., Daza, M.R.H., Hayazawa, N., Inouye, Y., Kawata, S., 2002, Appl. Phys. Lett. **80**, 3400.
Van Bladel, J., 1995, Singular Electromagnetic Fields and Sources, IEEE, Oxford.
Wurtz, G., Bachelot, R., H'Dhili, F., Royer, P., Triger, C., Ecoffet, C., Lougnot, D.J., 2000, Jpn. J. Appl. Phys. **39**, 98.
Yin, X., Fang, N., Zhang, X., Martini, I.B., Schwartz, J., 2002, Appl. Phys. Lett. **81**, 3663.
Zayats, A.V., Smolyaninov, I., 2003, J. Opt. A: Pure Appl. Opt. **5**, S16.
Zeleni, J., 1932, Science **79**, 269.

Chapter 8

Apertureless near-field microscopy of second-harmonic generation

by

A.V. Zayats

Nano-optics and Near-field Spectroscopy Group, Centre for Nanostructured Media, IRCEP, The Queen's University of Belfast, Belfast BT7 1NN, UK

Tip Enhancement
Advances in Nano-Optics and Nano-Photonics
ISSN: 1871-0018
S. Kawata and V.M. Shalaev (Editors)
© *2007 Elsevier B.V. All rights reserved.*
DOI: 10.1016/S1871-0018(06)01008-9

Contents

§ 1. Introduction

In addition to conventional scanning near-field optical microscopy (SNOM) with coated and uncoated fibre tips, continuous attempts to achieve truly nanoscale optical resolution have been recently focused on the so-called apertureless (or scattering-type) SNOM (for the recent reviews see, e.g., Richards and Zayats, 2004; Patane et al., 2004). One of the main motivations for apertureless SNOM (ASNOM) was to avoid the use of a subwavelength aperture which has small optical throughput and is difficult to fabricate in a controllable and reproducible manner at the scale of a few tens of nanometers. However, it has later appeared that the use of apertureless techniques has an additional bonus and allows to access a new class of optical phenomena dubbed as tip-induced effects.

In ASNOM, light interaction with a nanoscale object placed in the near-field vicinity to the surface under investigation is used to image its optical properties when the object is scanned across the surface (fig. 1). In addition to the improved spatial resolution, such an approach provides new conditions under which the optical processes are investigated since the probe tip introduces, in general, strong perturbation of the electromagnetic near-field around the studied object. This provides the possibility to enhance sensitivity to various optical processes by modifying the local electromagnetic field which is responsible for these processes. Tip-induced effects in apertureless SNOM include the enhanced Rayleigh

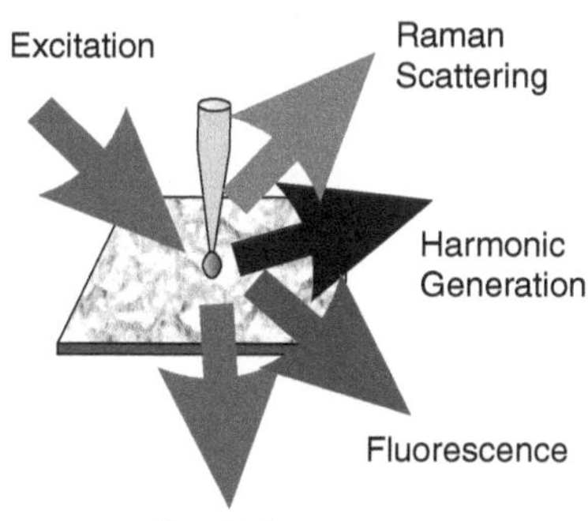

Fig. 1. Apertureless scanning near-field microscopy: possible detections modes.

and Raman scattering, one- and two-photon fluorescence, harmonic generation, modification of the relative multipole contribution in the optical properties of the atom and molecules, quantum electrodynamic effects in the optical properties that takes place in the strong electromagnetic fields.

The initial approaches to appertureless SNOM were based on the enhanced local scattering (Zenhausern et al., 1994; Bachelot et al., 1995 Keilmann and Hillenbrand, 2004) and local fluorescence (Lewis and Lieberman, 1991; Michaelis et al., 2000). In the former case, a nanoscopic scatterer at the extremity of a tip is employed to probe the distribution of the electromagnetic field in the near-field proximity to the surface under investigation, and the light that is locally scattered in the far-zone is detected. The small size of the scatterer required for achieving a good resolution results, however, in a low polarisability and, as a consequence, a low scattering signal. Metallic tips, silicon cantilevers, and metal-coated fibre tips have been used as probe tips. A major drawback of this technique is the poor signal-to-background ratio since the global scattering of the illumination light by the surface and the tip shaft is many orders of magnitude larger than the local tip-induced scattering of interest. Spatial resolution of scattering-type SNOM has been reported of the order of tens of nanometers. Optical contrast and image–object relation are determined by the electromagnetic interaction between the tip and sample and have a complicated nature (Zayats, 1999) for practically used systems because the field enhancement effects (Moskovits, 1985) that have nontrivial, complex dependencies on surroundings, are usually used to achieve better signal-to-background ratio. To eliminate the background scattering it is very helpful to use an optical signal at a wavelength different from that of the excitation light as is the case for probes made of nanoscopically confined fluorescent materials. A fluorescent nanoprobe can be implemented using dyes, semiconductor quantum dots, fluorescent polymers, single molecules, etc. but these are in many cases affected by bleaching.

To overcome some of these problems it has been recently suggested to take a nonlinear approach to ASNOM based on two-photon excited fluorescence (Kawata et al., 1999; Sanchez et al., 1999), second-harmonic generation (SHG) (Zayats and Sandoghdar, 2000, 2001) and Raman scattering (Stöckle et al., 2000; Hayazawa et al., 2001; Hartschuh et al., 2003). These techniques make use of the electromagnetic field enhancement at a probe tip to achieve a local enhancement of nonlinear processes in the vicinity of the tip. The signal related to nonlinear optical processes is determined by the super-linear dependencies on the intensity of the driving field (the fourth power of the local fundamental field in the case of SHG). This makes such techniques much more sensitive to the field distribution over a surface than apertureless SNOM based on linear optical processes. Not only the magnitude of the signal but also the spatial resolution of ASNOM is related to the field enhancement, since it determines the effective surface area most strongly contributing to the measured signal. As discussed by Kawata et al. (1999), in contrast to the scattering or conventional fluorescence, the light associated with nonlinear processes can be truly confined (falling faster than $1/r^3$ with a distance r from a tip apex) even without localised surface plasmon excitation.

An approach based on SHG is advantageous because there are almost no restrictions on tip and sample materials since all materials exhibit surface SHG. Only one of a tip or a sample needs to possess strong SHG to provide second-harmonic ASNOM imaging. If the nonlinear susceptibility of the sample is higher than that of the probe tip, the technique of apertureless second-harmonic SNOM allows one to probe the local second-harmonic field generated at the surface (a nonlinear background-free analogue of apertureless SNOM based on local scattering). In the opposite case, when the nonlinear susceptibility of the sample is smaller than that of the tip, one can achieve a strongly confined second-harmonic light source at the tip apex with properties dependent on the topographic and dielectric properties of the surface under investigation. From an experimental point of view, an apertureless SHG SNOM avoids some of the technical problems of both fluorescent and scattering-type ASNOMs: it will maintain advantage of a fluorescent ASNOM as the signal is spectrally isolated from the excitation light while not affected by bleaching and aging problems and tip formation as, e.g., metallic tips commonly used in scanning tunnelling microscopy or metal-coated cantilevers used in atomic force microscopy can be employed.

This paper presents an overview of apertureless probing of near-field SHG and its applications. After introducing the SHG and SNOM approaches to its studies in Section 2, the principles of the tip-induced SHG will be discussed in Section 3. The special cases of the SHG at the tip in the near-field proximity to the surface and the tip-enhanced surface SHG will be considered. Experimental configuration of the apertureless SHG SNOM is presented in Section 4 where the surface-related SHG probing is discussed. Section 5 presents the studies of the SHG enhancement at individual conically shaped objects, and Section 6 is devoted to the investigations of the properties of the local SHG source at the metallic tip apex. Finally, in Section 7, the applications of SHG ASNOM in imaging of magnetic and ferroelectric domains are described.

§ 2. Second-harmonic generation imaging with SNOM

Optical SHG is a nonlinear process leading to a partial transformation of the incident electromagnetic field at the frequency ω in the second-harmonic field at the frequency 2ω after interaction with material system (Shen, 1984):

$$E_i(2\omega) = \chi^{(2)}_{ijk} E_j(\omega) E_k(\omega) + \cdots , \tag{2.1}$$

where $\chi^{(2)}_{ijk}$ is the second-order nonlinear susceptibility of the material, and $\mathbf{E}(\omega)$ is the electric field of the incident or generated wave. Symmetry considerations show that this process is forbidden in centrosymmetric media in the dipole approximation. In this case, the only remaining source of nonlinearity is a surface interface where inversion symmetry is broken. Thus, SHG is very sensitive to surface states down to atomic scales, including topography variations, surface defects and adsorbates, magnetic and ferroelectric domains (see, e.g., McGilp, 1995, and references therein).

The recent advances in SNOM has opened up the possibility to study various optical phenomena with a resolution well below the diffraction limit (Richards and Zayats, 2004). The studies of near-field nonlinear processes were proposed by Vigoureux et al. (1994) and Zhao and Kopelman (1995). The first experimental observations of SHG using near-field optical microscopy was demonstrated by Smolyaninov et al. (1997a, 1997b) followed by numerous applications of this technique for studies of metal surfaces and functional materials, in chemistry and biology (for the recent review see Zayats and Smolyaninov, 2004, and references therein).

Most frequently a SHG SNOM with a fibre tip is used in the so-called collection mode when the SH excitation is achieved by direct illumination of a surface and the SH light is collected with a fibre tip (Zayats et al., 1999; Zayats and Smolyaninov, 2004) as shown in fig. 2(a). This mode is often used because the SHG requires high-intensity excitation light, which is achieved by a focused laser beam. In many experiments, femtosecond Ti:sapphire lasers are used, which provide high excitation power at low pulse energy, thus effectively generating SHG but introducing low heating and sample damage. Shear-force distance regulation is conventionally used to keep a tip in a near-field proximity and at a constant distance from the surface, thus providing simultaneous topographic imaging of the surface under investigation.

In most aperture-based SHG SNOM experiments, SH light is collected with uncoated fibre tips. The main disadvantage of a metal-coated fibre

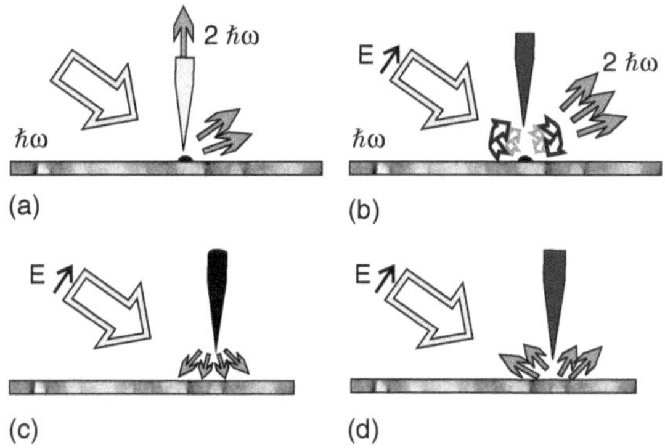

Fig. 2. Schematics of the near-field SHG measurements (a) in SH collection mode, (b) apertureless SHG SNOM: the electromagnetic interaction between the tip and surface leads to the SHG modification in the presence of the tip, (c) SHG from the tip apex, and (d) surface SHG enhanced by the probe tip.

tip in nonlinear optical measurements is a strong perturbation introduced by the tip in the local electromagnetic field distribution and SHG from the metal coating. Uncoated fibre tips introduce a much smaller perturbation since the dielectric constant of a fibre is much less than the one of a metal, and the effects related to the modification of the SHG by a probe tip can be neglected in many cases. However, even for fibre (glass) tips there can be some degree of the electromagnetic field enhancement associated with a tip apex (Kawata et al., 1999). This is crucial in near-field second-harmonic measurements for which the signal depends on the fourth power of the local excitation field. Owing to the field enhancement, the surface area just below the tip dominates the detected SH signal. As a result, the resolution of the SHG SNOM can be theoretically estimated to be as small as ~ 10 nm for 50 nm radius of a fibre tip apex if used to image materials with high dielectric constant and specific form of the nonlinear susceptibility tensor (Smolyaninov et al., 2001a). In this sense, one can consider this situation as "apertureless" SNOM with tip-enhanced optical processes and the signal scattering into the fibre instead of the free-space far-field (Smolyaninov et al., 2000, 2001b).

In fact, more severe limitations on the resolution of a SHG SNOM are imposed by the small number of SH photons collected by the tip and the related shot-noise and thermal drifts due to long-signal acquisition times. The high repetition rate lasers allow the image acquisition time to be kept in a reasonable range even at a low number of detected SH photons per pulse. The characteristic SH photon count rate in SNOM measurements is on average of the order of one SH photon count per 100–300 laser pulses. However, a long-signal averaging time which would help to register more photons is not desirable due to possible drifts in the experimental apparatus. For this reason, the resolution of the SHG SNOM is in many cases limited by a low detected SH signal rather than the geometry of a probe tip and determined by a trade off in the choice of SH signal acquisition time.

In the SHG measurements with uncoated fibre tips, the experimentally achievable spatial resolution is usually about 80–100 nm, however, in the case of the polarisation measurements, the SNOM resolution with bare fibre tips down to 40 nm was reported which is related to the strong field enhancement in the vicinity of piezo-electric materials (Smolyaninov et al., 2001a). The resolution and perturbation introduced by a fibre tip depends on the sharpness and size of the taper via the tip-induced field enhancement.

Apertureless second-harmonic SNOM makes total use of the effect of the field enhancement at a metallic tip, which is significantly stronger than in the case of a fibre tip, since permittivity of metal is much higher than that of glass. This leads to much stronger light confinement and better spatial resolution (Zayats and Sandoghdar, 2000, 2001). In this case, the tip-induced field enhancement effects at both fundamental and second-harmonic frequency result in SHG being strongly dependent on the tip presence. The scattered SH light in this case is detected in the far-field (figs. 2(b)–(d)). The origin of the SHG can be the nonlinear properties of the tip that are modified by the near-field electromagnetic interaction with the surface, the nonlinear optical properties of the surface that are modified by the tip, or the nonlinearities related to both the tip and the surface that are coupled via near-field interactions.

§ 3. SHG in the presence of a probe tip

To understand imaging mechanisms of the apertureless second-harmonic SNOM, let us consider SHG processes in the a probe-tip/surface system (fig. 2(b)). The generation of the second-harmonic signal in the apertureless SHG SNOM can be considered taking into account the electromagnetic coupling between the surface and probe tip at both fundamental and second-harmonic frequencies (Zayats and Sandoghdar, 2000, 2001). The electromagnetic interaction between the tip and the surface results in the establishment of the local field at the fundamental frequency. This local field generates the second-harmonic field due to the second-order nonlinear susceptibility of the surface and/or tip. The SH field in turn induces the coupling between the tip and the surface field at the frequency 2ω that leads to the local SH field re-distribution. The local SHG field is responsible for the SH radiation detected in the far-field.

Before discussing the self-consistent numerical calculations, let us understand the physical principles of second-harmonic apertureless SNOM. Although the tip as well as the sample could be made of SH-active material, for the sake of simplicity, we will assume that the SH signal discussed in the following considerations is generated by a layer of randomly oriented nonlinear molecules that cover the probe and/or the sample. If the near-field interaction between the surface inhomogeneities and the probe tip plays the dominating role, and if the dimensions of the tip–surface junction are small compared with the distance to the observation point, we can neglect retardation effects, and the SH intensity measured in the far field can be expressed as (Zayats and Sandoghdar, 2000, 2001)

$$I(2\omega) \propto \left| \sum_i \chi_i^{(2)} \alpha_i^{\text{eff}}(2\omega)(\alpha_i^{\text{eff}}(\omega))^2 + \chi_T^{(2)} \alpha_T^{\text{eff}}(2\omega)(\alpha_T^{\text{eff}}(\omega))^2 \right|^2 E^4(\omega), \quad (3.1)$$

where $\chi_i^{(2)}$ and $\chi_T^{(2)}$ are the second-order nonlinear susceptibilities of the surface structure and the tip, respectively, and $\alpha^{\text{eff}}(\omega) = f(\alpha_T, \alpha_i, d_{ij}, \mathbf{E}(\omega)/E(\omega))$ is the linear effective polarisability that depends on the polarisabilities of the surface structure (α_i) and the tip (α_T) as well as their separations d_{ij} and the polarisation of the fundamental light. This effective polarisability takes into account the field enhancement effects at the ω and/or 2ω frequencies related to the field concentration around highly curved surfaces (the so-called lightning-rod effect (Landau and Lifshits, 1984)) as well as possible localised surface plasmon resonances when the incident electromagnetic field is in a resonance with the electron plasma oscillations in a metallic particle (Zayats et al., 2005).

It is common to use instead of the effective polarisability α^{eff}, the so-called field enhancement factor $L(\omega) = L_{LR}L_{SP} = |E_{\text{loc}}/E_0| \sim \alpha^{\text{eff}}(\omega)$ which shows the enhancement of the local electric field, E_{loc}, compared to the incident one, E_0. The field enhancement factor has a complex dependence on the dielectric properties of the surface and probe tip, tip shape, surface topography, and their mutual geometrical configuration, leading to nontrivial object–image relations (Zayats, 1999). It should be noted that such a complex behaviour exists in any kind of apertureless SNOM where the tip not only scatters (or produces) the near fields but also modifies the local surface polarisability due to the near-field electromagnetic interaction.

First, we will consider the second-harmonic intensity distribution around the apex of a perfectly conducting metallic conical tip (Zayats and Sandoghdar, 2000). The corrections due to the finite dielectric constants of the tip and sample materials as well as the tip–sample interaction can also be taken into account (Kawata et al., 1999). The geometrical field enhancement which is associated with the field concentration at the highly curved surfaces, is only weakly dependent on the excitation wavelength via the dielectric constant of the material as long as the electrostatic approximation is valid. However, one should take into account the contribution of the resonant field enhancement if the tip or the surface can support the excitation of surface plasmons. Owing to the fundamental and second-harmonic field enhancement at the extremity of a probe tip, the main source of SH light is situated either at the apex of the tip or at the surface region just underneath it. If a probe tip exhibits stronger second-order nonlinearity than a surface (or vice versa), the situation is simplified considerably. In the former case, the tip of ASNOM can be considered as a strongly confined source of second-harmonic light (nanoscale light source which properties and, thus, radiation characteristics) depends on the surface structure underneath it. In the latter case, SHG from the surface under investigation is modified by the probe tip.

3.1. SHG from a probe tip: a localized light source

In the case, when the nonlinear susceptibility of the tip is much higher than the one of the surface, $\chi_i^{(2)} << \chi_T^{(2)}$, the signal of SHG ASNOM is determined by the second term in eq. (3.1):

$$I(2\omega) \propto |\chi_T^{(2)} \alpha_T^{\text{eff}}(2\omega)(\alpha_T^{\text{eff}}(\omega))^2|^2 E^4(\omega), \tag{3.2}$$

and the tip acts as a SH light source (fig. 2(c)). This situation is analogous to apertureless SNOM based on fluorescence from the tip where the SH light generated at the tip plays a role of nanoscopic light source. The field enhancement at the metallic conical tip dictates the spatial variation of the strength of the nonlinear response and, therefore, the attainable resolution.

Figure 3 shows the SH as well as fundamental intensity distributions when the tip apex is assumed to have a radius of 10 nm. For the tip with an opening angle of 10°, the SH intensity decreases by 10^3 times at the distance of 100 nm from the apex, at the same time, the fundamental light intensity decreases only by 10 times. For a 60° tip, the field confinement is

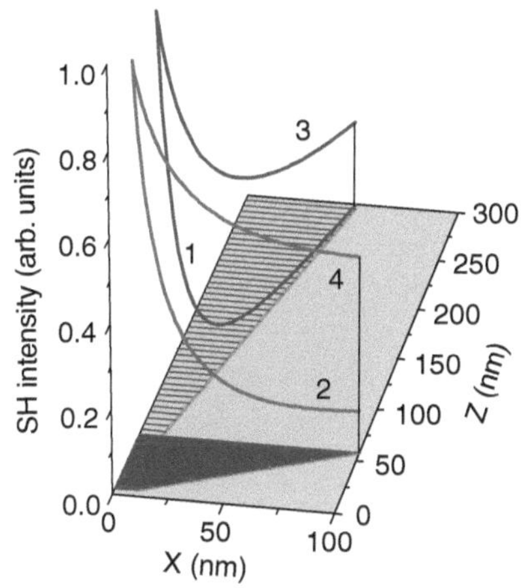

Fig. 3. Spatial variations of the SH (1,2) and fundamental (3,4) light intensities around a conical tip of 10° (1,3) and 60° (2,4) opening.

smaller: the SH and fundamental light intensities decrease at the 100 nm distance by 10 and 2 times, respectively. Thus, the SH field is much stronger confined to the probe tip than the fundamental field and, therefore, allows the higher spatial resolution to be achieved in optical studies. Labardi et al. (2004) experimentally imaged the fundamental scattering and SH radiation from metallised AFM tips and confirmed the predicted strong confinement of the SH light from the free-standing metallic tips. Bouhelier et al. (2003) observed strong SHG confinement during the near-field interaction of the tip with the surface. In the latter case, depending on the polarisation of the fundamental light, the SH field distributions around the tip has complex structure since both transverse and longitudinal fields generated at the tip can be detected in this case.

Since the effective polarisability at both fundamental and second-harmonic frequencies $\alpha_T^{\text{eff}}(\omega)$ are sensitive to the variations of the topography and the dielectric properties of the surface under the tip, the second-harmonic light generated at the tip strongly depends on the properties of the surface and can be used to image and study the surface without the need of illuminating it. This can be advantageous for the studies of materials and objects that can not withstand high intensities of the illuminating light.

3.2. *Tip-enhanced surface SHG*

In the case when the tip nonlinearity is much lower than the sample nonlinearity, $\chi_i^{(2)} >> \chi_T^{(2)}$, the first term in eq. (3.1) plays the dominating role:

$$I(2\omega) \propto \left| \sum_i \chi_i^{(2)} \alpha_i^{\text{eff}}(2\omega)(\alpha_i^{\text{eff}}(\omega))^2 \right|^2 E^4(\omega). \tag{3.3}$$

The SH light in this case is generated at the surface (fig. 2(d)), and the tip is used to modify and examine this SH field distribution (via modifications of α_i^{eff} in the presence of the tip). This imaging mode is based on a similar principle as conventional scattering apertureless SNOM or the tip-enhanced two-photon fluorescence or Raman scattering: the optical response of the surface is enhanced due to the local electromagnetic field modification by the tip. However, this enhancement maybe stronger in the case of SHG than in some of the other techniques since both fundamental and SH field can be enhanced by the tip.

As shown in fig. 4 for different opening angles of a probe tip, the SH light generated at the surface covered with nonlinear molecules is confined under the tip for the all tip parameters studied. The same as in the previous case, the tip provides stronger confinement of SH field compared to the fundamental field. Even for the tips with 60° opening, the SH light is strongly confined, while the half-width of the fundamental light intensity distribution is already twice the size of the tip.

An increase of the separation between the tip and the surface influences the absolute value of the enhancement but has a little effect on the relative spatial distributions of the light. This is inline with the near-field SHG in a mesoscopic nonlinear pad which has been shown to be much stronger confined than the fundamental light (Li et al., 1999).

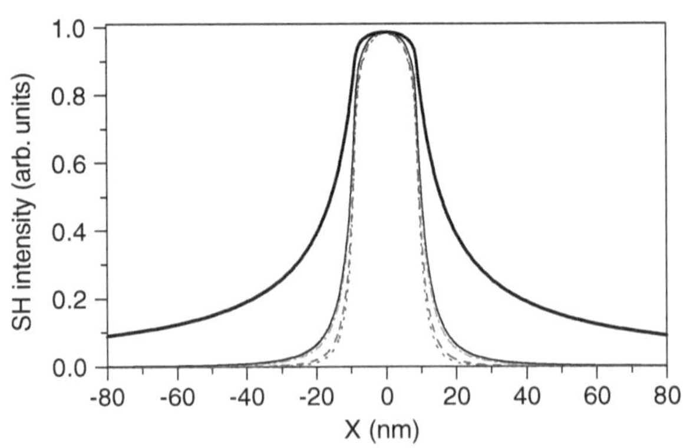

Fig. 4. Surface SHG intensity variations in the presence of a conical metallic tip of 10°, 20°, 45°, and 60° opening (dot, dash, dash-dot, and thin solid lines) placed 10 nm above the surface. Thick solid line shows the fundamental light intensity variation across the surface induced by the 60° tip.

3.3. *Self-consistent model of second-harmonic ASNOM*

Having discussed some aspects of the field confinement for the fundamental as well as the SH field components and two operational modes of apertureless SHG SNOM, let us now present a self-consistent model of near-field SHG. The generation of the second-harmonic signal in the apertureless SH SNOM can be formulated in a self-consistent manner using the dipole approximation and by taking into account the electromagnetic coupling between the surface and the probe tip (Zayats and Sandoghdar, 2000, 2001) similar to the calculations of SHG from randomly structured surfaces (Shalaev, 2000). Although the dipole approximation has its restrictions in the near-field, its modification allows the multipole contributions to be included effectively in the electric dipole formalism (Shalaev, 2000). In this model the fundamental light illuminating the tip–surface region induces the electric dipoles related to the surface objects as well as at the tip. A local field at the fundamental frequency is established due to the electromagnetic interaction among all the dipoles involved. Second-harmonic is generated by other dipoles associated with the particles driven by the local fundamental local field at the sites of the dipoles. Finally, taking into account the electromagnetic coupling between all the SH dipoles, one obtains the local second-harmonic field distribution which is the origin of the far-field SH signal detected in the experiment.

For simplicity we restrict ourself to the case of a metal surface and a metallic probe tip. We consider a structured surface as a set of spheroidal metallic particles placed on an ideally flat surface and another metallic spheroid as a probe tip with a major axis along its shaft. A spheroid has three electric dipoles associated with its axes. The one parallel to the tip shaft and perpendicular to the surface is most important as it interacts strongly with a surface. The local fundamental field at the dipole sites can be calculated using the well-known self-consistent formalism (Shalaev, 2000):

$$\mathbf{E}_{\rm loc}^{(\omega)}(\mathbf{r}_i) = \mathbf{E}_0^{(\omega)}(\mathbf{r}_i) - \varepsilon_0\omega^2 \sum_j \overleftrightarrow{G}_\omega(\mathbf{r}_i, \mathbf{r}_j)\left[\overleftrightarrow{\alpha}_j^{(\omega)}\mathbf{E}_{\rm loc}^{(\omega)}(\mathbf{r}_j)\right], \tag{3.4}$$

where ε_0 is the vacuum permittivity, ω is the fundamental light angular frequency, $\mathbf{E}_0^{(\omega)}(\mathbf{r})$ is the field which would prevail in space if the scattering sources were absent, and $\overleftrightarrow{G}(\mathbf{r}_j, \mathbf{r}_i)$ is the Green function which describes the field propagation from the ith dipole at location $\mathbf{r}_i$ to the point $\mathbf{r}_j$ taking into account also reflection from a surface. Finally, $\overleftrightarrow{\alpha}_i^{(\omega)}$ is the linear polarisability of the ith dipole which bears a local field $\mathbf{E}_{\mathrm{loc}}^{(\omega)}(\mathbf{r}_i)$. The solution of eq. (3.4) determines a local fundamental field at the dipole sites. This is a local fundamental field that gives rise to the signal detected in apertureless SNOM with a tip positioned in point $\mathbf{r}_i$.

The local fundamental field $E_{\mathrm{loc}}^{(\omega)}$ drives the nonlinear dipole to radiate second-harmonic field:

$$\mathbf{E}_0^{(2\omega)}(\mathbf{r}_i) = \varepsilon_0 \overset{\leftrightarrow}{\chi}{}_i^{(2)} : \mathbf{E}_{\mathrm{loc}}^{(\omega)}(\mathbf{r}_i)\mathbf{E}_{\mathrm{loc}}^{(\omega)}(\mathbf{r}_i), \tag{3.5}$$

where $\chi^{(2)}$ is the second-order nonlinear susceptibility of the metal or nonlinear molecules adsorbed on a sphere surface. Taking into account that "second-harmonic" dipoles interact via second-harmonic field, we can self-consistently calculate local second-harmonic field at the sites of the dipoles in the same way as for the fundamental field (eq. 3.4):

$$\mathbf{E}_{\mathrm{loc}}^{(2\omega)}(\mathbf{r}_i) = \varepsilon_0^{(2\omega)}(\mathbf{r}_i) - \varepsilon_0(2\omega)^2 \sum_j \overset{\leftrightarrow}{G}_{(2\omega)}(\mathbf{r}_i, \mathbf{r}_j)\left[\overset{\leftrightarrow}{\alpha}{}_j^{(2\omega)} \mathbf{E}_{\mathrm{loc}}^{(2\omega)}(\mathbf{r}_j)\right]. \tag{3.6}$$

By solving this equation analogously to eq. (3.4), we can obtain the second-harmonic field distribution that takes into account both fundamental and second-harmonic interactions in the system. This local second-harmonic field gives rise to a signal detected with an apertureless SHG SNOM.

To model a simple experimental situation, we consider a glass surface with a silver spherical particle as a test sample (fig. 5). The tip employed to probe second-harmonic field generated from the sample is treated as a silver sphere covered with randomly oriented SH active molecules. Second-harmonic generation from silver is assumed to be negligible compared to the second-harmonic from nonlinear molecules, $\chi_{\mathrm{Ag}}^{(2)} << \chi_{\mathrm{mol}}^{(2)}$. The excitation light of $\lambda = 740\,\mathrm{nm}$ which is typical wavelength of a Ti:sapphire laser is p-polarised with respect to the surface. To avoid the interplay between different polarisation components of the fundamental field, we consider the illumination parallel to the sample surface so that

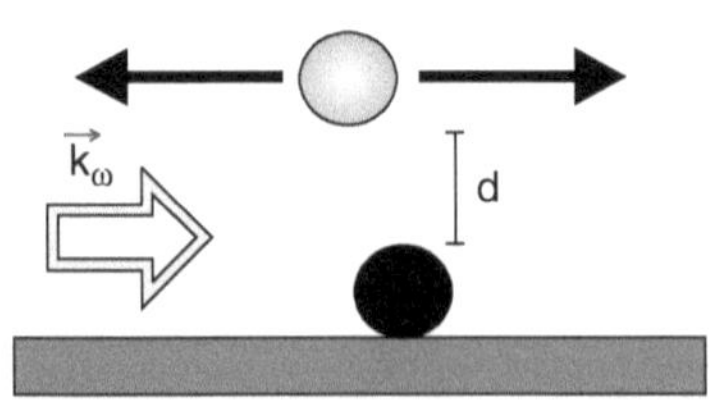

Fig. 5. Schematic of the numerical simulations: the sample is modelled with a silver sphere placed onto a dielectric surface; the tip is a silver sphere covered with nonlinear molecles.

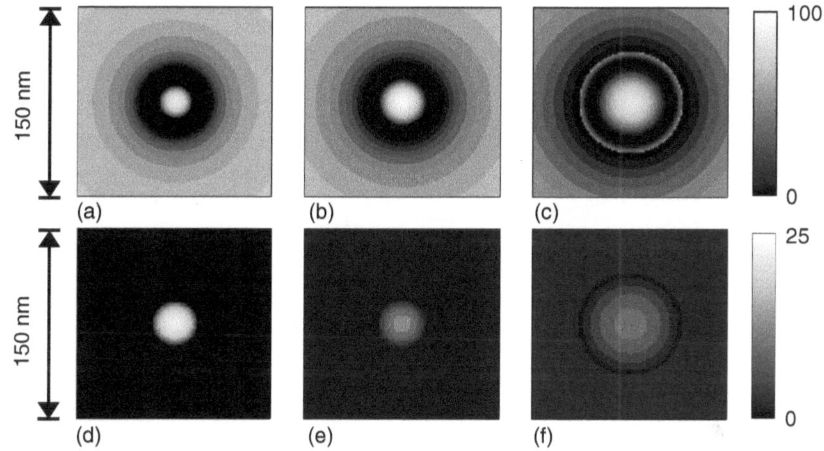

Fig. 6. (a,b,c) Second-harmonic and (d,e,f) fundamental light intensity distributions for the tip–surface separations of (a,d) 1 nm, (b,e) 7 nm, and (c,f) 20 nm. The object is a 10 nm radius Ag sphere placed on a glass surface, and the tip is a 10 nm Ag sphere covered with nonlinear molecules (fig. 5). The distances between spheres are measured from their surfaces. Excitation light ($\lambda = 740$ nm) is p-polarised with respect to a surface.

there is no electric field component parallel to the surface (to avoid shadowing effects, this can be achieved in the experiment, e.g., with illumination under the total internal reflection conditions). The dielectric constants of silver at the fundamental and second-harmonic frequency are $\varepsilon_{Ag}(\omega) = -22.8 + 1.35i$ and $\varepsilon_{Ag}(2\omega) = -2.06 + 0.48i$, respectively.

The images in fig. 6 are calculated for the constant-distance scanning mode so that the probe tip follows the surface topography. As one can see from the images, the fundamental and second-harmonic intensity distributions both reflect the structure of the sample. With an increase of the tip–surface distance the fundamental light intensity (and the contrast of the images) drops significantly, and respective images become broader. At the same time, the maximum of SH intensity is observed at the scanning distance of about 7 nm from the surface when the excitation of second-harmonic surface plasmon related to the tip-defect system is achieved (fig. 7). The SH images are subjected to widening much less than the fundamental light images due to the confinement of the SHG mediated by a probe tip. The SH image contrast variations are about 4 times greater than those of the images obtained with the scattered light. It should be noted that in a real experiment the contrast variation of the image obtained with scattered fundamental light will be additionally affected by the background due to light scattering from the surface and tip shaft. Under the excitation with s-polarised light, both SH

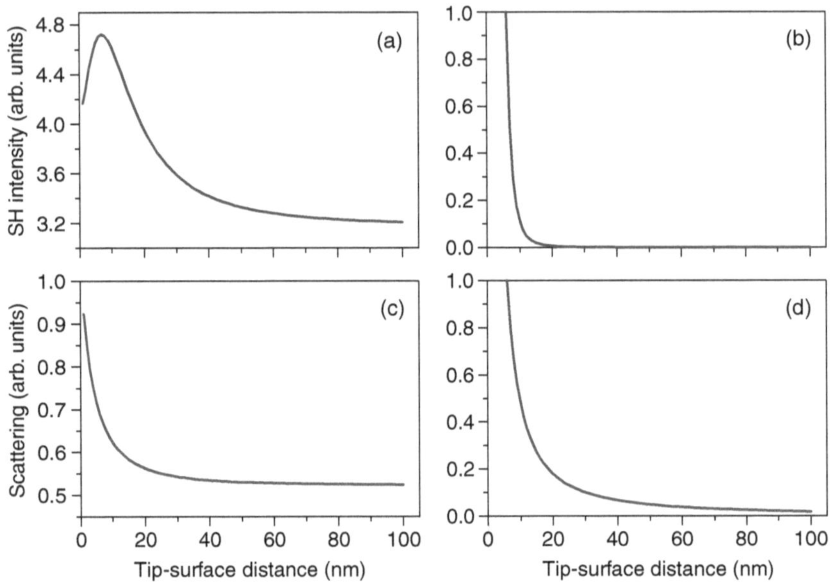

Fig. 7. Tip–surface distance dependencies of the intensity of (a,b) SHG and (c,d) scattered fundamental light obtained with the tip positioned above the object. (a,c) The model configuration is the same as in fig. 5. (b,d) The perfectly conducting ($Re\varepsilon = \infty$) conical tip with 20° opening. The intensity scales are internal, not to be compared.

and fundamental light images have complex structures related to an interplay between the components of the excitation and SH fields parallel and perpendicular to the tip-defect axis during scanning. This results also in the reverse contrast of the SH images because the s-polarised excitation light has no components of the electric field in the direction of axis between the sphere centres, and therefore, there is no SH enhancement when a probe tip is positioned above a defect. When a tip is displaced from the central position, the excitation field has such a component leading to the SH enhancement for certain scanning directions. The role of fundamental light polarisation on the SHG images in the system of interacting metallic objects was studied in detail by Bozhevolnyi et al. (2003).

The tip–surface approach curves in the case of p-polarised excitation (figs. 7(a) and 7(c)) are calculated in the configuration when the tip is positioned above the object. The SH signal reveals distinct resonant behaviour related to the excitation of localised surface plasmon in the system of coupled spheres used as the tip and defect in the model calculations. For complex surface structures consisting of ensembles of particles, the resonant electromagnetic field enhancement will play an even more significant role because the value of an enhancement can be stronger and resonant conditions can be reached more easily. This, together with the more complex nature of the interplay between the components of excitation and SH field parallel and perpendicular to the tip-defect axis, will create more difficulties for interpreting the object–image correspondence (Zayats, 1999).

If the system does not exhibit the resonant enhancement, as can be in the case of perfectly conducting metallic tip ($Re\varepsilon = \infty$), the distance dependence of the SHG and fundamental signal is determined by the geometrical enhancement at the tip apex and between the defects and/or tip. For the tip of a conical shape placed over a surface covered with nonlinear molecules such distance dependence is governed by a power law (Zayats and Sandoghdar, 2001) and can be steeper (depending on the angle of a cone opening) than the one for the resonant conditions but the absolute value of the SH enhancement is smaller (figs. 7(b) and 7(d)). The distance dependencies of the fundamental light scattering and SHG confirm a stronger confinement of the SH light in this case as is to be expected.

The distance dependencies of the field confinement at the metallic tips have been numerically studied in detail by Demming et al. (2005) in the context of Raman scattering and by Laroche et al. (2005) for both fundamental and SH light. Both models have shown the distance dependence of the localised surface plasmon resonance. It was also shown that the behaviour of the fundamental and SHG enhancement factor strongly wavelength dependent and can exhibit opposite tendencies (the increased or decreased scattered fundamental and SH intensities) with the increased tip–surface interaction.

§ 4. Second-harmonic ASNOM: experimental realisation

A typical experimental setup for near-field SHG studies with ASNOM is shown in fig. 8. To achieve local excitation/detection of SHG, a near-field optical microscope is built on top of an inverted optical microscope. A silver-coated adiabatically tapered fibre tip is fixed on a quartz tuning fork attached to the piezo-tube which is used for shear-force distance regulation. This arrangement provides constant tip–surface separation during scanning and, as the result, topography imaging. The coated tip is usually fabricated by exposing the fibre tip to 5 nm size colloidal gold and then to the silver enhancer solution (the technique commonly used in electron microscopy). The sample is fixed on a 2D-piezoscanner and scanned against the tip. The probe tip is positioned in front of an

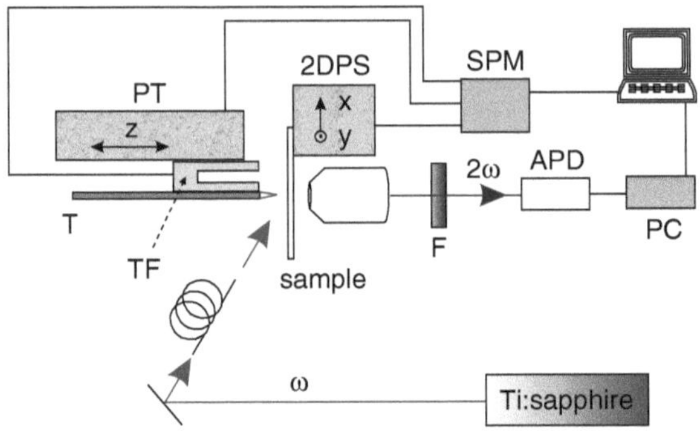

Fig. 8. Experimental setup of an apertureless SH SNOM in the direct illumination mode: (T) silver-coated adiabatically tapered fibre tip, (TF) quartz tuning fork, (PT) piezo-tube, (2DPS) 2D-piezoscanner, (F) interference filter, (APD) avalanche photodiode, (PC) photon counter, and (SPM) control electronics.

achromatic microscope objective with numerical aperture of 0.8 corresponding to an acceptance angle of about 106°. The SH light is separated by the interference filter and measured with the avalanche photodiode and the photon counter.

Light of the picosecond Ti:sapphire laser ($\lambda = 790$ nm, 82 MHz repetition rate) is coupled in the polarisation preserving single-mode fibre and used for the SH excitation at approximately 45° angle of incidence. Pulse duration after the fibre is about 5 ps. The excitation intensity on a sample surface is about 0.4–1 MW/cm^2. The polarisation of the excitation light can be controlled by rotating the output end of the fibre. The SH signal at every point of the image is averaged over 30 ms and typical count rates are of the order of one SH photon count per 100 excitation laser pulses. The measured SH intensity depends quadratically on the fundamental light power.

A metal film was used as a sample in order to ensure the local field enhancement in the metal-tip/metal-surface junction. The sample was the 30-nm-thick gold film thermally evaporated onto a glass substrate. The film surface reveals random topographic structure with height variations less than 20 nm (fig. 9).

The lateral distribution of second-harmonic signal measured with the presence of a silver-coated tip under the excitation with light having an electric field component parallel to the tip axis reveals complex behaviour depending on the defect structure but, in general, reflects the local topography (fig. 9). In this experimental realisation, the second-order nonlinear susceptibilities of the tip and sample are of the same order of magnitude, and, therefore, the observed SH light is generated at both the tip and the surface. Lateral variation of the SH signal is related to the

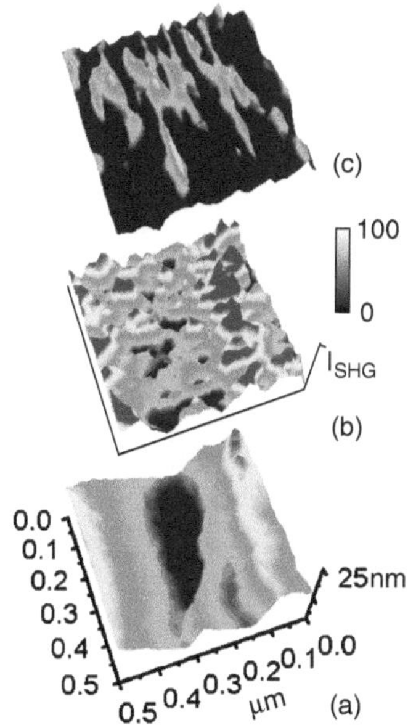

Fig. 9. The images of the rough gold film surface measured using apertureless second-harmonic SNOM: (a) topography, (b) SH light distribution measured with p-polarised excitation light, and (c) SH light distribution measured with s-polarised excitation light.

different degree of the electromagnetic field enhancement due to the interaction between the tip and surface defects.

In the areas of the film where the defect structure is simple, the SH images show good correlation with the field enhancement at the individual defects, e.g., with the groove on the surface. At regions with complex surface structure (an ensemble of ~ 50-nm-size defects at the distances of ~ 50 nm between each other), the SH image does not represent individual surface features but rather the local field distribution in the system of interacting defects (cf., eqs. (3.4)–(3.6)). It has been difficult to discriminate between geometrical and resonant field enhancement in the experiments (this can be done, e.g., by varying the excitation light wavelength). Although localised surface plasmons have not been especially excited in these experiments, at a rough surface where defects of all sizes and shapes are present, the excitation and SH light can always find a defect (or ensemble of defects) and a tip position where localised surface plasmons can be excited in the tip–surface junction (Zayats, 1999). These defects will dominate in the recorded SH images. The field enhancement depends dramatically on the geometry, size, and mutual position of the tip and surface defects. Slight variations of the defect shape may result in the

Table 1. Typical SH signal (in counts per 30 ms) measured using the silver-coated tip and gold surface for different polarisations of the excitation light with the probe tip in the near- and far-field. Background signal (BG) measured without sample originates due to SHG in the fibre and optical elements.

	NF	FF	BG
s-pol	6,000	5,000	3,000
p-pol	13,000	5,500	3,000

significant variations of the enhancement and, as a result, the observed SH signal. For this reason, the experiments involving resonant field enhancement (including scattering and fluorescent ASNOM) are difficult to interpret even for virtually "regular" surface structures that are indeed not regular on the nanometer scale.

To check the source of the nonlinear response and possible topographic artifacts, the polarisation and distance dependencies of the SH signal were studied. By changing the polarisation of the excitation light and the tip–surface distance, the electromagnetic interaction between the tip and the surface is modified and, hence, the SH intensity is also changed. It turns out that the contrast of the SH images disappears under the s-polarised excitation or when the tip is retracted in a far-field zone. Moreover, the overall intensity of the SHG depends strongly on the excitation configuration (Table 1). When the tip is far from the surface, the diffuse SH light from the metal surface has been detected as is expected for rough surfaces. In contrast, when the tip is in the near-field proximity from the metal surface, the change of polarisation from p to s results in a strong decrease of the SHG signal. An increase of the SH intensity with the tip approaching a surface is of the same order of magnitude as calculated for the model system (fig. 7).

Williame et al. (2005) have recently repeated these experiments on the visualisation of the field enhancement at rough metal surfaces using the apertureless SHG SNOM setup with metallised AFM cantilevers. Their results are in a very good agreement with the model calculations presented in Section 3.3 demonstrating much better resolution and higher sensitivity of the SHG measurements to the field variations across the surface compared to the fundamental light scattering.

§ 5. SHG enhancement at conical objects

Electromagnetic field enhancement and light confinement are closely related phenomena. It is the confinement that leads to the strong field

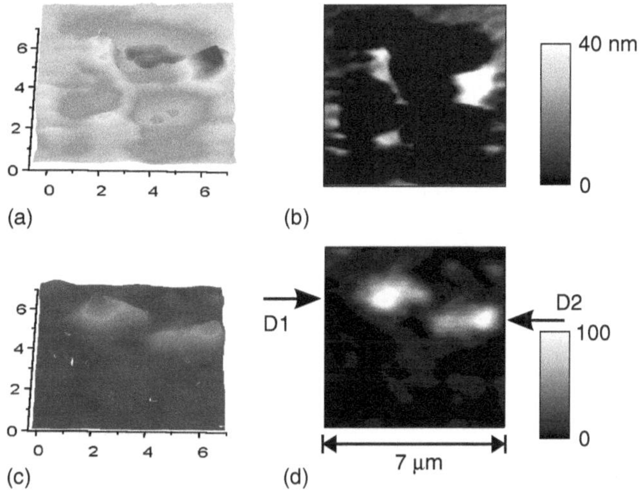

Fig. 10. (a,b) Topography and (c,d) corresponding SH images of the conical protrusions on the 30 nm gold film. 3D (a,c) and plane-view (b,d) images are presented. Arrows indicate the defects shown in fig. 11.

enhancement due to a small volume of the related electromagnetic modes. The enhancement of the incident field at a tip apex is one of the examples of a configuration leading to a strongly localised light source. The confinement can be estimated from the localisation size of surface plasmons at a tip apex and/or lightning-rod effect. Nonlinear techniques make use of the enhancement of the electromagnetic field at a probe tip to achieve a local enhancement of the second-order nonlinear processes at or in the vicinity of the conical tip.

The spatial variation of the SHG process across a conical object was studied using the protrusions on the surface of the thin gold film (Zayats et al., 2000). SHG was locally excited using the fibre-based SNOM, and the SHG enhancement dependent on the relative position of the SNOM tip with respect to the cone apex was studied. Figure 10 shows the SH intensity distribution over the area of the film with several pronounced conical protrusions of different size. The surface topography exhibits large smooth areas of the film with the height variations less than 10 nm. The height of the protrusions is about 30 nm while they have micron-sized bases. Such cones with large opening angle were chosen to facilitate SNOM measurements across the cone surface (SNOM measurements of smaller cones are limited by the size of the fibre tip aperture).

SHG is significantly stronger at these conical objects compared to the "flat" surface on which they are placed. Other surface defects also reveal themselves in the SHG images but the related SHG enhancement is much lower. The average SHG enhancement at this surface region recalculated from the data of local measurements is of about 1.2 times while the local enhancement reaches value up to 10 times even at the such "blunt" cones.

The dipolar localised surface plasmon resonances related to the protrusions are at about 0.9 eV, which is much lower than the energy of fundamental ($\hbar\omega \approx 1.57$ eV) and second-harmonic photons. For such objects, only the geometrical field enhancement effect is important in the wavelength range of interest. The local electric field close to the apex of a conical protrusion can be estimated in this case using the electrostatic approximation as

$$E_{\rm loc} \sim 1/r^{1-\nu} E_z^{(0)}, \tag{5.1}$$

where $E_z^{(0)}$ is the component of the electric field parallel to the cone axis near the plane surface far from the defect, r is the distance to the cone apex, and ν is the first root of the equation $P_\nu(\cos(\alpha)) = 0$, where P_ν and α are the Legendre function of the first kind and the cone opening angle, respectively. Depending on the opening angle, the parameter ν can be in the range 0–1 for an ideal metal (Landau and Lifshits, 1984). The corrections due to a finite dielectric constant of a material as well as the influence of surroundings have been extensively discussed by Kawata et al. (1999). This electrostatic model gives a good agreement with the experimentally measured SH intensity distributions (fig. 11) for the parameters of the cones obtained from the topographic images. For the cone D2 with the opening angle slightly larger than that of the cone D1, the enhancement and, thus, local SHG is smaller, as should be expected. It should be noted that the finite resolution of the SNOM leads to averaging of the signal over the excitation area and, therefore, to the subsequent reduction of the measured enhancement at a cone apex compared to its true value.

§ 6. SHG from a metal tip apex

Let us consider the confinement and polarisation properties of the SHG light generated at the apex of a conical metal tip. Takahashi and Zayats (2002) studied these properties under the evanescent near-field fundamental illumination of the tip. Bouhelier et al. (2003), Labardi et al. (2004) and Neacsu et al. (2005) investigated these properties under the far-field illumination. A simple electrostatic model of the SH field confinement and properties based on the considerations of Zayats and

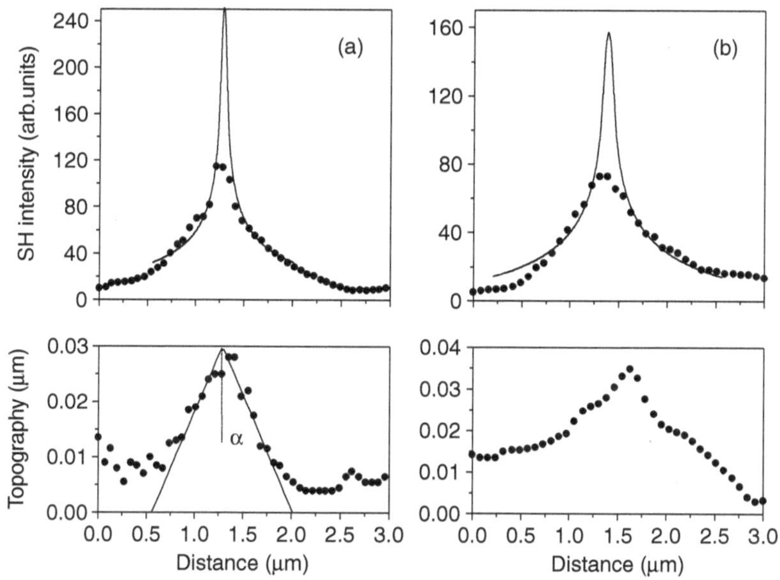

Fig. 11. Topography and SH intensity cross-sections indicated by arrows in fig. 10: (a) D1 and (b) D2 defects. Symbols represent the experimental data, solid lines are the model calculations for the lightning-rod effect.

Sandoghdar (2000, 2001) was briefly described in Section 3.1. Laroche et al. (2005) carried out detailed numerical modelling of the SH generation at the ASNOM tip and its polarisation and distance dependencies. These are the important questions since the field enhancement at the apex of a metallic tip will dictate the spatial variation of the strength of the nonlinear response and, therefore, attainable confinement and resolution. In the following, we will concentrate on polarisation and near-field interaction dependencies of the second-harmonic and the scattered fundamental light. In general, different topological features of a probe tip are responsible for scattering and SHG in the far- and near-field. The near-field interaction between a probe tip and a surface significantly modifies the electromagnetic field enhancement at the tip and results in the different distance dependencies of near-field scattering and SHG. The polarisation contrast obtained with second-harmonic signal significantly exceeds polarisation contrast of linear scattering from the tip.

The apertureless scanning near-field microscope for investigation of SHG properties of a metal tip apex is shown in fig. 12. Light of a femtosecond Ti:sapphire laser ($\lambda = 800\,\text{nm}$, $\tau = 300\,\text{fs}$) has been used for the SH excitation. The polarisation of the excitation light has been controlled with a $\lambda/2$-plate. By appropriate positioning of the prism, the gold tip apex can

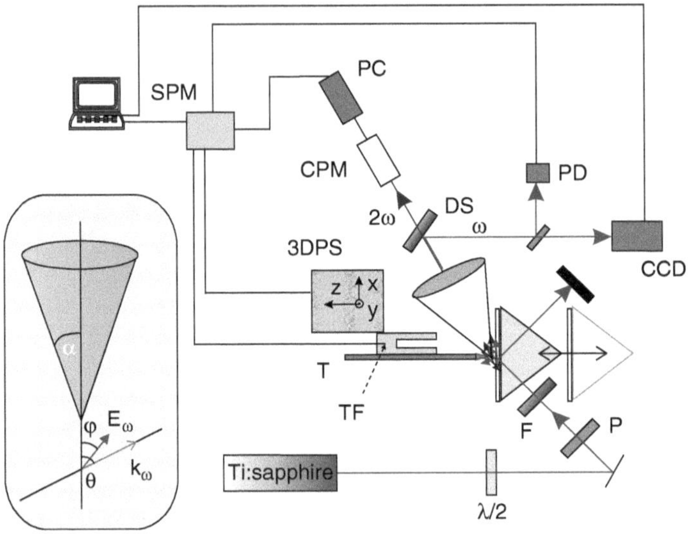

Fig. 12. Experimental setup of an apertureless SH SNOM in the evanescent field illumination mode: ($\lambda/2$) waveplate, (F) interference filter, (T) sharpened Au tip, (TF) quartz tuning fork, (P) polariser, (3DPS) 3D-piezoscanner, (DS) dichroic beam-splitter, (CPM) photon counting photomultiplier, (PC) photon counter, (PD) photodiode, (CCD) CCD camera, and (SPM) control electronics. The insert shows the geometry of the orientation of the fundamental light polarisation and wavevector with respect to the tip.

be illuminated either directly by laser light or through the glass prism in the total internal reflection configuration. In this way, far- or near-field excitation of SHG can be achieved at the same angle of incidence. In the near-field mode, the tip is kept close to a surface with a shear-force distance regulation based on a quartz-tuning fork. In this mode, only evanescent field (exponentially decaying from a surface) is present over the prism. This reduces the effects of light scattering by the tip shaft as only the part of the tip brought into the evanescent field is illuminated. The scattered fundamental and second-harmonic light are collected in the direction of about 70° to the direction of the tip axis with a lens of N.A. = 0.35. The dichroic beam-splitter is used to separate the fundamental and SH light which are then measured with the photo-diode and the photon counting photomultiplier, respectively. Thus, both scattered fundamental and second-harmonic light can be simultaneously measured.

The gold probe tips have been sharpened using conventional electrochemical etching. The use of a gold tip and a glass surface ($\chi_T^{(2)} >> \chi_s^{(2)}$) has ensured negligible nonlinear response of the surface compared to the tip. The measured SH signal from the gold tips has been of the order of

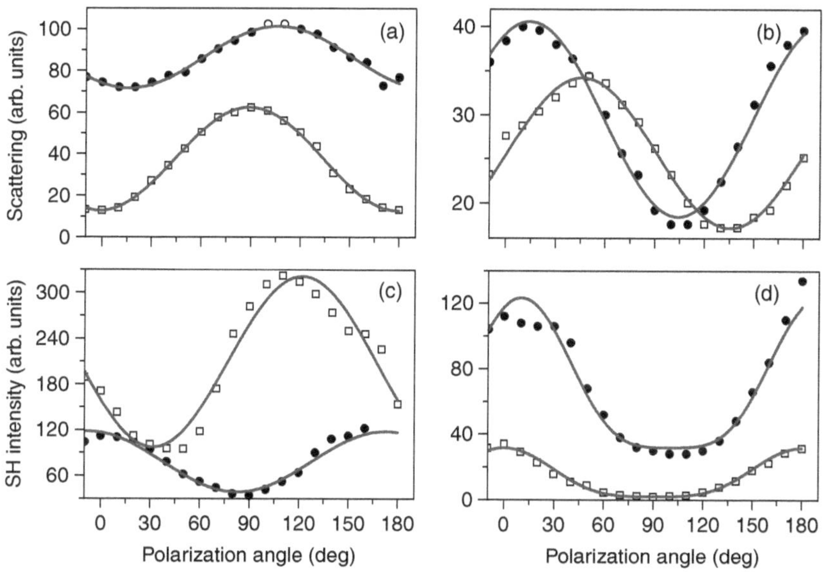

Fig. 13. Polarisation dependencies of the scattering (a,b) and SHG (c,d) from the tip apex under the far-field (a,c) and near-field (b,d) excitation: (symbols) experiment, (solid lines) model fits. The polarisation angle is measured with respect to the plane of incidence (fig. 12).

5,000 counts/s, the same order of magnitude as a signal measured in the experiments with apertureless SHG SNOM and from individual surface defects (Zayats et al., 2000; Zayats and Sandoghdar, 2001).

The typical polarisation dependencies of the scattering and SHG from a set of gold tips measured under the excitation in far- and near-field are presented in fig. 13. They reveal only two types of dependencies ($\cos^2\theta$ and $\cos^4\theta$) of the detected fundamental and SH intensities on the orientation of the fundamental light polarisation with respect to the tip axis. The optical properties of a sharp metal tip near a surface can be phenomenologically described in terms of the effective polarisabilities taking into account the electromagnetic field enhancement at the tip and tip–surface interaction (Section 3.1). Rigorous treatment of the electromagnetic field near a tip–surface junction has been recently achieved using the reciprocity theorem (Porto et al., 2003). This approach has been successful in explaining polarisation as well as frequency dependencies of linear scattering in apertureless SNOM. Adopting a model of a conical tip, the electromagnetic field distribution near the tip apex can be asymptotically expressed similar to eq. (3.2) as $\mathbf{E}(\omega) \sim \mathbf{L}(r, d, \phi, \alpha)\cos\theta$, where θ is the angle between the electric field of illuminating light of and the plane of incidence. $\mathbf{L}$ describes the field enhancement at a metal tip apex and depends on the angle of incidence ϕ, the distance from a cone apex r, the distance d from the tip apex to the surface, and the opening angle α of a cone. The tip–surface interaction results in the additional weak dependence on the polarisation angle but does not modify significantly the polarisation dependence of the scattered light intensity. Thus, the intensity of the far- and near-field scattering from a tip apex follows $I(\omega) \sim \cos^2\theta$ polarisation dependence. The measured polarisation dependencies of the scattering from the tip under the near- and far-field excitation are in good agreement with this model for all measured tips (figs. 13(a) and 13(b)). The shift of the angular position of the maxima is due to a slight inevitable misorientation of the tip with respect to the plane of incidence which is different for different tips in the experiment.

In the same model framework, SHG at a tip is induced by the tip-enhanced incident field $\mathbf{E}(\omega)$ followed by the enhancement of the second-harmonic field $\mathbf{E}(2\omega)$ at a cone apex: $E_i(2\omega) \propto \chi^{(2)}_{ijk} E_j(\omega) E_k(\omega)$, where $\chi^{(2)}_{ijk}$ is the second-order nonlinear susceptibility of the tip. The polarisation dependencies of the SH intensity are, in general, more complex than those of linear scattering as they depend on the details of the SHG at a metal tip and, therefore, on the form of $\chi^{(2)}_{ijk}$ tensor and the details of the tip shape (Neacsu et al., 2005). Nevertheless, even for an arbitrary $\chi^{(2)}_{ijk}$ tensor, the dominating SHG (intensity enhancement $L^4_\omega L^2_{2\omega}$) has $I(2\omega) \sim \cos^4\theta$ dependence on the polarisation of the excitation light, where L_ω and $L_{2\omega}$ are the field enhancement factors at the fundamental and SH frequencies, respectively. However, for some of the $\chi^{(2)}_{ijk}$ components, SHG with lower enhancement factors, for example, such as $L^2_\omega L^2_{2\omega}$ can also exhibit $\cos^2\theta$ and more complex polarisation dependencies.

In the experiments with the isolated tips (far-field excitation), the $\cos^4\theta$ dependence of the SH intensity has been observed for some of the investigated tips. However, for some tips, in addition to a phase shift of the SH polarisation dependencies from the respective scattering dependencies (different for different tips), the SHG exhibits $I(2\omega) \sim \cos^2\theta$ polarisation dependence (fig. 13(c)). This suggests that the dominating sources of the second-harmonic and scattering light are related to different dipoles with different orientations with respect to the tip axis (please note that the model polarisation dependencies have been discussed above for an ideal cone). The physical reason for this can be either complex shape of the tip apex or nanoscale protrusions (roughness) on the surface of the tip differently contributing to the field enhancement and SHG (Labardi et al., 2004; Neacsu et al., 2005).

The situation is simplified when the tip is in the near-field proximity to the surface. In this case, the strong electromagnetic coupling between the tip and the surface takes place. As one can see from figs. 13(b) and 13(d), the fundamental light scattering and SHG polarisation dependencies reveal the $\cos^2\theta$ and $\cos^4\theta$ behaviour, respectively, suggesting that the same dipole is responsible for both scattering and SHG in this case. These polarisation dependencies differ from the far-field dependencies due to the different dipoles dominating in the far- and near-field optical processes (in the latter case this probably is the one closest to a surface and bearing strongest interaction with it).

To investigate the origin of the observed tip–surface interaction, the distance dependencies of the scattering and SHG has been measured (fig. 14). Negligible variations of the scattering and SHG intensities have been observed under the s-polarised excitation (the electric field is perpendicular to the tip axis), while for the p-polarised excitation, the optical

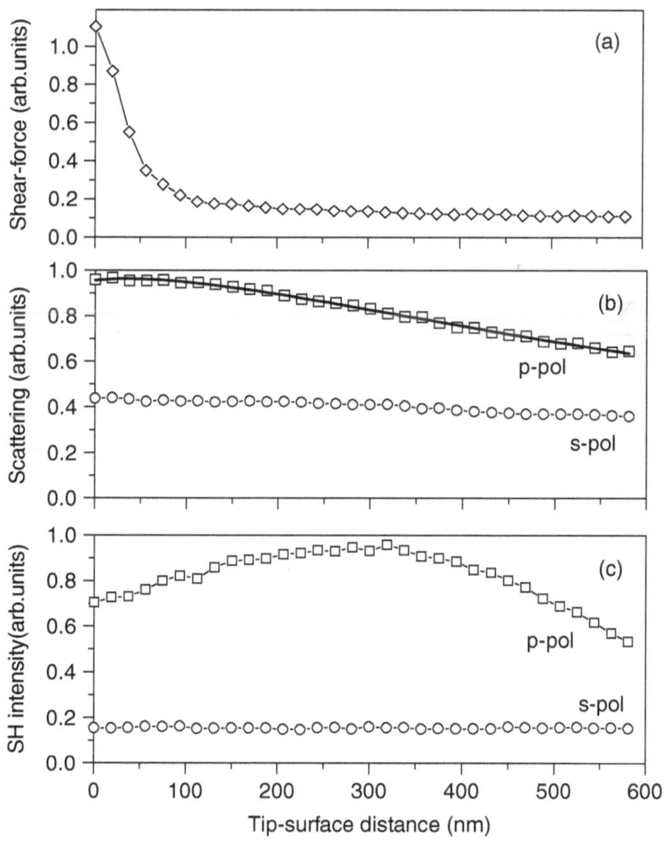

Fig. 14. Tip–surface distance dependencies of (a) shear-force signal, (b) scattering, and (c) SHG intensity for s- and p-polarisation of the excitation light. Thick solid line in (b) represents a fit in the dipole-image model.

signal depends significantly on the distance to the surface. This is expected because the influence of the field enhancement effects are most important for p-polarised light. The distance dependence of the scattering intensity (fig. 14(b)) is in agreement with the model taking into account the exponential decrease of the excitation field $E(\omega)$ from the surface and the changes of the tip polarisability resulting from the interaction with an image dipole induced by the surface presence. Such near-field interactions lead to the enhanced or suppressed scattering that depends on the dielectric constants of a surface and a tip as well as the distance between them. As a result, the distance dependence of the scattering intensity deviates from the exponential behaviour which is expected in the total internal reflection configuration. This deviation is observed at the distances smaller than 100 nm from the surface.

The near-field SHG distance dependencies are more complex, exhibiting maxima at about 250 nm from the surface which are absent in the scattering distance dependencies (fig. 14). Nonmonotonic dependencies of the near-field scattering as well as SHG have been predicted at much shorter distances from a surface ($\lesssim 100$ nm) related to the modification of the localised surface plasmon resonant conditions due to the interaction between the tip and surface (Zayats and Sandoghdar, 2001; Demming et al., 2005). On the another hand, the retardation effects are important at longer distances. The exact numerical simulations of Laroche et al. (2005) show that such dependence can be observed due to the interplay between the surface plasmon resonance modification and scattering efficiency at different tip–surface distances. Indeed, at some wavelengths the intensity of the scattered SH light drops when the metallic tip approaches the glass surface from 200 to 10 nm. Another possible origin of the measured SHG distance dependencies may be a change of dominating SH dipoles during the tip approach.

At longer distances ($\gtrsim 800$ nm) from the surface the oscillations in the SH signal related to the retardation effects at the SH wavelength have been observed (fig. 15). At the same time the scattering fundamental field reveals the simple exponential decay. This example illustrates the superior (compared to linear scattering) sensitivity of local SHG to the density of the electromagnetic states above the surface. The oscillations are related to variation of the electromagnetic density of states due to the presence of the surface and the tip and multiple scattering between them.

Polarisation and distance dependencies of the near-field scattering and SHG from a metal tip apex shows that the electromagnetic interaction between a probe tip and a surface results in much less influence of the tip–surface quality on the near-field SHG in contrast to the far-field SHG.

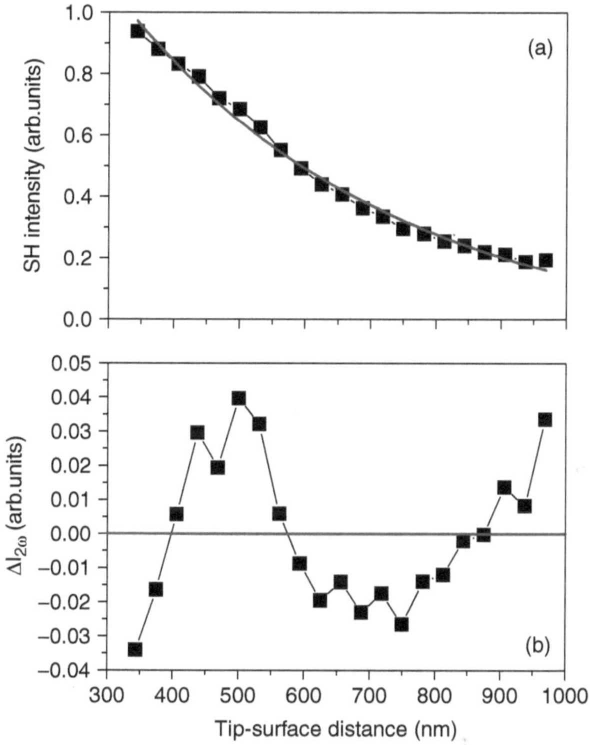

Fig. 15. (a) Distance dependence of the SH intensity in the middle-field from the surface. Thick solid line represents the dependence expected in the dipole-image model. (b) The deviation of the observed dependence from the dipole-image model: the SH intensity subtracted from the model calculations in (a). The fundamental light is p-polarised.

The near-field SHG is a relatively easy way (standard STM tip can be used) to achieve a reliable (no bleaching) and strongly confined (due to the field enhancement effects) light source close to the sample surface needed in numerous applications. The tip shape can be controlled by using either a single colloidal particle as a probe tip or specially nanofabricated probe tips of required shape and size to achieve the strongest field enhancement and desirable polarisation and distance dependencies for given materials of a tip and a surface. Characteristic polarisation dependencies of the near-field SHG show that the polarisation contrast (exceeding the contrast of linear scattering) which is important for apertureless near-field optical imaging of ferroelectric and magnetic domains as well as for nonlinear magneto-optics effects studies and applications on the nanoscale is possible to achieve.

Light confinement effects and related nanoscopic light sources are important from both a fundamental point of view and potential applications.

Nano-local light sources and the ability to manipulate strongly localised electromagnetic fields are required for imaging and characterisation of materials and devices on the nanoscale as well as for local optical and spectroscopic studies. They can be used for both photolithographic fabrication of nanostructures as well as for optical and magneto-optical high-density data storage. Closely related to this is photochemistry and photobiology on the nanoscale. Using localised light sources, photochemical and photobiological transformations and photo-chemical reactions can be induced locally on the level of individual molecules. SHG-based nanoscopic sources provide a wide spectral tuneability range and excellent polarisation properties.

§ 7. SHG ASNOM applications for functional materials characterisation

Magnetic and ferroelectric domain imaging and characterisation are the backbone for the development of new technological materials and structures for numerous functional applications, most importantly, for high-density data storage. It is both the size and behaviour (in the external magnetic or electric field) of the domains that determine the performance of new materials and devices. The understanding of the formation of the domain structure and its evolution are of particular significance on mesoscopic and nanoscopic scales where local physical behaviour may differ significantly from the bulk. Optical approaches occupy an important place among the techniques used for functional materials imaging. They are nondestructive, allow measurements in ambient conditions and applied external fields as well as imaging under protective layers.

Linear optical techniques with subwavelength spatial resolutions providing magneto-optical contrast (i.e., the ability to discriminate between differently oriented magnetic domains) are based on various SNOM approaches with polarisation resolution. These approaches rely on the magneto-optical Kerr or Faraday effects in reflection or transmission, respectively. Using linear apertured or apertureless SNOM, domain imaging has been achieved for polar (magnetisation perpendicular to the surface) as well as in-plane magnetised materials and thin films (Dickson et al., 2005a, and references therein). Similarly, several optical techniques exist with which it is possible to image ferroelectric domain structures in the near-field using apertured and apertureless SNOM (Dickson et al., 2005b, and references therein). Many linear optical techniques in this case are, however, limited by the necessary application of electric fields either perpendicular or parallel to ferroelectric sample surface. Apertureless

SNOM may additionally require the implementation of interferometric signal analysis to recover the phase information on domain orientation variations.

Nonlinear optical techniques based on SHG are advantageous for functional materials imaging due to a much higher sensitivity to magnetisation (or macroscopic polarisation) variations in the magnetic (or ferroelectric) domain structures. High surface and interface sensitivity of SHG in combination with the high spatial resolution provided by SNOM has been validated as a powerful tool for imaging of magnetic and ferroelectric domains (Zayats and Smolyaninov, 2004, and references therein). The spatial resolution in the collection mode SH SNOM, when the second-harmonic photons are collected with a SNOM tip, has been shown to be typically about 100 nm with uncoated fibre tips usually used in SHG measurements, reaching about 30 nm in the case of high-refractive index materials, such as piezo-electric films and crystals, with particular polarisation properties (Smolyaninov et al., 2001a; Zayats and Smolyaninov, 2004). In the latter case, although the detection of the SH light is done with a bare fibre tip, the field enhancement effects are similar to those relevant to apertureless SHG SNOM discussed in Section 3.2. As was proposed by Smolyaninov et al. (2001b), the fibre tip can be partially coated with metal to provide additional field enhancement and scattering of the SH field into the fibre as well as to apply external electric field to modify the ferroelectric domain structure under investigation.

SHG polarisation analysis which is in many cases essential for domain imaging, although in principle possible with an aperture-based SNOM, is not always feasible due to a weak SHG signal passing through the sub-wavelength aperture of a fibre tip. More importantly, due to the effects caused by depolarisation at the tip apex and in the fibre, it is usually difficult to routinely achieve the polarisation extinction better than 10 with apertured SNOM tips. As has been shown above, a probe tip of apertureless SNOM acts as an effective local polariser due to polarisation properties of the field enhancement and, thus, polarisation analysis of the detected light can be efficiently achieved in the apertureless configuration.

Thus, for magnetic and ferroelectric domain imaging, apertureless SHG SNOM has advantages in providing high sensitivity to domain orientation, stronger electromagnetic field enhancement and, therefore, higher SH signal, better spatial resolution (stronger light confinement) as well as good polarisation properties enabling polarisation analysis of SH light in the near-field. In addition, the apertureless configuration provides more flexibility in illumination/detection configurations needed to address various orientations of the domains and different components of the second-order susceptibility tensor that is needed for absolute determination of the domain orientation.

Applications of ASNOM to the observation of domain-orientation-dependent SHG has nontrivial consequences for imaging contrast mechanisms. For comparison, we consider both a collection mode SHG SNOM when a fibre tip is used to detect SHG signal in the near-field of a domain and a probe tip of an apertureless SNOM when the near-field scattering induced by the tip is observed in the far-field. The presence of the tip significantly modifies both the local fundamental field and generated SH field. These effects are most important for metallic probe tips of an apertureless SNOM. However, even for uncoated fibre tips, the probe tip influence can be important. Fibre tip effects are, in many cases negligible in linear SNOM measurements, but cannot be neglected in SHG studies since the SH intensity is proportional to the fourth power of the local fundamental field.

Following the analysis of Section 3.2, the SH field measured with SNOM (scattered into the fibre tip or scattered into the far-field zone by a metallic tip) above magnetic or ferroelectric domain can be expressed as (Dickson et al., 2004; 2005a)

$$\mathbf{E}(2\omega) \sim \mathbf{L}(2\omega)\overset{\leftrightarrow}{\chi}{}^{(2)}_{s}(\mathbf{M}) : [\mathbf{L}(\omega)\mathbf{E}(\omega)][\mathbf{L}(\omega)\mathbf{E}(\omega)]. \tag{7.1}$$

Here $L(\omega) = f(\varepsilon_T, \varepsilon_s, d, a, a/b)$ is the local electromagnetic field enhancement in the vicinity of the SNOM tip that depends on the permittivity of the surface $\varepsilon_s(\omega)$ and the tip $\varepsilon_T(\omega)$, their separation d and the size (a) and shape (a/b) of the tip. In the first approximation, this enhancement is only important for the z-components of the fundamental and second-harmonic fields that are parallel to the tip axis $\mathbf{L} = (1, 1, L)$.

In apertureless SHG SNOM, all components of the scattered local second-harmonic field contribute to the measured signal. In contrast, assuming that an SNOM fibre tip collects only dipole radiation and the SH dipoles oriented in the z-direction at the aperture do not radiate towards the tip, the SHG signal detected with SNOM will be $I_{\mathrm{SH}} \sim E_x^2(2\omega) + E_y^2(2\omega)$. This assumption works in SHG SNOM imaging of ferroelectric domains in piezo-electric materials ($\varepsilon \sim 100$) for which the enhancement is important even with bare fibre tips. Nevertheless, a small part of the $E_z(2\omega)$ field can be collected by the realistic fibre tip in the experiment due to possible shape irregularities and roughness of the tip apex. This makes the latter experimental situation somewhat similar to ASNOM measurements.

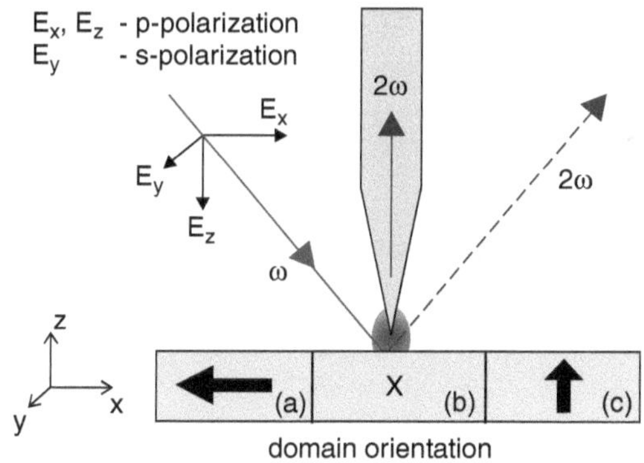

Fig. 16. Three principal domain orientations (a) longitudinal, (b) transverse, and (c) polar.

Three principal domain orientations should be considered for both ferromagnetic and ferroelectric materials (fig. 16). In the polar case, the domain polarisation is perpendicular to the sample surface, while the in-plane domain orientations constitute longitudinal and transverse configurations with the in-plane domain polarisation parallel and perpendicular to the plane of incidence, respectively. Taking into account the second-order nonlinear susceptibility tensor of magnetic metals (Kirilyuk, 2002), in the transverse configuration ($\mathbf{M}\|y$), (xz) is the plane of incidence, for the case of s-polarised fundamental light, $\mathbf{E}(\omega) = (0, E_y, 0)$, the SH field generated at the magnetic domain has both magnetic-related component

$$E_x(2\omega) \sim \pm\chi^{(2)}_{xyy}(\pm M)E^2_y(\omega) \tag{7.2}$$

as well as nonmagnetic one

$$E_z(2\omega) \sim L(2\omega)\chi^{(2)}_{zyy}E^2_y(\omega). \tag{7.3}$$

Both these components of the SH field are probed in the ASNOM configuration, providing the SH intensity sensitivity to magnetic anisotropy since both magnetic and nonmagnetic SHG contributions are p-polarised. These are the same components that are detected in the conventional far-field SHG measurements, but the contribution of different susceptibility tensor elements to the total SH field is different in the presence of the ASNOM tip. The nonmagnetic SHG $E_z(2\omega)$ in this configuration is enhanced at the probe tip, that leads to a decrease of the ratio of the magnetisation-related signal ($\Delta I_M(2\omega) \sim L(2\omega)\chi^{(2)}_{zyy}\chi^{(2)}_{xyy}$) to the total SHG signal ($I_T(2\omega) \sim L^2(2\omega)$) compared to the far-field measurements. The situation is different in the apertured SNOM measurements, when the SH field collected with a probe tip is dominated by $E_x(2\omega)$. Although the

SHG intensity $I_{SH} = |E_x(2\omega)|^2$ measured in this configuration is directly related to the magnetisation, the aperture-based SNOM measurements would not provide magnetic contrast if there is no other SHG signal (not related to magnetisation of the domain) collected by the tip to serve as a reference signal.

In the longitudinal configuration ($\mathbf{M}\|x$)) for s-polarised incident light, only polarisation contrast can be achieved in the apertureless SNOM measurements, since in addition to the s-polarised magnetisation-related contribution

$$E_y(2\omega) \sim \pm\chi^{(2)}_{yyy}(\pm M)E_y^2(\omega), \tag{7.4}$$

the nonmagnetic component of the p-polarised SH light

$$E_z(2\omega) \sim L(2\omega)\chi^{(2)}_{zyy}E_y^2(\omega) \tag{7.5}$$

is probed. This leads to the polarisation rotation, rather than the intensity variations, with magnetisation $\mathbf{M}$ in the measured signal. Thus, to image magnetic contrast, the scattered SH light should be observed through an additional polariser. In contrast, in apertured SNOM, the dominating SHG field component detected with the SNOM fibre tip is $E_y(2\omega)$, therefore, magnetic contrast is possible in the intensity measurements in collection mode SNOM in the presence of the "reference" nonmagnetic-origin SH light of the same polarisation as discussed above.

In the case of p-polarised fundamental light, the situation is more complicated due to the larger number of susceptibility tensor elements involved (Kirilyuk, 2002). However, the overall analysis is similar to the one presented above and generally allows ASNOM to detect the same components of the second-harmonic field as in the far-field measurements but with possibly different relative intensities due to the field enhancement effects on a probe tip which are differently important for different susceptibility tensor components. For example, in the case of transverse magnetisation, the SH intensity related to $\pm\chi^{(2)}_{zzx}(\pm M)$ component of the susceptibility will bear the enhancement $L^2(\omega)L^2(2\omega)$. In the case of the longitudinal configuration, the SH intensity related to $\pm\chi^{(2)}_{yxx}(\pm M)$ and $\pm\chi^{(2)}_{yzz}(\pm M)$ components will differ in the case of ASNOM detection by $L^4(\omega)$ compared to the conventional far-field detection. The enhancement or reduction of the observable magnetisation-related contrast depends on both the relative values of the magnetic and nonmagnetic susceptibility tensor elements and the field enhancement effects at the ASNOM tip.

The apertureless SNOM tip oriented perpendicularly to the film surface changes the relative contribution of the second-order susceptibility tensor components to the total SH field but does not mix the polarisation state of SH light and, as in far-field measurements, only polarisation contrast is possible in the longitudinal magnetisation configuration with apertureless SNOM. However, if the tip is oriented at some angle to the surface and is not in the plane of incidence, it will enhance the electric field components parallel to the tip axis in both s- and p-polarised SH light effectively acting as a local polariser. In this case, longitudinal magnetic contrast in SHG can be also observable without additional polarisation analysis in apertureless SNOM with metallic tips. However, such a configuration with a tilted SNOM tip might reduce achievable spatial resolution.

If a SHG-capable tip is used with the nonlinear susceptibility comparable to the magnetic one ($\chi_s^{(2)} \sim \chi_T^{(2)}$), it can provide the SHG intensity contrast in all magnetisation and polarisation configurations via interference term in eq. (3.1), removing the need for additional polarisation analysis of the detected SH light.

Similar considerations apply to the imaging of ferroelectric domains. The second-order susceptibility tensor for ferroelectric materials has more nonzero components, and, thus, the data analysis is more cumbersome than for magnetic materials but straightforward. The optical second-harmonic response of ferroelectric material in the far-field is given by eq. (2.1). Symmetry considerations show that the second-order susceptibility tensor $d_{ijk}^{(2\omega)}$ has the same form as the piezoelectric tensor. Information on the local ferroelectric polarisation may thus be recovered using second-harmonic SNOM. It is possible, knowing the nonzero nonlinear susceptibility tensor components, to determine the local domain direction by comparing the near-field SH response obtained with different fundamental light polarisations and different angles of incidence. Such polarisation dependencies are strongly influenced by the field enhancement effects (Dickson et al., 2005b).

For example, piezo-electric materials exhibit high dielectric constant and, thus, even an uncoated fibre tip provides significant degree of the enhancement allowing to achieve the spatial resolution significantly better than with linear SNOM measurements. Utilising the above-described polarisation properties of the SHG and assuming that domains are related to the grain structure of the film, the SHG variations can be directly correlated to the topography of the film. This is illustrated in fig. 17. The SHG polarisation difference image fig. 17(d) has been obtained by subtraction of the raw SH signal measured at different fundamental light polarisations. Grain/domain boundaries have been drawn as a visual guide. The area designated A shows almost no change in SHG as the fundamental light polarisation is rotated from s to p. This may indicate a domain oriented in the longitudinal direction for a tip–surface

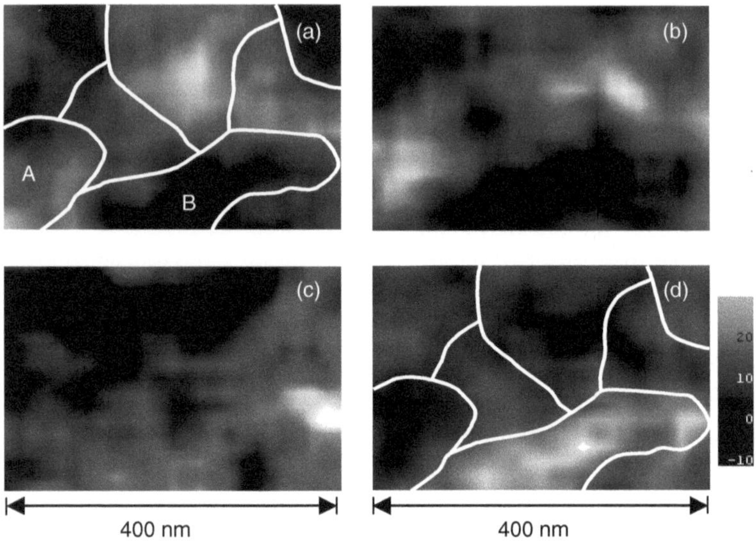

Fig. 17. (a) Shear-force topography and (b,c) corresponding SHG SNOM images of the PZT film obtained with (b) p- and (c) s-polarised excitation light. The polarisation-dependent change in the SH intensity is shown in (d). The grey-scale variations in (d) from black-to-white correspond to an increase in the SH intensity with the fundamental light polarisation rotated from s- to p-state. Possible grain boundaries have been traced on (a) and (d) to assist with visual interpretation.

combination exhibiting a small enhancement factor. The area marked B shows a strong increase in SHG for p-polarised fundamental light, indicating either transverse or polar orientation of the domain.

Thus, the advantages of the apertureless SH imaging of magnetic and ferroelectric domain is that the SHG contrast related to domain orientation is similar to that in the far-field measurements but with different contributions from the susceptibility tensor components. Nevertheless, in the apertureless SNOM configurations with a tilted metallic tip or a tip exhibiting second-order nonlinearity comparable to that of the sample, the SH intensity contrast can be obtained for different domain orientations without the need for additional polarisation analysis. The technique also provides better spatial resolution and stronger SHG signal due to the tip-induced field enhancement.

§ 8. Conclusion

We have presented the recently developed technique of apertureless near-field SHG microscopy based on the tip-induced enhancement of the electromagnetic field at fundamental and second-harmonic frequencies. This approach allows one either to probe the surface-related SHG by modifying it with a probe tip in the near-field proximity to the surface under investigation or to achieve a nanoscopic light source confined to the apex of the probe tip with the radiative properties that depends on the dielectric and topographic properties of the surface under investigation. In both cases, the probe tip acts as a local polariser due to the polarisation dependencies of the field enhancement and provides polarisation resolution which is needed in many applications.

Apertureless SHG SNOM holds a promise of higher spatial resolution than achievable with conventional apertured SNOM and linear scattering-type apertureless near-field techniques since the detected SH intensity depends quadratically on the local fundamental field. Owing to the same reason, in addition to the intrinsic sensitivity of SHG to surface states and other inhomogeneities of the surface, this nonlinear optical technique provides also higher sensitivity to variations of the electromagnetic fields for the applications in imaging and local optical characterisation of various materials and structures, including functional materials as well as chemical and biological species. Further extension of the technique for third-harmonic generation and four-wave mixing on the nanoscale can also be developed.

Nonlinear nano-optics in general and apertureless SHG in particular have opened up fascinating opportunities in studies of the electromagnetic field enhancement effects and their applications for achieving nanoscale light sources. Implementation of novel nano-fabrication technologies provide new possibilities for tailoring properties of nanostructures and probe tips in order to emphasise and target one or another effect in the nonlinear optics related to the field enhancement and light localisation. An understanding of nanoscale nonlinear processes and their realisation at low excitation light intensities might provide the possibility to develop novel nonlinear optical devices on the subwavelength scales needed in many areas of modern photonics and optoelectronics.

Acknowledgements

The author is indebted to all his colleagues and collaborators who contributed in one or another way to the progress of this work, among them C. C. Davis, W. Dickson, V. Sandoghdar, I. I. Smolyaninov, S. Takahashi and many others.

References

Bachelot, R., Gleyzes, P., Boccara, A.C., 1995, Opt. Lett. **20**, 1924.
Bouhelier, A., Beversluis, M., Hartschuh, A., Novotny, L., 2003, Phys. Rev. Lett. **90**, 013903.
Bozhevolnyi, S.I., Beermann, J., Coello, V., 2003, Phys. Rev. Lett. **90**, 197403.
Demming, A.L., Festy, F., Richards, D., 2005, J. Chem. Phys. **122**, 184716.
Dickson, W., Takahashi, S., McHugh, D., Atkinson, R., Pollard, R., Zayats, A.V., 2005a, J. Appl. Phys. **97**, 033505.
Dickson, W., Takahashi, S., Moros, C., Bowman, R.M., Gregg, J.M., Zayats, A.V., 2005b, Phys. Rev. **B72**, 094110.
Dickson, W., Takahashi, S., Pollard, R., Zayats, A.V., 2004, Appl. Phys. Lett. **85**, 6341.
Hartschuh, A., Sanchez, E.J., Xie, X.S., Novotny, L., 2003, Phys. Rev. Lett. **90**, 095503.
Hayazawa, N., Inouye, Y., Sekkat, Z., Kawata, S., 2001, Chem. Phys. Lett. **335**, 369.
Kawata, Y., Xu, C., Denk, W., 1999, J. Appl. Phys. **85**, 1294.
Keilmann, F., Hillenbrand, R., 2004, Phil. Trans. R. Soc. Lond., Ser. **A362**, 787.
Kirilyuk, A., 2002, J. Phys. **D35**, R189.
Labardi, M., Allegrini, M., Zavelani-Rossi, M., Polli, D., Cerullo, G., De Silvestri, S., Svelto, O., 2004, Opt. Lett. **29**, 62.
Landau, L.D., Lifshits, E.M., 1984, Electrodynamics of Continuous Media, Pergamon, London.
Laroche, T., Baida, F.I., van Labeke, D., 2005, J. Opt. Soc. Am. **B22**, 1045.
Lewis, A., Lieberman, K., 1991, Nature **354**, 214.
Li, Z.-Y., Gu, B.-Y., Yang, G.-Z., 1999, Phys. Rev. **B59**, 12622.
McGilp, J.F., 1995, Progr. Surf. Sci. **49**, 1.
Michaelis, J., Hettich, C., Mlynek, J., Sandoghdar, V., 2000, Nature **405**, 325.
Moskovits, M., 1985, Rev. Mod. Phys. **57**, 783.
Neacsu, C.C., Reider, G.A., Raschke, M.B., 2005, Phys. Rev. **B71**, 201402.
Patane, S., Gucciardi, P.G., Labardi, M., Allegrini, M., 2004, Riv. Nuovo Cimento **27**, 1.
Porto, J.A., Johansson, P., Apell, S.P., López-Rıós, T., 2003, Phys. Rev. **B67**, 085409.
Richards, D., Zayats, A.V., (Eds.), 2004, Nano-optics and near-field microscopy, Phil. Trans. R. Soc. Lond., Ser. A **362**, 699–919.
Sanchez, E.J., Novotny, L., Xie, X.S., 1999, Phys. Rev. Lett. **82**, 4014.
Shalaev, V.M., 2000, Nonlinear Optics of Random Media, Springer, Berlin.
Shen, Y.R., 1984, The Principles of Nonlinear Optics, Wiley, New York.
Smolyaninov, I.I., Liang, H.Y., Lee, C.H., Davis, C.C., Aggarwal, S., Ramesh, R., 2000, Opt. Lett. **25**, 835.
Smolyaninov, I.I., Liang, H.Y., Lee, C.H., Davis, C.C., 2001a, J. Appl. Phys. **89**, 206.
Smolyaninov, I.I., Liang, H.Y., Lee, C.H., Davis, C.C., Nagarajan, V., Ramesh, R., 2001b, J. Microsc. **202**, 250.
Smolyaninov, I.I., Zayats, A.V., Davis, C.C., 1997a, Phys. Rev. **B56**, 9290.
Smolyaninov, I.I., Zayats, A.V., Davis, C.C., 1997b, Opt. Lett. **22**, 1592.
Stöckle, R.M., Suh, Y.D., Deckert, V., Zenobi, R., 2000, Chem. Phys. Lett. **318**, 131.
Takahashi, S., Zayats, A.V., 2002, Appl. Phys. Lett. **80**, 3479.
Vigoureux, J.M., Girard, C., Depasse, F., 1994, J. Mod. Opt. **41**, 49.
Williame, L., Lecaque, R., Le Moal, E., Fort, E., Gresillon, S., Rivoal, J.-C., Boccara, C., 2005, Linear and non-linear near field behaviour on random metal films and surfaces, in Kawata, S., Shalaev, V.M., Tsai, D.P. (Eds.), Plasmonic Nano-imaging and Nano-fabrication, Proc. SPIE, vol. 5928, pp. 1–8.
Zayats, A.V., 1999, Opt. Commun. **161**, 156.
Zayats, A.V., Kalkbrenner, T., Sandoghdar, V., Mlynek, J., 2000, Phys. Rev. **B61**, 4545.
Zayats, A.V., Sandoghdar, V., 2000, Opt. Commun. **178**, 245.

Zayats, A.V., Sandoghdar, V., 2001, J. Microsc. **202**, 94.
Zayats, A.V., Smolyaninov, I.I., 2004, Phil. Trans. R. Soc. Lond., Ser. A **362**, 843.
Zayats, A.V., Smolyaninov, I.I., Davis, C.C., 1999, Near-field microscopy of second-harmonic generation, in: Chikishev, A., Zadkov, V.N., Zheltikov, A.M. (Eds.), Novel Trends in Laser Spectroscopy and Optical Diagnostics, Proc. SPIE, vol. 3732, pp. 81–92.
Zayats, A.V., Smolyaninov, I.I., Maradudin, A.A., 2005, Phys. Rep. **408**, 131.
Zenhausern, F., O'Boyle, M.P., Wickramasinghe, H.K., 1994, Appl. Phys. Lett. **65**, 1623.
Zhao, X., Kopelman, R., 1995, Ultramicroscopy **61**, 69.

Chapter 9

Resonant optical antennas and single emitters

by

B. Hecht, P. Mühlschlegel, J.N. Farahani, H.-J. Eisler, D.W. Pohl

Nano-Optics group, National Center of Competence for Research in Nanoscale Science, Institute of Physics, University of Basel, Klingelbergstr. 82, CH-4056 Basel, Switzerland, e-mail: bert.hecht@nano-optics.ch

Tip Enhancement
Advances in Nano-Optics and Nano-Photonics
ISSN: 1871-0018

S. Kawata and V.M. Shalaev (Editors)
© *2007 Elsevier B.V. All rights reserved.*
DOI: 10.1016/S1871-0018(06)01009-0

Contents

§ 1. Introduction

Light is an electromagnetic phenomenon. This notion is as old as are Maxwell's equations. As such light belongs to the broad variety of electromagnetic disturbances that exhibit self-contained propagation in various media. Our everyday life and modern technology have been revolutionized by the use of propagating electromagnetic waves. Radio waves, for instance, with wavelengths ranging from meters to hundreds of meters are used to transmit information over large distances. Mobile phones operate at wavelength between 15 and 30 cm using microwaves. Slightly shorter waves are used in radio detection and ranging (RADAR) to determine velocities and distances of moving objects like cars and clouds. On the short-wavelength side of the electromagnetic (em) spectrum, for example, x-rays with wavelength in the sub-nanometer range are used to determine structures of proteins and materials.

In this broad spectrum, visible light is of particular interest since the energy of optical and near-infrared photons match the electronic transitions in larger molecules which is, by-the-way, the reason we are able to see. Also vibrational transitions fall into this regime and albeit having smaller transition energies they can still be detected using visible light (Raman effect). In addition, infrared spectra provide detailed chemical fingerprint information. The optimal photon energy and momentum of visible and near-infrared light with respect to electronic transitions in matter led to the use of light for manipulation and modification of matter, e.g. in optical tweezers and in optical data storage. Another intriguing property of visible and near-infrared light is that it can be transmitted without noticeable losses over huge distances using waveguides. This is a feature which makes light extremely attractive for telecommunication purposes. Finally, the energy of visible photons is sufficiently larger than the thermal energy at room temperature to enable the detection of single photons via the photo-electric effect.

From a classical point of view electromagnetic waves, and also light, are created by time-dependent current distributions (Balanis, 1997; Lee, 1984). The *efficient* creation and detection of electromagnetic waves is a field of research with immediate and profound relevance for technology and applications. Specifically shaped conductors used in this context are usually called *antennas*. All antennas are based on the fact that a time-dependent current distribution will emit electromagnetic radiation and that, vice versa, charges are accelerated in the conducting material by incoming electromagnetic waves thus locally generating strong oscillating electric (near-)fields. Such efficient interconversion of propagating and localized, enhanced fields with wavelengths in the optical regime is instrumental for advances in optical characterization (Kneipp et al., 1997; Knoll and Keilmann, 1999; Hartschuh et al., 2003; Jackson et al., 2003; Haes et al., 2004), manipulation (Pohl, 1994; Novotny et al., 1997; Xu and Käll 2002; Hallock et al., 2005), and (quantum) optical information processing (Krenn et al., 1999; Michler et al., 2000; Bozhevolnyi et al., 2001; Barnes et al., 2003; Maier et al., 2003) on the nanometer-scale.

Antennas at optical frequencies require antenna dimensions in the sub-μm range. The realization of such antennas requires micro and nano fabrication techniques. First steps in this direction were made for antennas optimized for infrared radiation (Wilke et al., 1994; Fumeaux et al., 1999). The question whether concepts of antenna theory may be applied to the optical wavelength regime has been addressed by Pohl (1991, 2000) and Grober et al. (1997). First experimental studies on optical antennas were following some time after (Crozier et al., 2003; Fromm et al., 2004; Farahani et al., 2005; Mühlschlegel et al., 2005).

This chapter is intended as an introduction to important elements of antenna theory and their application to optical antennas. We start out with a review of classical antenna theory with emphasis on simple thin linear dipole antennas. We give plausible arguments for the field confinement in the antenna feed gap as well as for the occurrence of antenna resonances at uneven integer multiples of half the wavelength for an emitting antenna. We then discuss possible deviations from classical antenna theory that may occur at optical frequencies: Surface plasmon resonances and finite conductivity of metals. Field simulations are used to investigate basic properties of optical antennas made from plasmonic and nonplasmonic metals, gold and aluminum, respectively. Similarities and discrepancies to classical antenna theory are discussed. We then discuss briefly the fabrication and characterization of gold antennas that are resonant at near-infrared wavelengths. Finally we discuss the interaction of single dipole emitters with resonant optical antennas. We introduce a basic framework in which the dipole emitter is considered as a classical point dipole and the antenna as an arbitrary scatterer to show that the coupled system of emitter and antenna may be considered as an effective super-emitter with properties inherited from both the nano-antenna and the single quantum system.

§ 2. Antenna basics

The efficiency of interconversion between propagating electromagnetic fields and local (near-)fields strongly depends on the geometry and the material properties of the respective antenna. It is clear that a good antenna material is one that has a high conductivity to avoid energy dissipation to heat (Ohmic losses) in the antenna. Choosing a clever geometry allows us to engineer the emission pattern and the total emissivity of an antenna by taking advantage of the fact that the contributions of different oscillating current elements can be arranged to interfere constructively or destructively in the far field (Balanis, 1997; Lee, 1984). As a matter of fact depending on the geometry there are very efficient as well as very inefficient antennas. In inefficient antennas there is little or even no far-field emission of radiation although high-amplitude oscillating currents occur. At optical frequencies an extended noble metal halfspace is an example of a very inefficient antenna. The geometry is sketched in fig. 1(a). Although surface plasmon polaritons exist at the metal/vacuum interface which exhibit strong locally oscillating currents and strongly enhanced optical near fields, no far-field emission occurs for such a mode since the adjacent surface current elements are out of phase and therefore cancel in the far field. Another example of a bad antenna is an open two-wire waveguide, shown in fig. 1(b), which consists of closely spaced parallel wires that are thin compared to the wavelength. In such a system a standing current wave exists on both wires with a node at the open end of the waveguide. Each current element in the wires faces its anti-phase counterpart in the other wire. Therefore the total far-field radiation of a two-wire waveguide vanishes although strongly oscillating currents exist. Notably, the open two-wire waveguide of fig. 1(b) can be converted into an efficient dipole antenna just by bending up its ends as sketched in

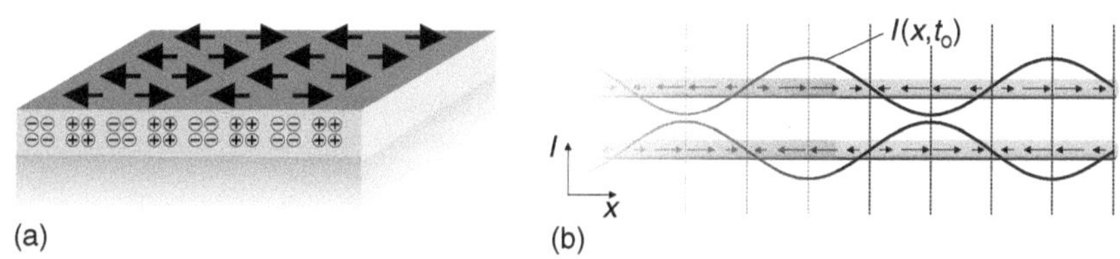

Fig. 1. Inefficient antennas: (a) surface plasmon polariton resonance; (b) Two-wire waveguide.

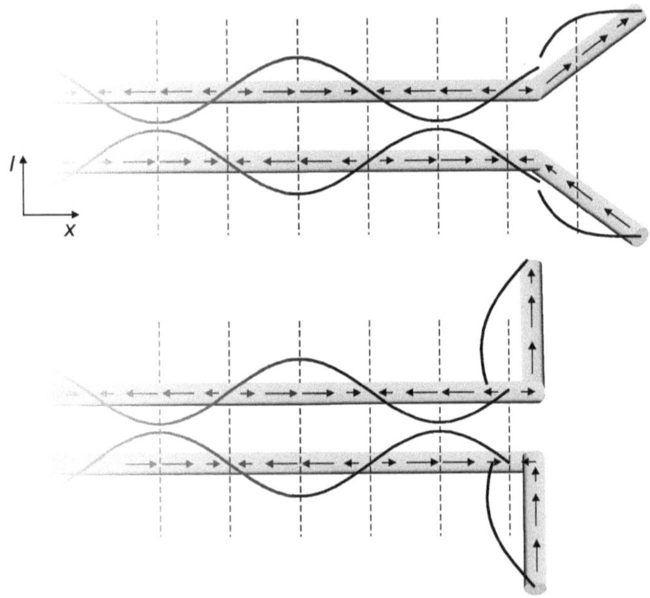

Fig. 2. Efficient antenna created by bending up the ends of the two-wire waveguide in fig. 1 (Balanis, 1992).

fig. 2. The bending in opposite directions results in in-phase current elements yielding effective emission of electromagnetic waves into the far field.

In summary, efficient antennas maximize the amount of far-field radiation by avoiding Ohmic losses in the antenna itself and by avoiding totally destructive interference of far fields of the oscillating current elements on the antenna.

Since the current distribution along the two-wire transmission line is a sinusoidal standing wave, we may assume as a first approximation that the current along the bent-up arms of the dipole antenna follows a sinusoidal behavior as well. Actually, for radio waves, this approximation is known to be good for thin antennas (wire radius $a<\lambda/20$ (Balanis, 1997)) and even for thicker antennas which do not exceed $\lambda/2$ in length (King and Harrison, 1943). The one-dimensional sinusoidal current distribution in the antenna arms is given by (Balanis, 1997; Lee, 1984)

$$j(z) = j_{\max}(l)\sin\left[k\left(\frac{l}{2} - |z|\right)\right], \tag{2.1}$$

where $j_{\max}(l) = j_0/\sin\left(\frac{1}{2}kl\right)$ is the current amplitude as a function of the current at the feed gap j_0, l the total antenna length, and k the wave vector. The antenna is oriented in z-direction with the origin of the coordinate system at the center of the antenna feed gap. Note that the harmonic time dependence of the one-dimensional current (eq. (2.1)) and a possible phase shift with respect to the driving voltage have been omitted for now.

2.1. Field enhancement in resonant dipole antennas

This first approximation is very useful to study e.g. the field enhancement or radiation pattern as a function of the antenna length l for a thin antenna. We concentrate here on the field enhancement in the antenna feed gap and its dependence on the antenna length. As a starting point we consider the continuity relation $\dot{\rho} = -\nabla j$ between the charge density $\rho(z)$ and the current density j (eq. (2.1)) on the isolated dipole antenna that is directly connected to a high-frequency AC voltage source at the feed gap.[1] For a time-harmonic one-dimensional current in the antenna the continuity relation is evaluated as

$$\rho(z) = -\frac{1}{\omega}\frac{\mathrm{d}}{\mathrm{d}z} j(z), \tag{2.2}$$

which shows that for fixed frequency ω the charge density $\rho(z)$ is proportional to the gradient of the current. We may now consider the current distribution for dipole antennas of different lengths. For the $\lambda/2$ dipole antenna in fig. 3(b) we find that the current has its maximum at both sides of the gap. From this maximum value it drops to zero since no charges are transported over the gap. The gradient of the current is therefore maximal at the gap and charges of opposite signs face each other over a very short distance. This accumulation and concentration of charges leads to very strong electric fields which can be identified with the quasi-electrostatic near field of a resonant $\lambda/2$ dipole antenna. Additional but much weaker charge accumulation occurs at the ends of a linear dipole antenna. When Hertz performed first antenna experiments in 1886 it was the accumulation of opposite charges near the feed gap that resulted in an electric discharge spark in a receiving antenna which caught Hertz's

[1] An isolated dipole antenna is an antenna for which the two-wire waveguide that delivers the energy from the AC voltage source has been made shorter and shorter such that it may be safely neglected. From another point of view the infinitely thin antenna feed wires are spaced by a small distance that is much less than the length l of the antenna. In this case their effect cancels already at distances that are small compared to the antenna dimensions (Lee, 1984; Balanis, 1997).

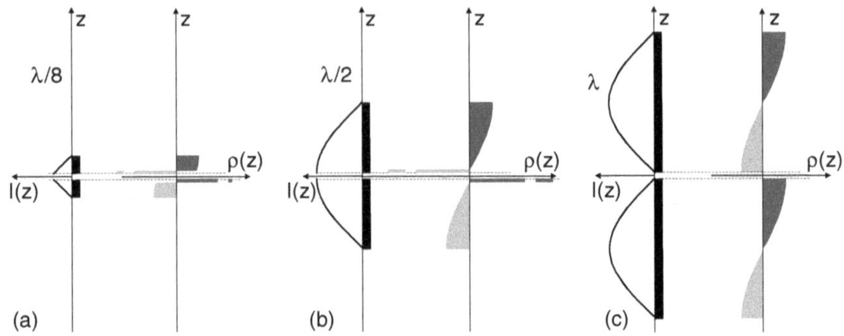

Fig. 3. Charge distribution in dipole antennas of different lengths according to eq. (2.2). On each antenna arm, the total charge cancels limiting the height of the charge peaks near the feed gap.

attention (Hertz, 1971). For dipole antennas of shorter length, as illustrated in fig. 3(a), a similar charge accumulation at the feed gap occurs, however the current gradient is much less pronounced. For antennas longer than $\lambda/2$ the current gradient at the gap is reduced again until it reaches the same low value in the feed gap as at the ends of the antenna for an antenna of length λ (see fig. 3(c)). For even longer antennas it is obvious that another field maximum in the gap will be found for $l = 3\lambda/2$ and for further uneven multiples of $\lambda/2$ as long as the sinusoidal approximation holds.

2.2. *Emission of radiation from dipole antennas*

Now that we have a zero-order estimate for the current distribution on the antenna we could consult an antenna theory text book and calculate the power emitted into the far field as a function of the antenna length. We may use the fact that the vector potential can be calculated from the current distribution (Novotny and Hecht, 2006)

$$\mathbf{A}(\mathbf{r}) = \mu_0 \mu \int_{\mathrm{V}} \mathbf{j}(\mathbf{r}')\, G_0(\mathbf{r}, \mathbf{r}')\, \mathrm{dV}', \tag{2.3}$$

where $G_0(\mathbf{r}, \mathbf{r}') = \mathrm{e}^{\pm \mathrm{ik}|\mathrm{r}-\mathrm{r}'|}/(4\pi|\mathrm{r}-\mathrm{r}'|)$ is the Green's function of the inhomogeneous Helmholtz equation and V is the volume in which the current density $\mathbf{j}(\mathbf{r}')$ is different from zero. Once the vector potential $\mathbf{A}(\mathbf{r})$ is determined the electric and magnetic fields as well as the Poynting vector can be derived from it. Using the Poynting vector the radiated power and the emission pattern can be calculated straightforwardly. All would be settled now if the *amplitude* of the current j_o, which entered our considerations as a free parameter, would be known. It is our goal in the following to determine the current amplitude as a function of the length. This will lead us to the notion that not only the current amplitude is a function of the antenna geometry but also the current distribution.

2.2.1. Antenna equivalent circuit

Why this must be the case is best understood by considering the equivalent electric circuit representation of the antenna and the driving generator. Figure 4(a) shows how a series resonant circuit consisting of a single-loop inductor, a capacitance, and a wire resistance can be converted into a driven dipole antenna. The respective equivalent electronic circuit is depicted in fig. 4(b). Assuming that the generator is ideal, this equivalent circuit is a series resonant circuit consisting of a resistor, a capacitor, and an inductor mimicking the properties of the antenna. In such a circuit the current j naturally depends on the driving frequency ω

$$j(\omega) = j_o(\omega)e^{i\omega t+\varphi(\omega)}, \tag{2.4}$$

with a frequency-dependent current amplitude $j_o(\omega)$ and a phase shift $\varphi(\omega)$ which depends on R, L, and C in a well-known fashion. The

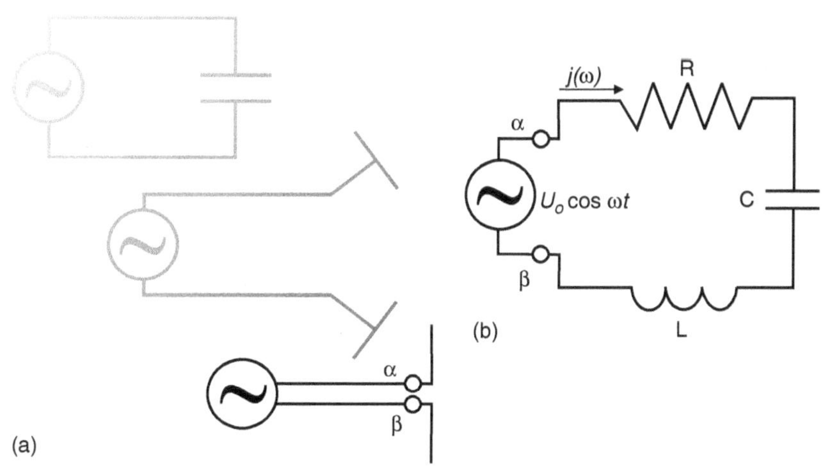

Fig. 4. Equivalent circuit representation of an antenna. (a) Dipole antenna created by bending a single wire loop. (b) Equivalent resonant circuit of a driven antenna.

resonance frequency of the circuit is given by

$$\omega_o = \sqrt{\frac{1}{LC} - \frac{R^2}{2L^2}}, \tag{2.5}$$

and its Q-factor by

$$Q = \sqrt{\frac{L}{CR^2}}. \tag{2.6}$$

In order to create a resonant circuit an alternative approach is to keep the frequency fixed and tune the antenna length. By doing so L and C change simultaneously such shifting the resonant frequency relative to the fixed excitation frequency. In this picture the antenna dimensions at which a resonance occurs can be determined roughly by a simple consideration. We assume that the capacitance of the antenna is described by a relation similar to that of a simple plate capacitor with a virtual characteristic dimension x that describes the extension of the capacitor

$$C(x) = \varepsilon_o x, \tag{2.7}$$

and that the inductance is given by an expression similar to the inductance of a single wire loop

$$L(x) = \mu_o x. \tag{2.8}$$

Using eq. (2.5) in the limit of small damping we easily find that x must be on the order of wavelength $\lambda/(2\pi)$ of the radiation used to fulfill eq. (2.5). On the other hand, using eq. (2.7) and eq. (2.8) in eq. (2.6) leads to

$$Q = \frac{Z_0}{R}, \tag{2.9}$$

where $Z_0 = \sqrt{\varepsilon_o/\mu_o} = 376.7\Omega$. This and in fact also eq. (2.6) shows that high Q-factors may be achieved by minimizing R. We turn to the meaning of R in the context of an antenna in the following section.

2.2.2. *Antenna impedance*

The equivalent circuit in fig. 4(b) shows that the antenna can be characterized by its input impedance Z measured between the two terminals α and β. The input impedance is defined by

$$Z = U_o/j_o, \tag{2.10}$$

and is a complex quantity:

$$Z = R + iX. \tag{2.11}$$

The real part R is a measure of how much real power is dissipated by the antenna. It can be decomposed into a contribution due to Ohmic losses, R_{Ω}, in the antenna material and a contribution due to radiation, R_{rad},

$$R = R_{\Omega} + R_{\mathrm{rad}}. \tag{2.12}$$

R_{rad} is called the radiation resistance of the antenna. This name is a bit misleading since a high radiation resistance means strong radiation. The imaginary part, X, is the so-called antenna reactance due to the contribution of both the capacitance and the inductance of the antenna. It is due to the fact that energy is stored in the resonant circuit. As such it can be connected to the energy stored in the nonpropagating near field of the antenna. A fundamental feature of a series resonant circuit is the fact that the current depends strongly on how well the resonance of the circuit is matched and that the current is maximal at the resonance frequency. To maximize the current, the total impedance of the circuit must be minimal which is achieved when the impedances of the inductor and the capacitor, $Z_{\mathrm{L}}(l, \omega)$, and $Z_{\mathrm{C}}(l, \omega)$, respectively, are equal and the total impedance becomes real. In other words, the antenna is in resonance whenever the imaginary part of the impedance vanishes and the antenna current is maximized. Coming back to the interpretation of eq. (2.9), it becomes clear that for fixed R_{Ω} the highest Q-factors are achieved for antennas that have the smallest radiation resistance.

2.2.3. *True current distribution in a thin dipole antenna*

We have seen that the current amplitude in the equivalent antenna circuit will vary as a function of the antenna length as both L and C will change. In addition the current acquires a phase shift with respect to the driving voltage. Our simple approximation of the current eq. (2.1) must therefore be extended to describe the true current in an antenna. The problem of determining the current distribution in a thin antenna is the key problem in antenna theory and has been treated extensively (see, for example, Lee, 1984; Balanis, 1997). Rather than calculating the values of L and C as a function of the antenna length it turns out to be much simpler to directly calculate the antenna current as the solution of an electromagnetic boundary value problem. As such the solution for the current is often given as an integral equation that can be solved numerically. Mainly two integral equations are used in antenna theory depending on the exact conditions of the problem: the Pocklington, (1897) and the Hallén integral equation (Hallén, 1938). Halléns integral equation for infinitely small antenna gaps has been used by King and Harrison (1943) to obtain a solution as a series expansion in powers of the small quantity $1/\Omega = 1/(2ln(l/a))$, where l is the overall antenna length and a the radius of the antenna wire.[2] Ω is large for thin antennas. The King/Harrison solution is therefore a good approximation for thin antennas. Furthermore, as a consequence of the assumptions used to derive Hallén's solution good conductors are required. Up to the first order in $1/\Omega$ the King/Harrison solution reads as (Lee, 1984)

$$j(l,z) = \mathrm{i}\frac{2\pi U_o}{\Omega\mu_o c}\left[\frac{\sin\left[k\left(\frac{l}{2} - |z|\right)\right] + b_1(l,z)\Omega^{-1}}{\cos\left[\frac{1}{2}kl\right] + d_1(l,z)\Omega^{-1}}\right], \tag{2.13}$$

where the complex functions $b_1(l, z)$ and $d_1(l, z)$ are defined in (King and Harrison, 1943). It is easy to see that in the limit of thin antennas, $\Omega^{-1} \to 0$, eq. (2.13) reduces to an expression similar to eq. (2.1) albeit with a modified current amplitude that is now complex and depends on l. Figure 5 shows the amplitude of the current along the antenna derived from eq. (2.13). It is interesting to note, that deviations to the sinusoidal approximation occur mostly for antennas that are longer than $\lambda/2$ near the gap. In particular for the λ-antenna the current at the feed gap is nonzero and a finite input impedance of the antenna according to eq. (2.10) can be defined. The input impedance is a measure for how well the current distribution along the antenna can be excited by coupling an AC voltage to the gap. The real and the imaginary part of the input impedance as a function of the antenna length of a thin antenna according to eq. (2.13) are plotted in fig. 6(a). The zero crossing points of the imaginary part of the impedance indicate resonant behavior of the antenna equivalent circuit and therefore for the antenna itself. As can be inferred from fig. 6(a) zero crossing points occur close to even and uneven integer multiples of $\lambda/2$ indicating resonant behavior for $l = \lambda/2$, λ, $3\lambda/2$ and so on. Maxima of the real part of the impedance occur close to multiple integers of the wavelength λ. While for these antenna lengths the constructive action of all current elements on the antenna is maximal (see fig. 2) the input impedance is very high indicating bad coupling to these

---

[2] Ω is chosen as symbol because it is common in the literature. It should not be confused with the unit of the resistance.

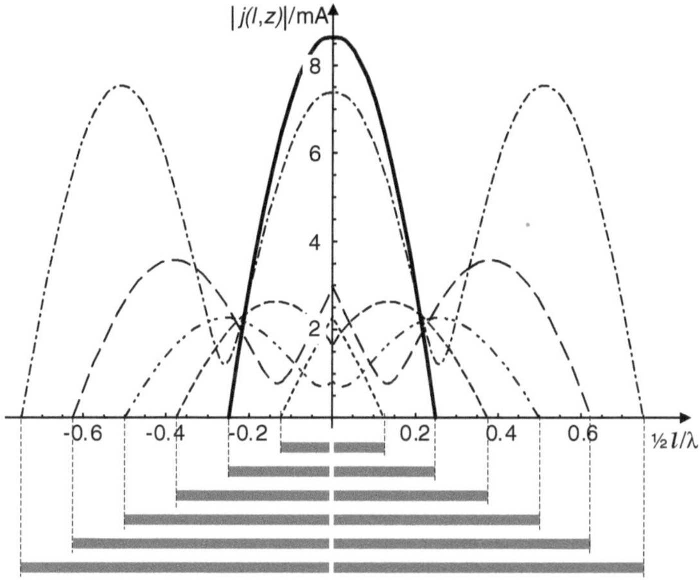

Fig. 5. Current distribution in a thin, ideally conducting antenna after King and Harrison (1943) as a function of the antenna length. A driving voltage amplitude of 1 V and $\Omega = 10$ have been used.

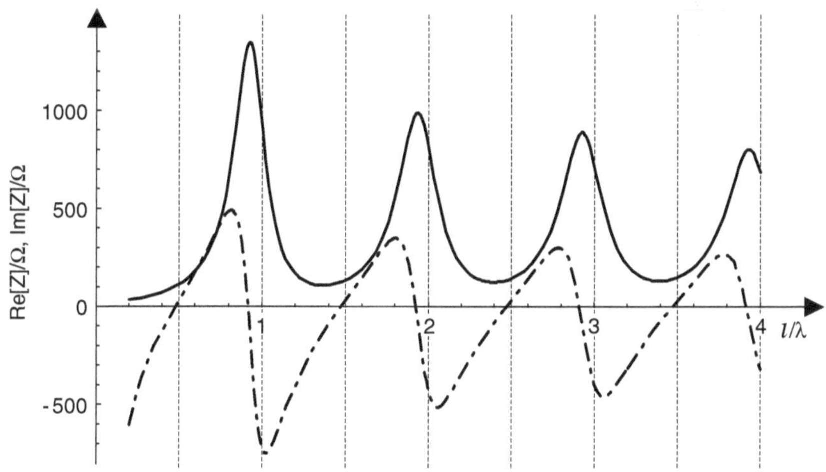

Fig. 6. Input impedance of a center-fed antenna. (a) Real part (solid line) and imaginary part (dash-dotted) of the antenna input impedance as a function of the antenna length. (b) Imaginary part of the impedance Z versus real part of Z as a function of the antenna length l. A driving voltage amplitude of 1 V and $\Omega = 10$ have been used.

current modes which finally results in very low current amplitudes in the antenna and thus to only weak emission. This idea is corroborated by an analysis of the power coupled to the antenna. As for any resonant circuit this power is a complex number. In the case of an antenna we identify the real part of the power with energy consumed by the antenna either to produce heat via Ohmic losses or to emit radiation. The imaginary part of the power corresponds to energy stored within the antenna's near field. The power delivered to the antenna can be written as $P = 1/2I^2Z$. Using eq. (2.10) this can be written as

$$P = \frac{1}{2}\frac{U_o^2}{|Z|^2}\,Z = \frac{1}{2}\frac{U_o^2}{|Z|^2}(R + \mathrm{i}X), \tag{2.14}$$

which is easily calculated using eqs. (2.10) and (2.11). Figure 7 shows the real and imaginary part of P as a function of the antenna length. For a lossless antenna all real power is converted into radiation. We see that Re(P) has maxima at antenna lengths that are uneven integers of $\lambda/2$ with the maximal power delivered for the $\lambda/2$ antenna. Im(P) shows a strongly dispersive behavior in the vicinity of the maxima of Re(P) with zero crossings for the resonant lengths. This behavior is indicative of peculiar behavior of the antenna near field close to the antenna resonances.

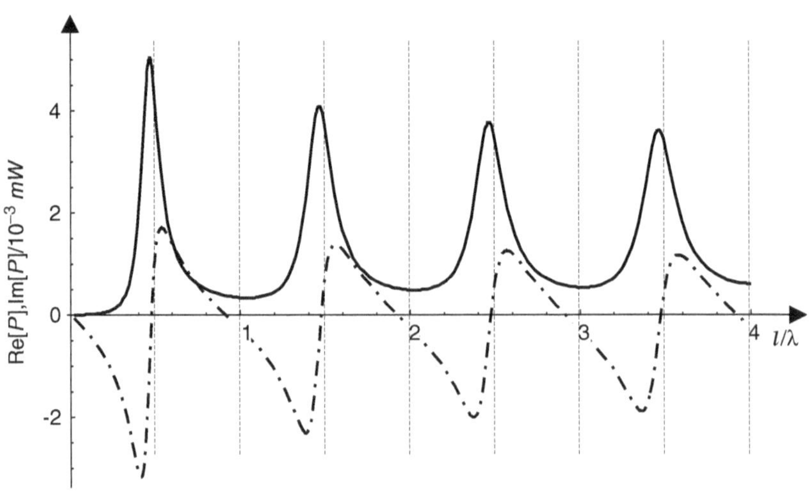

Fig. 7. Power delivered to a center-fed antenna. Real part (solid line) and imaginary part (dash-dotted) of the power delivered to the antenna terminals as a function of the antenna length. A driving voltage amplitude of 1 V and $\Omega = 10$ have been used.

§ 3. Antennas for light

3.1. Introduction

In our analysis of antennas we have so far reviewed some important aspects of classical antenna theory which essentially assumes that the antenna structure is made from very good conductors. The latter assumption is no longer true if we move toward the optical regime which we assume to include also the near infrared wavelength range. A consequence of the finite conductivity is the fact that the electromagnetic fields oscillating at optical frequencies will be able to penetrate metallic particles. The depth of penetration is defined by the skin depth of the material. Figure 8 shows a plot of the skin depth of gold as a function of the wavelength. We observe that while the wavelength of the radiation changes over many orders of magnitude between the visible and the radio wave range, the skin depth only changes by a few orders of magnitude. The skin depth at optical frequencies can be calculated from the refractive index of the metal as

$$\delta = \frac{\lambda}{2\pi \mathrm{Im}(n)}, \tag{3.1}$$

where Im(n) is the imaginary part of the refractive index. The latter can be expressed by the dielectric function as $\mathrm{Im}\,(n)^2 = \frac{1}{2}\sqrt{\mathrm{Re}(\varepsilon)^2 + \mathrm{Im}(\varepsilon)^2} - \mathrm{Re}(\varepsilon)$. For gold and aluminum in the visible and near infrared range the dielectric function is shown in fig. 9. Their skin depths at a wavelength of 830 nm are

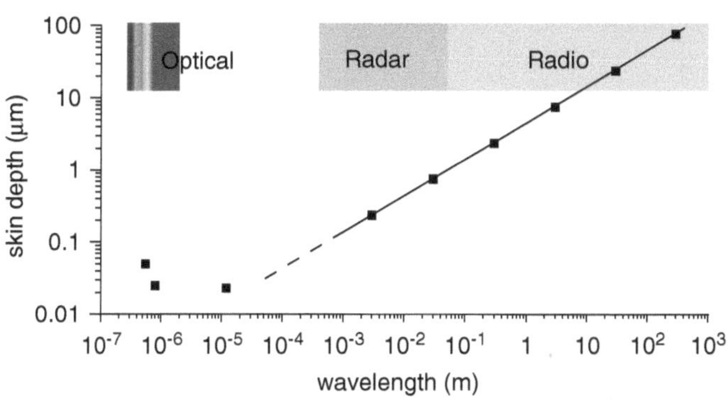

Fig. 8. Skin depth in gold as a function of the wavelength. Note that the skin depth does hardly change over the range of wavelength displayed. The skin depth is therefore negligible for radio waves but is very relevant at optical wavelengths.

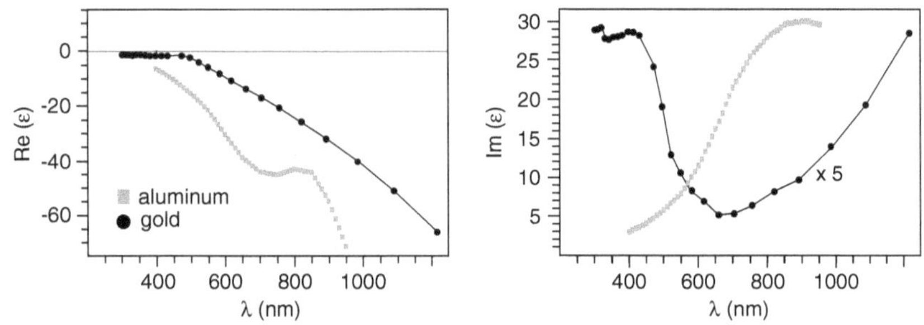

Fig. 9. Dielectric function of gold and aluminum (Schulz, 1954; Schulz and Tangherlini, 1954; Johnson and Christy, 1972).

26 and 19 nm, respectively.[3] Since the typical diameter of optical antennas will be less that $\lambda/20$ this means that the field is no longer confined to the surface of the antenna and the boundary conditions underlying classical antenna theory are significantly violated while still the current maximum is at the antenna surface. To get an idea as to what degree this behavior influences the operation of metallic antennas at optical frequencies we will apply finite-difference time-domain (FDTD) numerical simulations (Taflove and Hagness, 2005) to determine the near-field of such metal nano structures.

The finite skin depth is not the only parameter to be considered. A second property of metals at optical frequency is much more intriguing: it is the ability to sustain surface plasmon polariton oscillations at the boundary between a metal and a dielectric. These surface plasmons exhibit strongly enhanced local fields and may lead to resonantly enhanced polarizabilities of small noble metal particles (Novotny and Hecht, 2006). As we have seen before that the larger the current amplitude the more radiation is emitted by an antenna, it is well conceivable that the performance of optical antennas will benefit from surface plasmon resonances.

3.2. Light confinement by resonant dipole antennas

3.2.1. Nonplasmonic optical antenna

To stay as close to classical antenna theory as possible let us investigate the properties of optical dipole antennas made from aluminum.

[3] For long wavelengths $\delta = \sqrt{2\varepsilon_0 c^2/(\sigma\omega)}$, where σ is the conductivity (Feynman, 1989).

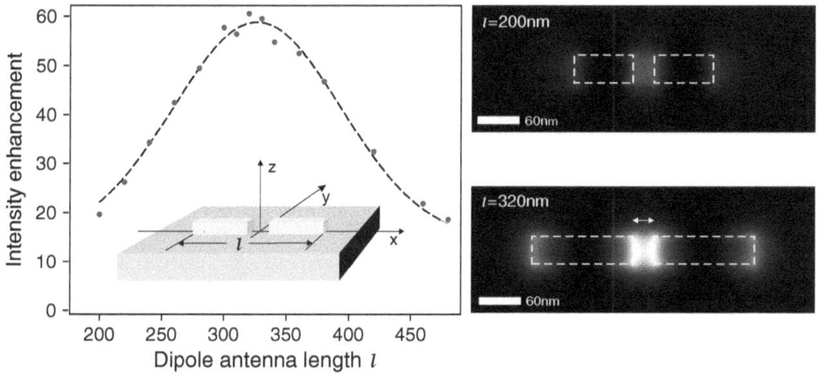

Fig. 10. FDTD simulation of aluminum dipole antennas. Left: Near-field intensity enhancement in the gap as a function of the antenna length. Right: Field distribution 10 nm above the antenna for a resonant antenna, $l = 320$ nm and an off resonant antenna, $l = 200$ nm. Maximal intensity enhancements are 23 and 7, respectively. Wavelength 830 nm, antenna height and width 30 and 40 nm, respectively. Gapwidth 30 nm.

Aluminum among all commonly used metals has the smallest skin depth at optical frequencies. Therefore aluminum antennas represent the best approximation to ideally conducting antennas at radio frequencies. The geometry of the antenna to be studied is sketched in the inset of fig. 10. The antenna is assumed to be supported by a glass half space with refractive index 1.5. The medium surrounding the antenna is assumed to have a refractive index of 1. For the simulation, we keep the wavelength fixed at 830 nm and change the length of the dipole antenna. The antenna length is measured from end to end and includes the constant gap width of 30 nm. In order to avoid artifacts and unrealistically large values for the intensity enhancement we measure the intensity enhancement in a plane 10 nm above the antenna. The maximum of the intensity enhancement occurs right above the center of the gap. A nice resonance curve is obtained when plotting the intensity enhancement factors as a function of the antenna length l. It has a maximum at $l = 320$ nm and a full width at half maximum of about 150 nm (length difference). The resonant length is clearly shorter than $\lambda/2$, however, this may be expected since the antenna is supported by the glass halfspace in which the wavelength is reduced by a factor of 1.5. It turns out to be a good approximation to assume that the antenna interacts with light of an effective wavelength which may be determined to a first approximation by the mean of the wavelengths in both media surrounding the antenna. We obtain $\lambda_m = 830\,\text{nm}/1.25 = 664$ nm and expect that the $\lambda/2$ resonant antenna should have a length $l = \lambda_m/2 = 332$ nm.This value is surprisingly close to the length we have found to produce the maximum field enhancement for the aluminum optical antenna. Note that also for the ideal antenna the resonance occurs at a length slightly smaller than $\lambda/2$ (see fig. 7).

What we have found is a surprising result in the sense that we have constructed a metallic nano structure that shows a resonant localized intensity enhancement at near infrared wavelength. However the structure does not sustain surface plasmons in the visible and near infrared wavelength range as we will see in the following. We therefore conclude that antenna resonances are an independent and alternative way to achieve resonant enhancement of confined optical fields in metallic nano structures.

3.2.2. Plasmonic optical antenna

Having said this we now consider optical antennas made from gold. The dielectric function of gold is shown in fig. 9. Gold nano particles and gold nano rods which make up an optical dipole antenna are well known for the surface plasmon resonances that govern their interaction with light. For spherical gold particles with diameters up to 20 nm and for gold ellipsoids (rods) with the same diameter and a long axis extension of up to 100 nm the plasmon resonance is very well described by the quasistatic approximation (Sönnichsen et al., 2002). The long-axis polarizability of a cigar-shaped ellipsoid (rod) for a fixed wavelength depends on the eccentricity $1-b^2/a^2$ which depends on the ratio of the short axis radius b to the long axes radius a (Bohren and Huffman, 1983; Kreibig and Vollmer, 1995). Figure 11 shows the absolute square of the long-axis polarizability of gold and aluminum rods as a function of their total length $l = 2a$. For gold rods of 20 nm radius at 830 nm illumination a sharp resonance is observed for a rod length of 190 nm while no resonance is observed for aluminum rods. Now, an optical antenna consists of two rods separated by a narrow gap. If two rods of a certain length that are resonant at a certain wavelength are brought in close proximity δ the long-axis resonance of the coupled rods is shifted to longer wavelengths (Rechberger et al., 2003). If we assume that the excitation wavelength is fixed this means that in order to make the coupled rods resonant again their length has to be reduced. In the extreme case of zero gap δ between the two rods, we recover again the problem of the single rod which should have again the original length to be on resonance. For a dipole antenna consisting of gold rods on a glass interface in air we therefore predict, that the plasmon

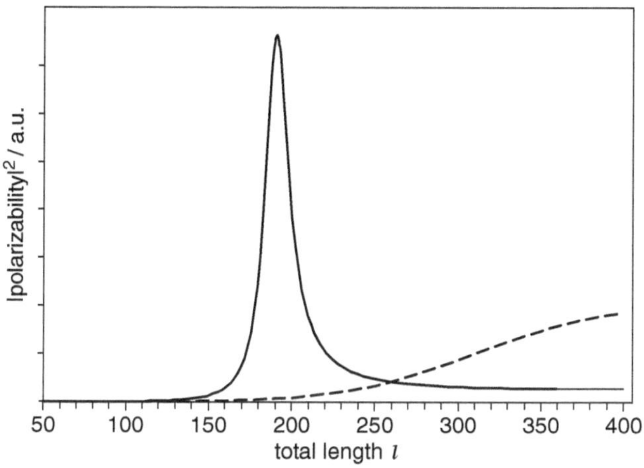

Fig. 11. Quasistatic long-axis plasmon resonance of a prolate ellipsoid with $b = c = 20$ nm, corresponding to a diameter of 40 nm as a function of the total length $l = 2a$. Solid line: gold, dashed line: aluminum. The dielectric functions for gold and aluminum are $\varepsilon_{\text{gold}} = -25 + 1.6i$ and $\varepsilon_{\text{alu}} = -46 + 29i$ at a wavelength of 830 nm.The dielectric constant of the environment models a glass/air system by assuming an average dielectric constant of 1.625.

resonance of the coupled rods should occur for an overall antenna length l somewhere between $2a$ and $4a + \delta$ where $\delta << 2a$ is an initial gapwidth. In our example this is a range between 190 and $\approx$400 nm. Interestingly, the length for which the $\lambda/2$ antenna resonance is expected to occur also falls within this range for $\lambda = 830$ nm.

We therefore conclude, that for an optical antenna made of gold we expected that both the antenna resonance and the plasmon resonance of the coupled antenna arms fall within the *same* rather narrow window of antenna lengths. We therefore expect for a gold antenna to observe contributions of both resonances at the same time. It is obvious that a future challenge will be to engineer the metal nano structure in a way that both resonances occur at exactly the same antenna length.

If we look at the result of a simulation of gold antennas for the same parameters as in fig. 12 we observe a notable shift of the resonance towards shorter wavelength as well as an asymmetry of the resonance line. A possible interpretation of this behavior is that the observed resonance line is the sum of two contributions, (i) the plasmon resonance (dashed line) and (ii) the antenna resonance (dash-dotted line). The position of the plasmon resonance is compatible with our previous considerations, however, a detailed analysis of the problem is need to come to definite conclusions.

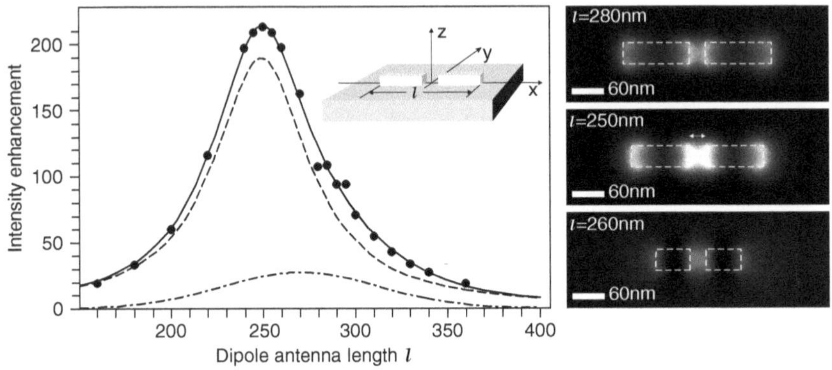

Fig. 12. FDTD simulation of gold dipole antennas. Left: Near-field intensity enhancement in the gap as a function of the antenna length. Right: Field distribution 10 nm above the antenna for three different lengths l, 280, 250 and 160 nm. Maximal intensity enhancements are 34, 65 and 6, respectively. Wavelength 830 nm, antenna height and width 30 and 40 nm, respectively. Gapwidth 30 nm.

3.3. *Light confinement by a resonant bowtie antenna*

It is interesting to note that resonance of gold bow tie antennas with dimensions given in fig. 13 is found at a total length of $l = 240$ nm only slightly shorter than for the dipole antenna. The advantage of the bowtie geometry is that the field is concentrated in single maximum since the charges at the long sides of the triangles are allowed to spread over a much larger spatial region.

3.4. *Fabrication and characterization of resonant optical antennas*

We now turn to the actual fabrication and experimental characterization of optical antennas. Precise structuring of metals in the nm-regime is a nontrivial task. This is true in particular for gold which has a high mobility at room temperature and tends to form clusters. The precision requirements for the fabrication of the simplest antenna structures (dipole antennas) are high. Fabrication of smooth metallic lines well below 100 nm wide with feed gap dimensions in the range of 10 nm is required. Only a few lithographic methods based on electron beam (e-beam) (Howard et al., 1980; Park et al., 1997; Fromm et al., 2004), focused ion beam (FIB) (Shedd et al., 1986; Hoffmann et al., 1993; Matsui and Ochiai, 1996; Kong et al., 2004) and atomic force microscopy (AFM) (Bouchiat and Esteve, 1996; Lüthi et al., 1999; Martin et al., 2005) are able to

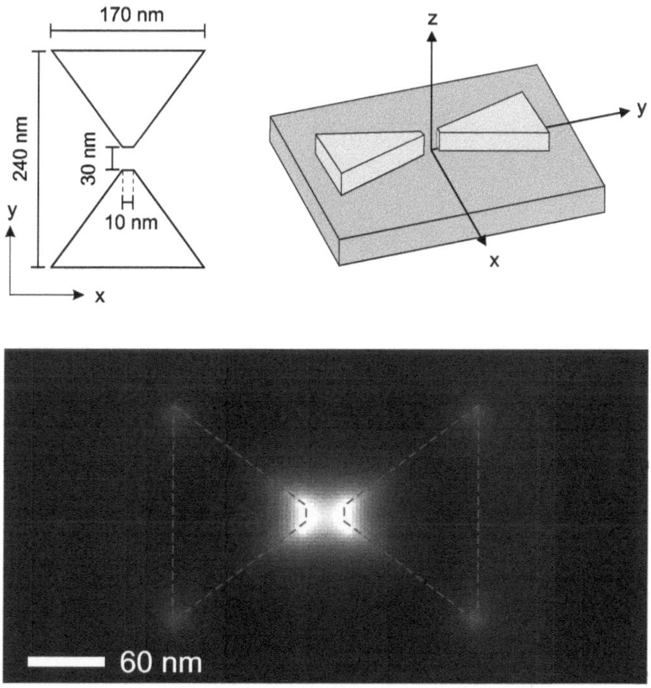

Fig. 13. Antenna geometry and intensity enhancement 10 nm above a resonant gold bowtie antenna. Maximum intensity enhancement is 75. The antenna thickness is 30 nm.

generate sub-100 nm metal lines. However, e-beam lithography is the by far most used and flexible method. A difficulty arises for the optical investigations of antenna structures since it is necessary to fabricate them on a transparent (e.g. glass) substrates. However e-beam lithography, FIB milling and scanning electron microscopy (SEM) characterization require a conductive sample. Therefore the glass substrate has to be coated with a thin layer of a transparent and conductive material. A natural choice for the coating material is indium-tin oxide (ITO).

Even after such a coating it was found that it is very difficult to fabricate high-definition antenna structures by applying a *direct* electron-beam lithography and lift-off procedure. We therefore first create a pre-structure consisting of an array of rectangular gold patches by electron-beam lithography which is later on refined by focused-ion beam milling. Figure 14 shows a scanning electron microscopy image of the pre-structured gold patches on an ITO-coated cover glass.

In a second fabrication step the FIB was used to remove material at the position of its focus. By using small ion currents a spot size in the

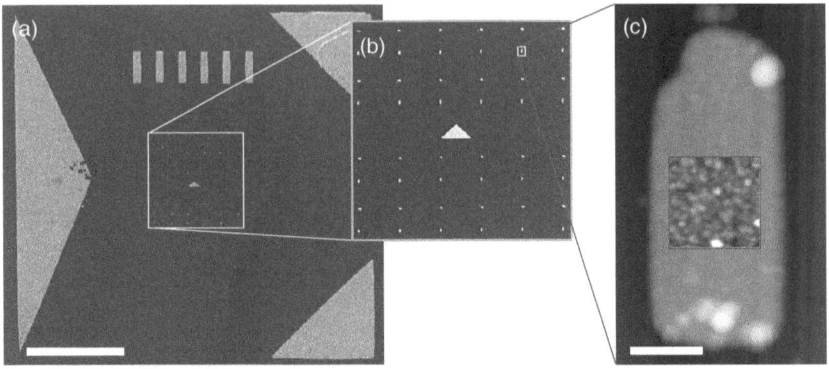

Fig. 14. Prestructuring of a gold film by electron beam lithography. The SEM image shows three zoom levels. (a) Scanning electron microscopy image of the gold patch array and triangular marker structures. (b) Zoom ($50 \times 50\ \mu m^2$) on the array of metal patches. (c) AFM-image of one gold patch ($400 \times 800 \times 40\,nm^3$) before FIB milling. Inset shows high-resolution AFM-image ($250 \times 250\,nm^2$) used for surface roughness measurement.

sub-10 nm is achievable. The resolution is, however, limited by the small depth of focus. The minimum line width actually depends on the depth of the milled line (Li et al., 2003). This limits, for example, the achievable line width of a 60 nm deep line to about 30 nm (Li et al., 2003). The drawback of a low ion currents are low milling rates, that make material removal time consuming and hence very expensive. This is one of the reasons that makes pre-structuring with electron-beam lithography necessary.

Figure 15 illustrates the strategy used in FIB structuring of optical antennas. Rectangular areas were milled away sequentuially (white boxes 1–5 in fig. 15(a)) to sculpture the optical antenna shown in fig. 15(b). Inspection of the antenna by AFM confirms the dimensions and the high quality of the structure.

In a first attempt to characterize the fabricated resonant optical antennas the field enhancement in the gap was exploited to promote non-linear effects like two-photon excited photoluminescence of gold. Upon excitation with 830 nm ps-pulsed light indeed green shifted luminescence was observed. This effect is well documented in the literature and is most efficient for rough gold films (Boyd et al., 1986; Beversluis et al., 2003; Schuck et al., 2005) and single gold particles (Beversluis et al., 2003). However, in addition to this effect which is known to occur inside the metal itself a second effect with a much stronger power dependence (power of 4) was found to contribute to the light emission of resonant dipole antennas. It is speculated that this power-of-4-effect is due to the formation of a plasma and/or dielectric breakdown in the antenna feed

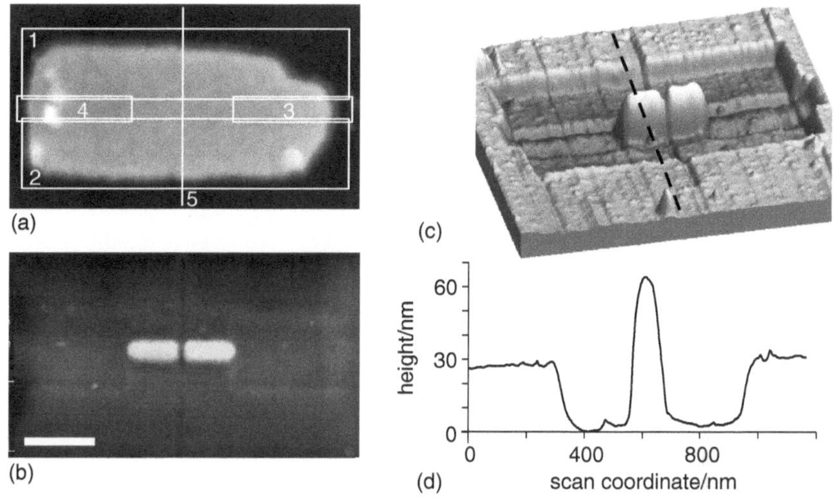

Fig. 15. FIB-assisted antenna fabrication. (a) SEM-image of gold patch. Patterns 1–5 (white boxes) where removed subsequently. (b) Resulting antenna structure. (c) 3D AFM-image, dashed line is indicating the position of the line profile shown in (d).

gap (von der Linde and Schueler, 1996). The process was denoted as white light continuum (WLC) generation in Mühlschlegel et al. (2005). Figure 16 shows SEM images alongside with scanning confocal images for excitation polarized along and perpendicular to the antenna main axis obtained in the same area. A clearly length dependent emission from antennas is observed which is strongly dependent on the excitation polarization. Details are described elsewhere (Mühlschlegel et al., 2005). The occurrence of the high nonlinearity in any case shows that indeed very strong field enhancement exists in the feed gap of a resonant optical dipole antenna (Mühlschlegel et al., 2005).

§ 4. Single dipole emitters coupled to optical antennas

So far we have studied the field enhancement created in the gap of resonant optical antennas by external illumination. Analogous to the the radio frequency case we also would like to stimulate enhanced emission of light by an optical antenna through excitation by an oscillating electric field applied over the antenna feed gap. For optical antennas such an excitation cannot be delivered via galvanic contacts because of size constraints. However, one may think of a single dipole emitter, such as a single molecule or a quantum dot, placed inside the antenna feed gap

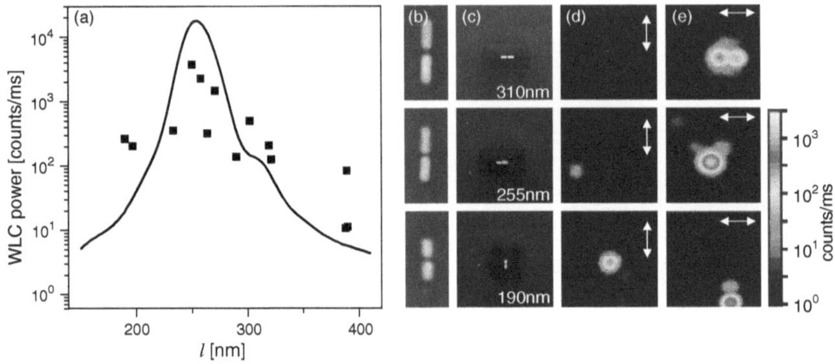

Fig. 16. Experimental results obtained for optical antennas. WLC emission of antennas of different lengths (dots) and corresponding numerical calculation (solid line) (Mühlschlegel et al., 2005). SEM images, zoom (b) and overview (c), respectively. (d) and (e): Confocal scan images of the WLC generated by vertically and horizontally polarized laser pulses, respectively (average power 110 μW, logarithmic color code). Dimensions: (b) $450 \times 180\,\text{nm}^2$; [(c), (d), and (e)] $2 \times 2\ \mu\text{m}^2$.

acting as sources for excitation. In this case the emitters quasistatic near field drives the electrons in the antenna material near the antenna gap thus exciting currents in the antenna due to capacitive coupling. In the following we will outline how the interaction between antenna and dipole emitter can be treated in the framework of classical electrodynamics and nano-optics (Novotny and Hecht, 2006). The emitters that suggest themselves for such experiment are displayed in fig. 17(a). Organic dye molecules and semiconductor nano crystals can be readily observed on the level of single emitters using advanced optical microscopy in conjunction with single photon counting (Novotny and Hecht, 2006). For a single emitter the detected number of photons per time interval can be easily deduced by assuming that the emitter behaves as an effective 2 or 3 level system with a ground state, a first excited state, and possibly a dark state, like the triplet state in organic molecules. The different levels are connected by transition rates as denoted in fig. 17(b). Solving the resulting system of rate equations in the steady state leads to the following photon detection rate expressed by the systems transition rates:

$$R = \xi R_\infty \frac{I/I_\text{S}}{1 + I/I_\text{S}}. \tag{4.1}$$

Here, ξ is the efficiency at which the emitted photons are detected. R_∞ and I_S are the emission rate at saturation and the saturation intensity, respectively, both depending on the transition rates. The constants

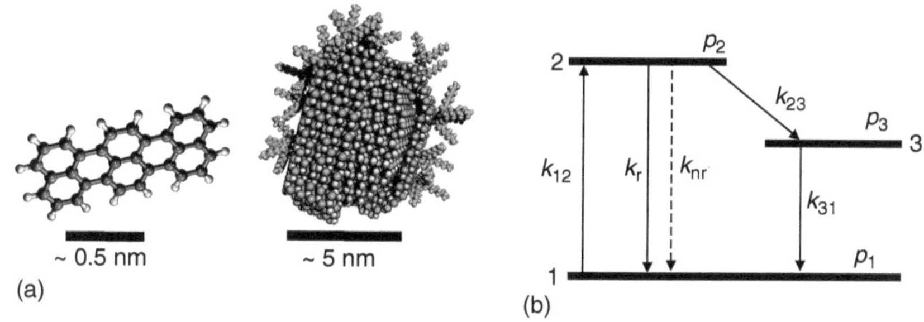

Fig. 17. Single emitters. (a) Organic dye molecule (terrylene) and semiconductor nano crystal. (b) Level scheme: Single emitters can be treated as 2 or 3 level systems.

R_∞ and I_S are defined as

$$R_\infty = \frac{k_{31}k_r}{k_{23}+k_{31}}, \qquad I_S = \frac{(k_r + k_{nr} + k_{23})k_{31}}{\sigma(k_{23}+k_{31})}. \tag{4.2}$$

At low intensities, far from saturation, the detected rate increases linearly with the excitation intensity as

$$R = \xi \frac{R_\infty}{I_S} I = \xi \frac{k_{31}k_r}{k_{23}+k_{31}} \frac{\sigma(k_{23}+k_{31})}{k_{31}(k_r+k_{nr}+k_{23})} \approx \xi \frac{k_r}{k_r+k_{nr}} \sigma I. \tag{4.3}$$

Recognizing that the ratio of radiative to the total decay rate is the quantum efficiency η we arrive at the following simple expression for the count rate detected from a single emitter below saturation as a function of the excitation intensity I

$$R = \xi \eta \sigma I. \tag{4.4}$$

The emission of light by a single emitter may be treated as an electromagnetic antenna problem (Pohl, 1991, 2000) in which the single emitter acts like a point dipole. It can be shown theoretically that although a single emitter emits single photons one by one the emission pattern of a point-dipole antenna will still be observable in experiments (Loudon, 1983). It is well known that a point dipole is a very bad antenna so that we may expect that the total photon emission rate of a single emitter is limited by its small extension to the values we typically observe. Since the inverse of the photon emission rate at saturation coincides with the lifetime of the excited state we may argue that apart from variations introduced by the density of photonic states, the lifetime of a single emitter is also determined by the fact that it is a badly matched antenna for the wavelengths it emits.

4.1. Properties of single dipole emitters near metal nano structures

If a single emitter is brought in close proximity of a metal nano structure the optical properties of the molecule and the optical properties of the metal nano structure become important for the behavior of the total system. Emitter and nano particle can be treated as a new entity with modified optical properties inherited from both the molecule and the metal nano structure.· We will show in the following how the parameters σ,η, and ξ are modified by the presence of a general metal nano structure. If a single emitter is placed near a metal nano structure in addition to the external illumination field $\mathbf{E}_o(\mathbf{r}_o)$ it will also see the scattered field $\mathbf{E}_S(\mathbf{r}_o)$ (see fig. 18). If the scattered field is enhanced by a factor of α then the total excitation intensity seen by the emitter is $(1+\alpha^2)I$. This can be accounted for by defining a modified absorption cross section $\sigma' = (1+\alpha^2)\sigma \approx \alpha^2\sigma$. The last step is a good approximation if the enhancement factor is sufficiently large. We conclude that in presence of the nano structure the absorption cross section of the single emitter is enhanced.

The other parameters in eq. (4.4) are also modified due to the presence of the nano structure. Lets first consider the quantum efficiency η. The nonradiative decay rate is modified since the emitter transfers energy to the electrons in the metals whose motion (current) is subject to Ohmic losses finally dissipating the energy to heat. The radiative rate is modified since the emitters own radiation field is reflected back to the emitters

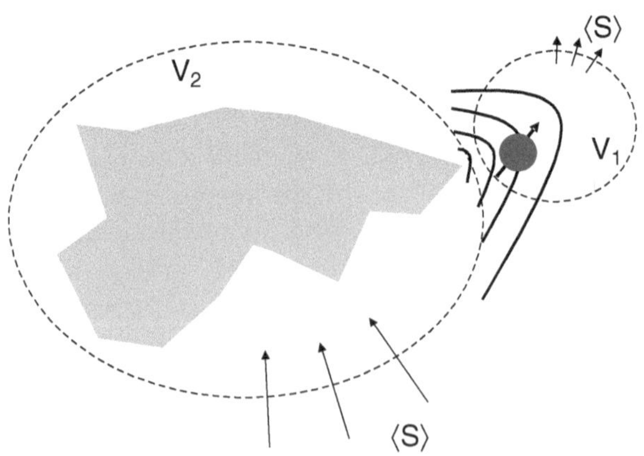

Fig. 18. Single emitter close to a metal nano structure.

position by the metal nano structure thus driving the emitter with a distance dependent phase delay. Both effects can be described by making use of Poynting's theorem (Novotny and Hecht, 2006) which states for a linear medium that in a closed volume V the electromagnetic energy can only decrease due to Ohmic losses inside V or due to radiation of energy through the surface of the volume ∂V:

$$-\frac{1}{2}\frac{\partial}{\partial t}\int_V [\mathbf{D}\cdot\mathbf{E}+\mathbf{B}\cdot\mathbf{H}]\,\mathrm{d}V = \int_V \mathbf{j}\cdot\mathbf{E}\,\mathrm{d}V + \int_{\partial V}[\mathbf{E}\times\mathbf{H}]\cdot\mathbf{n}\,\mathrm{d}a. \tag{4.5}$$

We choose a volume V_1 that encloses the dipole emitter but not the metal nano structure. In the steady state the rate of change of the electromagnetic energy density inside this volume is zero. The only current density inside V_1 is the current density that characterizes the dipole emitter at the position $\mathbf{r}_0$, $\mathbf{j} = -i\omega\,\mathbf{p}\,\delta(\mathbf{r}-\mathbf{r}_0)$. We obtain an expression for the effective flow of energy out of the volume V_1 by introducing this current density into eq. (4.5). In the time-averaged version for harmonic time dependence this reads as

$$P = \frac{\mathrm{d}W}{\mathrm{d}t} = \int_{\partial V_1} \langle\mathbf{S}\rangle\cdot\mathbf{n}\;\mathrm{d}a = \frac{\omega}{2}\mathrm{Im}\{\mathbf{p}^*\cdot\;\mathbf{E}(\mathbf{r}_o)\}, \tag{4.6}$$

where $\langle\mathbf{S}\rangle = \frac{1}{2}\mathrm{Re}(\mathbf{E}\times\mathbf{H}^*)$ is the time-averaged Poynting vector. Since the field at the position of the molecule is a sum of the original and the scattered field we may write

$$\frac{P}{P_0} = 1 + \frac{6\pi\;\varepsilon_0\varepsilon}{p^2\;k^3}\,Im\{\mathbf{p}^*\cdot\;\mathbf{E}_S(\mathbf{r}_0)\} \equiv (k_r + k_{nr}), \tag{4.7}$$

where we have normalized by the energy dissipation rate of the free dipole $P_0 = p^2\;\omega\;k^3/(12\pi\;\varepsilon_0\varepsilon)$. Equation (4.7) describes the total decay rate $(k_r + k_{nr})$ of the dipole emitter in presence of the metal structure. It solely depends on the back-scattered field at the position of the dipole emitter. Since the emitter itself is assumed to be lossless, the only way that energy can be dissipated is inside the metal structure. By applying eq. (4.6) to a volume V_2 that encloses only the metal structure the rate of energy dissipation inside the volume may be calculated by integrating the Poynting vector flux *into* the volume V_2 over its surface. This rate must be identical to the nonradiative decay rate k_{nr} of the dipole emitter in presence of the metal structure. Finally let as note that naturally also the detection efficiency ξ will be modified by the presence of the metal nano structure. The currents in the metal particle will contribute to the far fields of the coupled system and thus change the emission pattern. If an optical system with a finite numerical aperture is used to collect the emitted light, a change in emission pattern will in general affect the detection efficiency.

4.2. Experimental realization: creating an antenna-based super-emitter

In the context of the previous discussion the question naturally arises: what would be the optimal shape of a metal nano structure to influence the properties of a single emitter in the most favorable way, e.g. enhancing the absorption cross section, increasing the radiative rates while keeping the nonradiative rate in bounds, and optimizing the radiation pattern in a way to send all the radiation towards the detector? According to the discussion of previous sections the answer to this problem is simple: one has to use an optical antenna! The problem of placing a single emitter in the feed gap of an antenna is not trivial. One possible solution is to employ scanning probe technology to position an optical antenna with nanometer precision close to a single, well-characterized dipole emitter fixed to a solid surface. The experiment is described in Farahani et al. (2005). To position the antenna it has to be fabricated at the apex of a metal-coated Si_3N_4-AFM tip using FIB milling. The fabrication steps are sketched in Fig. 19(a), (b). They involve coating of the bare tip and subsequent FIB milling with a pattern that is adapted to the symmetry of the tip. An SEM image and a zoom of the resulting bowtie antenna are shown in fig. 19(c).

Figure 20 shows the results of the following experiment: a single emitter (semiconductor nano crystal) was selected using confocal microscopy. Its

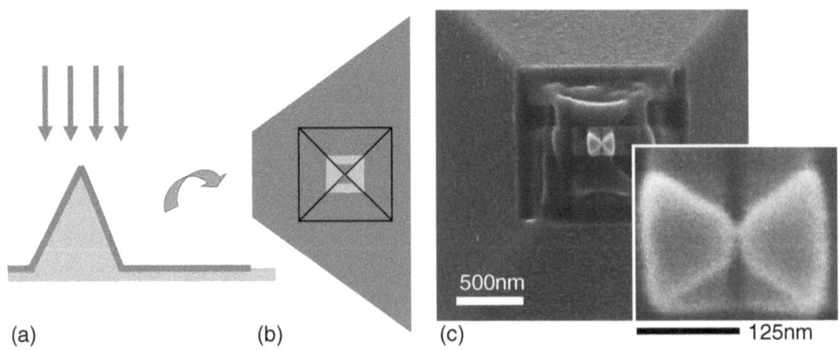

Fig. 19. Fabrication of antenna AFM tips: (a) metal coating; (b) FIB milling; (c) SEM picture: top view.

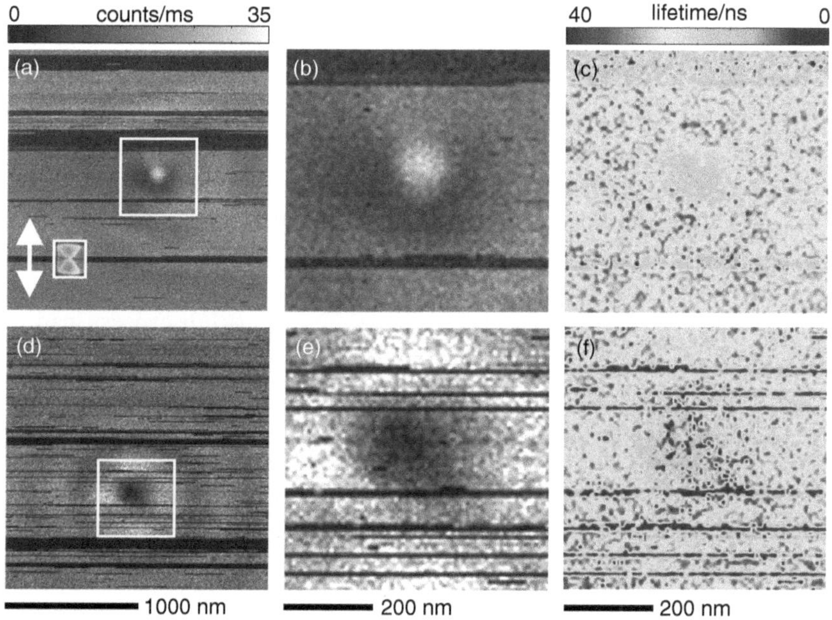

Fig. 20. Emission rate and lifetime modification of a single emitter by a scanning optical antenna. (a,b) Emission rate as a function of the antenna tip position. (c) Lifetime as a function of the tip position. Same are as (b). (d,e) Emission rate as a function of the tip position of a fully coated tip. (f) Lifetime as a function of the fully coated tip position.

fluorescence rate R and the lifetime of its excited state $1/(k_r + k_{nr})$ were measured as a function of time. Then the antenna tip was raster-scanned at the location of the emitter such that eventually it will move directly across the emitter. While raster-scanning the tip lifetime and rate are recorded continuously thus forming images of the single-emitter response as a function of the antenna position (Farahani et al., 2005). Figure 20(a) (b) and (c) show the fluorescence rate (and a zoom) as well as the lifetime of the emitters excited state as a function of the antenna tip position. Once the antenna is positioned above the emitter, its fluorescence is increased (fig. 20(b)) while its lifetime is decreased (fig. 20(c)). This is compatible with a coupling between antenna and single emitter that increases the overall emission of the single emitter. An effective super-emitter has been created. For control, the same experiment has been performed with fully metal-coated AFM tip. In this case the luminescence is quenched rather than increased while the lifetime is also reduced – in this case, however, due to an increase of non-radiative decay channels in the metallic tip.

§ 5. Conclusion

Resonant optical antennas behave similar to their much larger radiowave counterparts. However, at optical frequencies the metals susceptibility allows for surface plasmons to exist at the boundaries of metallic nano antennas. These resonant excitations play an important role for the behavior of resonant optical antennas made from noble metals. The fabrication of resonant optical antennas is a challenging task which applies modern micro fabrication techniques at the limit of their present capabilities. The rather wide tolerances during the fabrication process lead to antenna structures whose optical properties scatter to some extent around the expected behavior. Nevertheless, enhanced fields are created by resonant antennas that are strong enough to generate a power-of-4 nonlinear dependence of the emission observed from such structure. Dielectric breakdown and/or plasma generation in the feed gap suggest themselves as possible explanations.

Coupling a single emitter to a resonant optical antenna creates a new kind of emitting entity. The macroscopic antenna and the microscopic quantum emitter act as a single object which inherits properties of both constituents. Such a combination holds promise for applications in scanning near-field optical microscopy as well as photonics and quantum information processing.

Acknowledgments

The authors are grateful to H.J. Güntherodt for continuous support and to P. Gasser, S. Meier, P. Nellen, and U. Sennhauser (all EMPA) for FIB operation. Helpful discussions are acknowledged with A. Lieb, Y. Lill, J.Y.P. Butter, S. Karotke, and J. Toquant. Financial support by the KTI/CTI (project 7991.1;5) as well as by the Swiss National Science Foundation via the National Center of Competence in Research (NCCR) in Nanoscale Science, a grant "Optical antennas for nano-optics", and a professorship for one of the authors (B.H.) is gratefully acknowledged.

References

Balanis, C., 1997, Antenna Theory: Analysis and Design, 2nd edn., Wiley, New York, Chichester.

Balanis, C.A., 1992, Antenna theory: A review, Proc. IEEE **80**, 7–22.

Barnes, W., Dereux, A., Ebbesen, T., 2003, Surface plasmon subwavelength optics, Nature **424**(6950), 824–830.

Beversluis, M., Bouhelier, A., Novotny, L., 2003, Continuum generation from single gold nanostructures through near-field mediated intraband transitions, Phys. Rev. **B 68**, 115433–115510.

Bohren, C., Huffman, D., 1983, Absorption and Scattering of Light by Small Particles, Wiley Science Paperback Series, Wiley, New York.

Bouchiat, V., Esteve, D., 1996, Lift-off lithography using an atomic force microscope, Appl. Phys. Lett **69**, 3098–3100.

Boyd, G., Yu, Z.H., Shen, Y., 1986, Photoinduced luminescence from noble metals and its enhancement on roughened surfaces, Phys. Rev. B **33**, 7923–7936.

Bozhevolnyi, S., Erland, J., Leosson, K., Skovgaard, P., Hvam, J., 2001, Waveguiding in surface plasmon polariton band gap structures, Phys. Rev. Lett **86**(14), 3008–3011.

Crozier, K., Sundaramurthy, A., Kino, G., Quate, C., 2003, Optical antennas: Resonators for local field enhancement, J. Appl. Phys. **94**, 4632–4642.

Farahani, J.N., Pohl, D.W., Eisler, H.-J., Hecht, B., 2005, Single quantum dot coupled to a scanning optical antenna: A tunable super emitter, Phys. Rev. Lett **95**, 017402.

Feynman, R., 1989, The Feynman Lectures on Physics II, 1st edn, Addison-Wesley Publishing Company, MA.

Fromm, D.P., Sundaramurthy, A., Schuck, P.J., Kino, G., Moerner, W.E., 2004, Gap-dependent optical coupling of single "bowtie" nanoantennas resonant in the visible, Nano Letters **4**(5), 957–961.

Fumeaux, C., Alda, J., Boreman, G.D., 1999, Lithographic antennas at visible frequencies, Opt. Lett **24**, 1629–1631.

Grober, R., Schoellkopf, R., Prober, D., 1997, Optical antenna: Towards a unity efficiency near-field optical probe, Appl. Phys. Lett **70**, 1354.

Haes, A.J., Hall, W.P., Chang, L., Klein, W.L., Duyne, R.P.V., 2004, A localized surface plasmon resonance biosensor: First steps toward an assay for alzheimers disease, Nano Letters **4**(6), 1029–1034.

Hallén, E., 1938, Theoretical investigations into the transmitting and receiving qualities of antennae, Nova Acta Regiae Societatis Scientiarium Upsaliensis **11**(4), 1–44.

Hallock, A.J., Redmond, P.L., Brus, L.E., 2005, Proc. Natl. Acad. Sci. USA **102**, 1280.

Hartschuh, A., Sánchez, E., Xie, X., Novotny, L., 2003, High-resolution near-field raman microscopy of single-walled carbon nanotubes, Phys. Rev. Lett **90**, 095503.

Hertz, H., 1971, ber sehr schnelle elektrische Schwingungen: vier Arbeiten (1887 – 1889), vol. 251 of Ostwalds Klassiker der exakten Wissenschaften, 2nd edn, Thun, Frankfurt am Main.

Hoffmann, P., Ben Assayag, G., Gierak, J., Flicstein, J., Maar-Stumm, M., van den Bergh, H., 1993, Direct writing of gold nanostructures using a gold-cluster compound and a focused-ion beam, J. Appl. Phys **74**, 7588–7591.

Howard, R.E., Hu, E.L., Jackel, L.D., Grabbe, P., Tennant, D.M., 1980, 400-angstrom linewidth e-beam lithography on thick silicon substrates, Appl. Phys. Lett **36**, 592–594.

Jackson, J., Westcott, S., Hirsch, L., West, J., Halas, N., 2003, Controlling the surface enhanced raman effect via the nanoshell geometry, Appl. Phys. Lett **82**(2), 257–259.

Johnson, P., Christy, R., 1972, Optical constants of noble metals, Phys. Rev. B **6**, 4370–4379.

King, R., Harrison, C.W., 1943, The distribution of current along a symmetrical center-driven antenna, Proc. I.R.E **31**, 548–567.

Kneipp, K., Wang, Y., Kneipp, H., Perelman, L., Itzkan, I., Dasari, R., Feld, M., 1997, Single molecule detection using surface-enhanced raman scattering (sers), Phys. Rev. Lett **78**(9), 1667–1670.

Knoll, B., Keilmann, F., 1999, Near-field probing of vibrational absorption for chemical microscopy, Nature **399**, 134–137.

Kong, D.S., Varsanik, J.S., Griffith, S., Jacobson, J.M., 2004, Conductive nanostructure fabrication by focused ion beam direct-writing of silver nanoparticles, J. Vac. Sci. Technol. **B22**, 2987–2991.

Kreibig, U., Vollmer, M., 1995, Optical Properties of Metal Clusters, 1st edn, Springer Series in Materials Science, University of California Berkeley.

Krenn, J., Dereux, A., Weeber, J., Bourillot, E., Lacroute, Y., Goudonnet, J., Schider, G., Gotschy, W., Leitner, A., Aussenegg, F., Girard, C., 1999, Squeezing the optical near-field zone by plasmon coupling of metallic nanoparticles, Phys. Rev. Lett **82**, 2590–2593.

Lee, K., 1984, Principles of Antenna Theory, Wiley, Chichester.

Li, H.-W., Kang, D.-J., Blamire, M.G., Huck, W.T.S., 2003, Focused ion beamfabrication of silicon print masters, Nanotechnology **14**, 220–223.

Loudon, R., 1983, The Quantum Theory of Light, Oxford Science Publications, Oxfort University Press, Oxford.

Lüthi, R., Schlittler, R.R., Brugger, J., Vettiger, P., Welland, M.E., Gimzewski, J.K., 1999, Parallel nanodevice fabrication using a combination of shadow mask and scanning probe methods, Appl. Phys. Lett **75**, 1314–1316.

Maier, S., Kik, P., Atwater, H., Meltzer, S., Harel, E., Koel, B.E., Requicha, A., 2003, Local detection of electromagnetic energy transport below the diffraction limit in metal nanoparticle plasmon waveguides, Nat. Mater **2**, 229–232.

Martin, C., Rius, G., Borrise, X., Perez-Murano, F., 2005, Nanolithography on thin layers of PMMA using atomic force microscopy, Nanotechnology **16**, 1016–1022.

Matsui, S., Ochiai, Y., 1996, Focused ion beam application to solid state devices, Nanotechnology **7**, 247–258.

Michler, P., Kiraz, A., Becher, C., Schoenfeld, W., Petroff, P., Zhang, L., Hu, E., Imamoglu, A., 2000, A quantum dot single-photon turnstile device, Science **290**(5500), 2282–2285.

Mühlschlegel, P., Eisler, H.-J., Martin, O., Hecht, B., Pohl, D., 2005, Resonant optical antennas, Science **308**, 1607–1609.

Novotny, L., Bian, R.X., Xie, X.S., 1997, Theory of nanometric optical tweezers, Phys. Rev. Lett **79**, 645–648.

Novotny, L., Hecht, B., 2006, Principles of Nano-Optics, Cambridge University Press, Cambridge.

Park, Y.D., et al., 1997, Fabrication of nanometer-sized magnetiuc structures using e-beam patterned deposition masks, J. Appl. Phys **81**, 4717–4719.

Pocklington, H., 1897, Electrical oscillations in wires, Proceedings of the Cambridge Philosophical Society **9**(7), 324–332.

Pohl, D., 1991, Scanning Near-field Optical Microscopy (SNOM), vol. 12 of Advances in Optical and Electron Microscopy, Academic Press, New York, pp. 243–312.

Pohl, D., 1994, Photons and forces i: Light generates force, NATO ASI on Forces in Scanning Probe Methods, Black Forest.

Pohl, D., 2000, Near field optics seen as an antenna problem, in: Ohtsu, M., Zhu, X. (Eds.), Near-Field Optics: Principles and Applications, World Scientific, ISBN 981-02-4365-0, Singapore, pp. 9–21.

Rechberger, W., Hohenau, A., Leitner, A., Krenn, J., Lamprecht, B., Aussenegg, F.R., 2003, Optical properties of two interacting gold nanoparticles, Opti. Commun **220**, 137–141.

Schuck, P., Fromm, D., Sundaramurthy, A., Kino, G., Moerner, W., 2005, Improving the mismatch between light and nanoscale objects with gold bowtie nanoantennas, Phys. Rev. Lett **94**, 017402–017404.

Schulz, L., 1954, The optical constants of silver, gold, copper and aluminum. (1) the absorption coefficient *k*, J. Opt. Soc. Am **44**(5), 357–362.

Schulz, L., Tangherlini, F., 1954, The optical constants of silver, gold, copper and aluminum. (2) the index of refraction n, J. Opt. Soc. Am **44**(5), 362–368.

Shedd, G.M., Lezec, H., Dubner, A.D., Melngailis, J., 1986, Focused ion beam induced deposition of gold, Appl. Phys. Lett **49**, 1584–1586.

Sönnichsen, C., Franzl, T., Wilk, T., von Plessen, G., Feldmann, J., Wilson, O., Mulvaney, P., 2002, Drastic reduction of plasmon damping in gold nanorods, Phys. Rev. Lett **88**, 077402.

Taflove, A., Hagness, S., 2005, Computational Electrodynamics, 3rd edn, Artech House, Boston.

von der Linde, D., Schueler, H., 1996, Breakdown threshold and plasma formation in femtosecond laser–solid interaction, J. Opt. Soc. Am. **B13**, 216–222.

Wilke, I., Herrmann, W., Kneubühl, F.K., 1994, Integrated nanostrip dipole antennas for coherent 30 tHz infrared radiation, Appl. Phys. **B58**(2), 87–95.

Xu, H., Käll, M., 2002, Surface-plasmon-enhanced optical forces in silver nanoaggregates, Phys. Rev. Lett **89**, 246802.

Author Index

C

I

Subject Index

V

W

FSC
www.fsc.org
MIX
Papier | Fördert gute Waldnutzung
FSC® C083411

Kolibri 360 GmbH
Ferdinand-Jühlke-Straße 7
99095 Erfurt, Deutschland
produktsicherheit@kolibri360.de